The Kaiser's Flyers Macedonia 1914–1918

A Study of Military Aviation in South-East Europe

Falk Breuer & Walter Waiss

Translated by Pilar Elvira Wolfsteller

Special thanks for collaboration on this project goes to:

- Marton Szigeti
- Wolfgang Kaulertz
- Thorsten Pietsch
- Rolf Nagel
- Lance Bronnenkant
- Heinz Michael Raby
- Rainer Absmeier
- Stefan Kruse
- Matthias Hundt
- Reinhard Zankl
- Albin Denis
- Prof. Dr. Heinz Richter
- Jörg Kempf, Norwegen

Acknowledgements

Color aircraft profiles © Bob Pearson. Purchase his CD of WWI aircraft profiles for $50 US/Canadian, 40 €, or £30, airmail postage included, via Paypal to Bob at: bpearson@kaien.net

For our aviation books in print please see our website at: www.aeronautbooks.com.
I am looking for photographs of the less well-known German aircraft of WWI for future titles. For questions or to help with photographs please contact me at jherris@me.com

Interested in WWI aviation? Join The League of WWI Aviation Historians (**www.overthefront.com**), Cross & Cockade International (**www.crossandcockade.com**), and Das Propellerblatt (**www.propellerblatt.de**).

ISBN: 978-1-953201-15-7

© 2021 Aeronaut Books, all rights reserved
Text © 2021 Falk Breuer & Walter Waiss
Design and layout: Jack Herris
Cover design: Aaron Weaver
Digital photo editing: Walter Waiss & Jack Herris

Books for Enthusiasts by Enthusiasts
www.aeronautbooks.com

Table of Contents

Foreword by Falk Breuer	3
Thanks for Collaboration	

Part 1

Introduction by Dr. Heinz Richter	6
The Political and Military Developments Until the Beginning of the Fighting for Thessaloniki	9
The Balkan Wars 1912–1913	9
The Politics of the Balkan States on the Eve of the First World War	12
The Attack on Serbia. Start of the First World War in the Southeast	14
The Politics of the Other Balkan States Before the Beginning of the War	32
The Allied Expedition to Serbia October 1915–December 1915	59
Political Maneuvering in the Background	59
The Austro-German Campaign Against Serbia in Autumn 1915	71
The Allied Advance on Serbia	78
The Allied Occupation of Macedonia January–August 1916	84
Macedonia Under French Occupation	134
The Entrenched Camp of Saloniki	190
Romania's Entry into the War	212
Split of the Country August–December 1916	216
The Evacuation of the 4Th Army to Görlitz	216
The Governments of Zaimis and Kalogeropoulos	222
Increasing French Pressure	245
The Noemvriana (November Events)	258
The Provisional Government	281
Italy's Occupation of North Epirus	286
The Allied Autumn Campaign September 12Th–December 16Th 1916	287
The Deposition of Constantine and the New Regime	296s
The Conference of Rome January 1917	296
Blockade and Further Repression	316
Constantine's Last Interview as King	319
The Deposition of King Constantine	320
The New Regime	327
The Allied Spring Offensive 1917	329
The Further Military Developments	331
German-Greek Relations 1916–1917	334
The Collapse of the Macedonian Front 1918	338
The Spring Offensive	338
Building a New Greek Army	341
Ahead of the Big Offensive	343
Planning the Allied Offensive	347
The Attack	349
Interpretation and Legends Following Bulgaria's Capitulation	352
Macedonia in the Second World War	354

Part 2

Secondary War Theater Balkans	357
Illnesses in the Southeast	358
Zeppelins in the Southeastern Theater	359
Captive Balloons at Work on the Macedonian Front	360
Aviation Units	
Kampfgeschwader 1 in Macedonia	361
Feldflieger-Abteilung (Field Aviation Unit) 1	361
Flieger-Abteilung (Aviation Unit) 20 (Fa 20)	361
Flieger-Abteilung (Aviation Unit) 30 (Fa 30)	361
Crew of B- or C-Airraft	362
Stage Aircraft Park and Army Air Park	362
Air Force News Bulletins	363
Individual Reports from 1916 and 1917	363
Individual Reports from 1918	373

Part 3

Appendix	386
Photos, Documents	386
Personal Descriptions and Military Resumes	386
Logbook of Johannes Keller	388
Bibliography	390
Endnotes	393

Foreword

Last year numerous events celebrating the end of the First World War one hundred years ago have taken place across Germany as well as in neighboring countries. Presidents and Chancellors of former enemy nations reached out to shake each other's hands at the large soldiers' cemeteries in France.

Names that were synonymous with the so-called *"original catastrophe"* of the 20th century were once again spoken out loud: Verdun, the battleground of *In Flanders Fields*, the Somme and Champagne. Names that will always be connected with the first instance of industrial and technical war-faring practices, and through which, in the end, millions would become victims. One hundred years on, the killing fields of France and Belgium have still not lost their horror. Still no news from the western front…

In contrast, no one spoke of Drama, Xanthi or the Struma Valley. No one reminded the world that here, too, there was suffering and death. Who still knows today that there was a front far away from the great theatres of war, for example in European Macedonia? Embedded in a complex system of allies, the German empire was militarily obligated to support its Osmanian ally after its not very successful intervention in the Balkans. In addition to a stately number of ground forces and a small naval contingent, numerous air force sections were sent to the southeastern theatre. In their war diaries, the German soldiers wrote about the difficult road and traffic conditions, the searing heat and the violent sand storms that challenged humans, animals and material in a way they had never before experienced. In a completely new cultural environment, and with unfamiliar traditions and extreme climactic conditions, survival was key.

In their photo albums, the troops and officers documented the face of what was for them the true meaning of the "oriental" world. There was a particular fascination for the historical sites of Hellenic antiquity. Often the young German officers would engage in archaeological research in their free time, and dug for antique artifacts, likely with the words of Homer's *"Iliad"*, likely seared in their memories from their school days.

The relationship with the Bulgarian *"brothers in arms"* seems to have been friendly, but true respect was reserved for the military adversary, the British *Royal Flying Corps*. One engaged each other with the highest deference and communicated via written notes, that were thrown overboard by courageous pilots flying over enemy positions. In this way, information about the fate of downed or stranded crews which had not returned to their airfields, was exchanged. It was in this way that the series of photographs of the funeral of the legendary *"Eagle of the Aegean"*, Lt. Rudolf von Eschwege – delivered by British air officers - landed in the possession of his company comrade Johannes Keller, who kept them in his photo album.

His photo album and his flight log, as well as the comprehensive photo legacy of the former members of the German air force in Macedonia, Walter Schwabedissen, Wilhelm Grasmeher, Johannes Schaaf, Gerhard Fieseler, Fritz Kempf and the observation officer Walter Wehmeyer, are presented and analyzed in this publication about the secondary theatre of war, Macedonia, for the first time.

These extensive documents from the *Falk Breuer Collection* have the disadvantage, though, that we could only analyze small parts of the photographic legacy that pertain to Macedonia, even though the other parts are very exciting from a historical perspective. But we have to, and we want to, limit ourselves to this topic, and will bring only short periods of time before and after the fighting in the south-east into this study. It is also important to note that some individuals were ordered away and were also in Romania for a short time. These photographs, however, do not minimize the broad general overview of the South-East theatre, with Macedonia, but rather offer a deeper insight into the political and military operations.

The authors drew on the competent study[1] from Prof. Dr. Heinz A. Richter, who, as a profound expert on the political situation in Macedonia and its neighboring countries during the time of the First World War was very knowledgeable, and to whom we give a special thanks.

Therefore, I would like to say a few words here about why Falk Breuer and I were particularly and increasingly interested in this work. Falk Breuer is a passionate collector of historical photographs and documents depicting German aviation during the First World War. It is no surprise that the extensive aviation legacy is still awaiting a complete analysis. Older documentation has the drawback that the available photographic materials are usually difficult or impossible to utilize due to the poor quality of the print. Often the miserable quality of the images required additional treatment or improvements.

After two years of research, Falk Breuer and I decided to attack this big project. The decisive motivation was the publication of these extremely rare photographs of this south-eastern European theatre of war. In addition, we had the amazing luck to be able to win over Professor Dr. Heinz A. Richter to our project. Prof. Dr. Richter is an accomplished subject expert in this area and has already analyzed the German, British and Greek primary and secondary source literature. He will add his research through excerpts and some complete chapters, and bring the political aspects to the reader, which neither Falk Breuer nor I would have been qualified to do. As I mention, we look at the war in the southeast from 1915 through the end of the war in 1918, with a special emphasis on the allied opponents, with a political eye. Only in this way can we see a complete picture of the German aviation forces.

This book is written from numerous perspectives and is comprehensive, so that after a short introduction we jump into the core of the work: the political situation in Macedonia and its neighbors from 1914 until the end of the war in 1918. Biographies of individual actors, be they politicians or accompanying pilots, will be presented in the third section, following a chronological portrayal from Professor Dr. Heinz A. Richter in the second section, which has been assembled through a reconstruction of documented events. The third section also holds an appendix for documents, etc. The authors are always grateful for any additions or correction of any errors.

To Begin, A Short Introduction About the War's Events So Far

World War I began on August 1, 1914 and immediately developed into a conflict on several fronts. On the German side, the Austrian Empire, with its multiethnic state, Turkey as the Osmanian Empire and Italy, which immediately declared itself neutral. In 1914, the Austro-Hungarian army was not in a position to defeat the Kingdom of Serbia. This was the first indication that the army was in a miserable state. Only when German units came to support the Austrians in October 1915 in the "*Second Serbian campaign*" and with the help of the Bulgarian army that marched in from the east, were the combined forces able to push the kingdom of Serbia to the Greek border. This was called the secondary theatre "Makedonia" or "Macedonia". The forces entrenched themselves at the end of 1915, and the fruitless trench warfare began here also.

The 11th German Army here was supported by the aviation units (Artillery) [FFA] 1 (Westarp). FFA 28 under the command of von der Goltz, FFA 30 under Hptm. Martin Niemöller, FFA 57 under Heyder, FFA 66 under Liebermann and FFA 69 under Müller, which arrived from Russia on November 20, 1915.[2] A unified and targeted leadership was missing, and there was no clear focus, with too many tasks at hand. Some units were later switched out, re-established or renamed. According to Jörg Mückler, the well-known expert about aviation activities during World War I, in 1918 the German air forces were distributed across four regions:

- In the Chernobog: FFA 246 FFA 38 (ex FFA 69), Jasta 25
- In the Varda Valley: FA 30, FA 34 (ex FFA66), Jasta 38
- In the Struma Valley: FA20 (ex FFA 51)
- On the Aegean: Naval station Xanthi (SFA)

Bulgaria entered the war on the side of Germany and Austria on October 14th 1915. The ground link and supplies for the Osmanian Empire were thereby secured. The allied ground troops of the unsuccessful campaign for the Dardanelles were brought by ship into neutral Greece, and the harbor of Thessaloniki. We take the political developments from the descriptions[3] of Prof. Dr. Heinz Richter.

Special Thanks for Collaboration on This Project Goes To:

- Martin Szigeti
- Wolfgang Kaulertz
- Thorsten Pietsch
- Rolf Nagel
- Lance Bronnenkant
- Heinz Michael Raby
- Rainer Absmeier
- Stefan Kruse
- Matthias Hundt
- Reinhard Zankl
- Albin Denis
- Prof. Dr. Heinz Richter
- Jörg Kempf, Norwegen

Next Page, Above: Lithograph by Eugen Osswald, about 1915, with the title "*Ins Lazarett*" ("Into the hospital"). Eugen Osswald, Born January 22nd 1879 in Stuttgart, died February 17th 1960 in Munich.

Notes

1. Richter, Heinz: "*Der Krieg im Südosten*", 2nd volume, Macedonia 1915–1918, Rutzen-Verlag, Wiesbaden
2. Mückler, Jörg: "*Deutsche Flugzeuge im 1. Weltkrieg*", Motorbuchverlag Stuttgart, 2013
3. Richter, Heinz: "*Der Krieg im Südosten*", 2nd volume, Macedonia 1915-1918, Rutzen-Verlag, Wiesbaden, 2014

Facing Page, Lower: Map showing declarations of war on Germany. Austro-Hungary, Germany and its colonies in red.

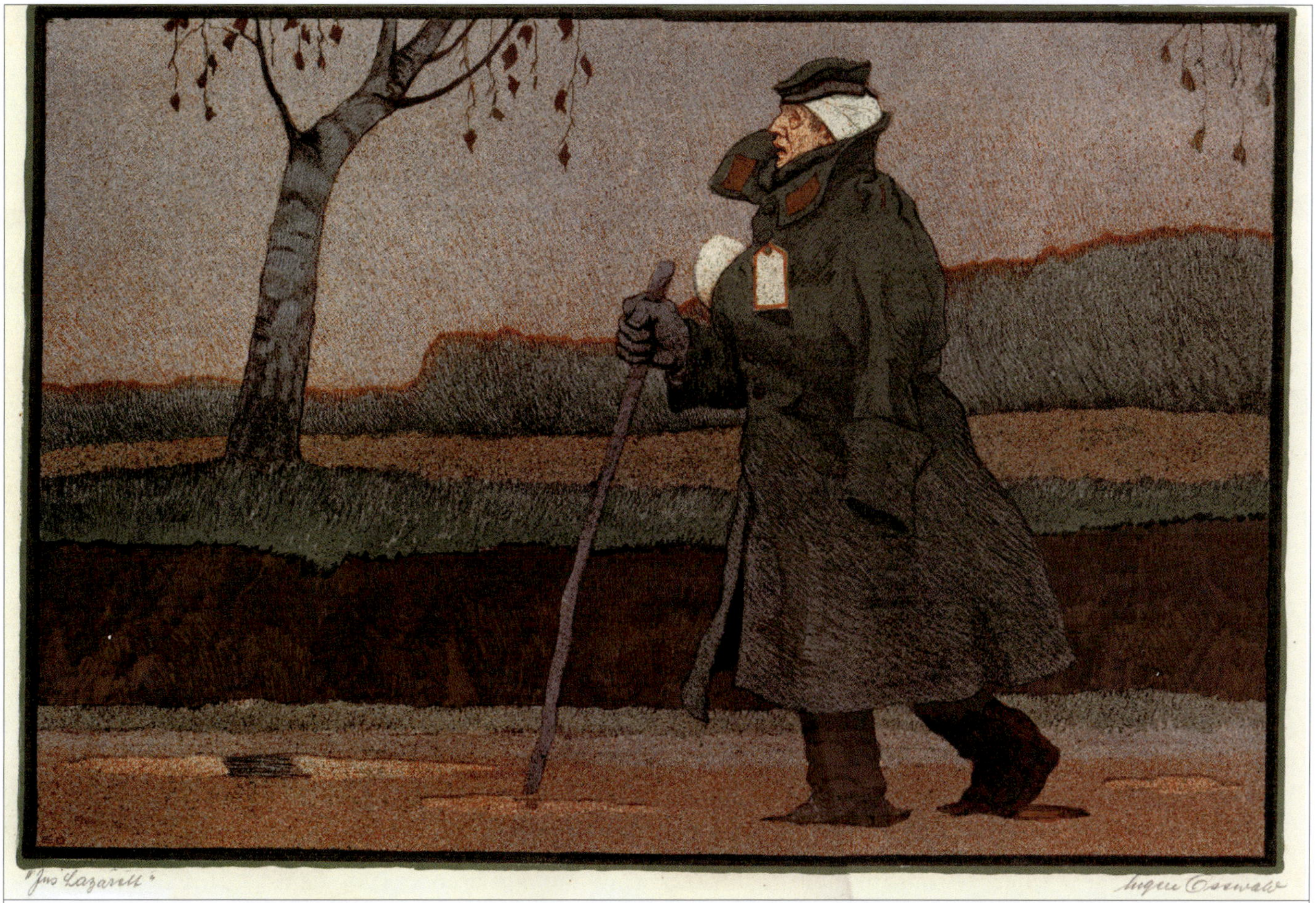

5

"Ins Lazarett" Auguts Oswald

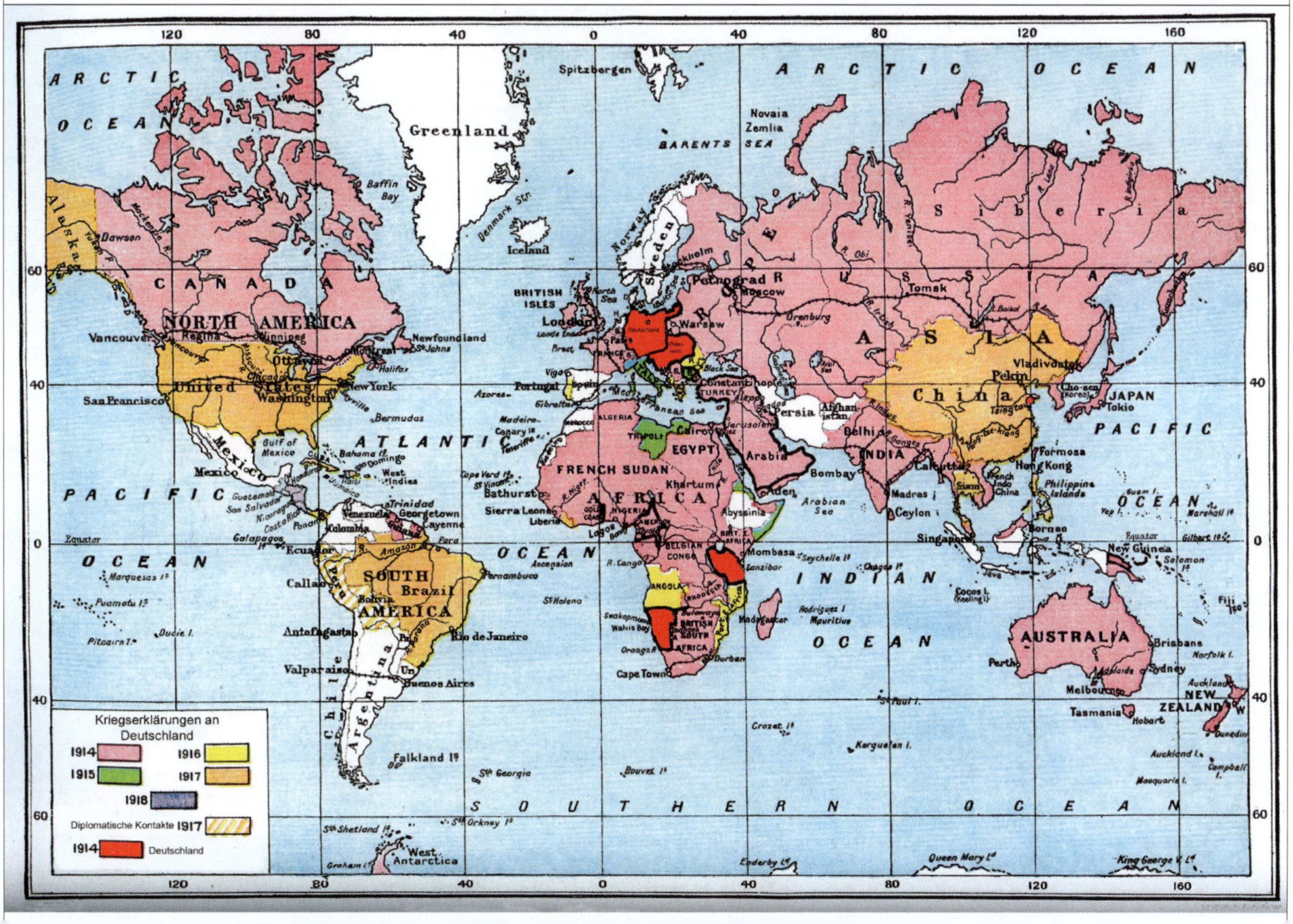

Kriegserklärungen an Deutschland
- 1914
- 1915
- 1916
- 1917
- 1918
- Diplomatische Kontakte 1917
- 1914 Deutschland

Part 1

Introduction from Prof. Dr. Heinz Richter:

"The southeast European theaters of World War I, so the Dardanelle peninsula and Macedonia, are no longer very present in – and in the case of Macedonia, probably completely disappeared from – the consciousness of many Germans and western Europeans. As we see in the *Wissenschaftliche Buchgesellschaft's* ("scientific book association's") commemorative volume marking the 100-year anniversary of the beginning of the First World War, published a short time ago, Gallipoli is worth a mention, but Macedonia and Saloniki do not even appear in the index. Apparently, the Macedonian front is no longer an important topic for historical analysis.

While a single book about Gallipoli has been published by German historiography since the end of the second World War[4], the most recent German publication about the fighting in Macedonia were two volumes from the Reichsarchives, in the middle of the 1920s.[5] More recent English-language literature about the *Saloniki front* is similarly sparse, as is the French. It is amazing that the fighting near Saloniki has been forgotten in Germany, since in the memoirs of von Hindenburg and Ludendorff the collapse of the Bulgarian front in October 1918 becomes the second legend after the "stab-in-the-back" legend, which was supposed to cover as the reason for the defeat in the West. The fact that English authors cite these arguments as their own in order to magnify the meaning of the victory in 1918, is baffling. In the framework of this study, then, we hope to prove von Hindenburg's and Ludendorff's claims, and find out why the British took on these theses.

The operations in Macedonia from 1915-1918 are, to a certain extent, the continuation of the Gallipoli operation. But they are very different from each other. The Gallipoli operation had a strategic goal: The opening of the Dardanelles Straits, in order to open the sea passage to Russia. Had this been successful, Russia would have been able to be supplied with all kinds of war material and munitions. Whether or not the revolution would have happened is doubtful. Conversely, the Allies would have had access to Russian grain. But the Gallipoli operation failed, and in 1916 the Allies left the peninsula. The reasons for this will be explained in the first half of the volume.[6]

In contrast to the Gallipoli operation, the Allied landing in Macedonia had a less lofty goal: They wanted to help the Serbs. Why this was not achieved is one of the aspects that we will examine in this study. The question of why the Allies remained in Saloniki until the end of the war, without having engaged themselves significantly militarily after the failure of this operation, is a central topic we explore in this study.

In contrast to the Dardannelles operation, that took place on the enemy's (Osmanian) area, with their landing in Saloniki in 1915 the Allies occupied a part of neutral Greece. While the first volume of this study focuses on the description of the fighting near Gallipoli and the political decision-making process in London and Paris, another aspect arises for this study, namely, the politics in Greece. How did the king and the government react to the occupation of a part of Greece? As a result of the occupation, in 1916 Greece was split, when Venizelos formed a competing government in Saloniki. In Greek history, this period is called *Ethnikos Diachasmos*, the national split. In Greek historiography, as well in England and France, King Constantine is portrayed as pro-German, with Venizelos as a supporter of the Allies. This study will address this.

There is a great deal of literature from the years of the First World War, and afterwards. But rarely is this objective, since all were influenced by the poison cauldron of the Allies' propaganda. Our picture of the development of Greece in these years is, to this day, more or less influenced by this propaganda.

A famous exception is the book of the English war correspondent George Frederick Abbott, which was published by the Methuen publishing house in 1920.[7] His description is probably the most objective on this topic, since he had no qualms about naming the allies' failures and atrocities. As early as 1916, Abbott wrote a book that dealt critically with the Allied nations' policy toward Turkey and Greece.[8] When he began to write a series of articles for *The National Weekly* about the situation in Greece, the censors stepped in after two installments and forbade further publication. Venizelos' provisional government in Saloniki promptly claimed that Abbott was in the employ of the Greek general staff, which was a smear.[9]

Therefore, it is laudable that it was made available in 1922, by Hamburg's Reprint-Verlag, Tredition Classics of Research. According to the *"Virtual*

Catalogue Karlsruhe" this book is only available in four libraries in Germany. The national catalogues in Great Britain account for just five copies, and in France only one. The astonishingly low number of copies in England and France suggest that someone may have "helped" make them disappear.

But censorship was not only common amongst the European allies. In 1914 the American diplomat Paxton Hibben became a war correspondent.[10] At first, he worked for *Collier's Weekly*, and later for the *Associated Press*. The AP sent him to Greece. Hibben interviewed all of the protagonists and watched the events critically. After King Constantine was deposed, he returned to the USA in 1917, and wanted to publish a book in which he would expose the Allies' intrigues, But, *"in certain quarters it was felt that its publication at that precise moment would embarrass our associates in the war. This intimation was conveyed to the publisher."* Therefore, the book could only be published in 1920.[11] In the introduction Hibben wrote, *"I felt that I should write down, in black and white, what I saw, and what I knew" […] in my thoughts every phase of the Greek tragedy in early 1917 was clear and present, and what I wrote was the truth."*[12]

Hibben's eyewitness account complements Abbott's description and includes many unknown details. Together, they show a very new view of the events in Greece in these years. Both authors complain about censorship, and describe how the Allies' propaganda machine worked. They also describe how the Allies' war correspondents allowed themselves to be used by this propaganda apparatus, and told lies in their newspapers, and how they had been supplied with Venizelist propaganda, and false and phony news. This study will also point out how the historical accounts were influenced through these reports, and to correct them. Further, it is necessary to reexamine the official documents, for example the Greek white books, for their authenticity. Is the evidence real or are some documents forgeries? The goal of this study is, therefore, a thorough revision of the Saloniki front operation, which was characterized by war propaganda, and of the internal developments in Greece during the time of the First World War.

Notes

(4) Klaus Wolf, *"Gallipoli 1915. Das deutsch-türkische Militärbündnis im Ersten Weltkrieg"*, (Sulzbach: Report Verlag, 2008)

(5) Reichsarchive (ed.) *"Herbstschlacht in Macedonien Cernabogen 1916"* (Oldenburg: Stalling, 1925). Reichsarchive (ed.) *"Weltkriegsende an der Mazedonischen Front"* (Oldenburg: Stalling, 1925).

(6) Heinz A. Richter, *"Der Krieg im Südosten"*, Volume 1, Gallipoli 1915, (Ruhpolding: Rutzen, 2014).

(7) George Frederick Abbott, *Greece and the Allies* (Hamburg: Tredition, 2008), reprint of the version from 1920

(8) George Frederick Abbott, *Turkey, Greece and the Great Powers: A Study in Friendship and Hate* (London: Robert Scott, 1916) The study can be downloaded here: http://en.youscribe.com/catalogue/books/literature/others/greece-and-the-allies-1914-1922-33170

(9) George B. Leon *Greece and the Great Powers 1914-1917*, (Saloniki: IMXA, 19745) p. 467.

(10) A short and impressive biography of Hibben's can be found in John Dos Passos *Neunzehnhundertneunzehn* (Stuttgart), Europäischer Buchklub, year unknown, p. 209'215.

(11) Paxton Hibben, *Constantine I and the Greek People* (New York: Holt, Century Company, 1920; reprint edition Memphis: General Books, 2012) p.4.

(12) Ibid, p. 2.

Above: The Austrian artist Theo Matejko (March 18th 1893–September 9th 1948) presented this sketch in his book *"Erlebnis Front"* for the first time.

Above: During his training, staff sergeant Johann Keller sits as a pilot in the front of an AEG C.I C.88/15 on a training flight. This aircraft was earlier in an aviation unit of the Army unit Gaede, see the stripes on the hull. Reconniassance was the most valuable function of WWI planes and was reflected in the fact that more than half of German warplanes were two-seat reconnaissance aircraft.

The Political and Military Developments Before the Beginning of the Fighting in Saloniki 1915

The Balkan Wars 1912–1913

In the first Balkan war in 1912, the Balkan states had come together for a few months and successfully fought against the Ottoman Empire. In the peace treaty of London on May 20th, 1913, Turkey renounced all areas west of the Enos-Midia Line from the Black Sea to the Aegean Sea and Crete. Macedonia was divided between Serbia, Greece and Bulgaria. Serbia received the inland with Skopje, Ohrid, Prilep and Bitola, Greece got the coastal region with Thessaloniki and Bulgaria received a smaller portion. Thrace went to Bulgaria. The major powers were to decide the fate of Albania and the Aegean islands.[13]

No sooner was the treaty signed, the victorious parties began squabbling about the distribution of the spoils. When the Serbo-Bulgarian alliance was forged, Serbia was allotted a part of western Macedonia, but this was now part of a newly-formed Albania. Therefore, the Serbs demanded a larger part of Macedonian territory, a demand that the Bulgarians of course rejected. Because Bulgaria feared a military confrontation with Serbia, it sent troops into the disputed areas from East Thrace. Due to the fact that the Greek-Bulgarian military alliance had no say in the distribution of the Macedonian areas, in Athens it was feared that the Bulgarians would challenge Greece for Saloniki. On June 1st 1913, the Greeks finalized a new military alliance with Serbia, that was clearly in opposition to Bulgaria. In the meantime, Romania also wanted a say in the matter, which demanded the separation of Dobruja, but wanted to stay out of the impending conflict. Montenegro hinted that in the case of war, it would enter on the side of Serbia. The Serbs contacted Turkey, which did not take sides, but made it understood that it would be prepared to intervene if necessary. In other words, a new war was imminent.

The only nations that could have hindered this were the European powers. But these too had different positions. Russia had demanded the formation of the Balkan alliance, in order to stop the further expansion of the Austro-Hungarian empire to the southeast, but condemned a Bulgarian expansion to the Aegean Sea and the straits. In the end, it also had ambitions in this region. They reminded Belgrade and Sofia that in the event of a conflict, the Serbo-Bulgarian contract was to fall under Russian arbitration. Austria-Hungary was naturally against the expansion of the southern Balkan states, because it also wanted access to the Aegean near Saloniki. Vienna knew that Serbia was dreaming of annexing Austrian Bosnia. Actually, it wanted to show stronger force. But since German support was missing, Vienna decided to force the Serbs to only clear out of the Albanian area.

Officially, Great Britain was allied with Russia, but did not want her on the Straits and in the Aegean Sea. Therefore, the British secretly supported Bulgarian ambitions in Thrace. London saw a similar role with the Greeks. The French, who were afraid of a war with Germany if Russia clashed with Austria in the Balkans, wanted to avoid a new Balkan war but did not make this clear to London. On the other hand, Germany had nothing against a pro-German Large Bulgaria.

It was only a question of time until a new war would break out. Politicians on both sides raised excessive demands on the Macedonian area. At the beginning of June, the Russian czar attempted to mediate, but to no avail. Bulgarian public opinion was against this. On June 28th, 1913, Czar Ferdinand ordered an attack on the Serbian and Greek positions in Macedonia. When the attack began on June 29th, without a declaration of war, Athens and Belgrade, joined by Montenegro, answered with that declaration. On July 10th, Romania declared war on Bulgaria, and the Turkish declaration of war followed on July 12th.

So Bulgaria - similar to the Ottoman Empire in the first Balkan war - was attacked from all sides. Romanian troops marched, almost without any notable opposition, to the suburbs of Sofia. The Ottoman units occupied undefended Adrianopel (Edirne). The main fighting took place in West-Macedonia, between the Greek and Serbian troops on the one side, and the Bulgarian troops on the other. A cease-fire was declared on July 31st.

In the peace treaty of Bucharest, signed on August 10th, Bulgaria lost almost all of the territory it had won in the first Balkan war. Greece received the so-called Aegeais-Macedonia, with the cities of Saloniki and Kavalla, so the entire coastline to the mouth of the Evros. Serbia won the so-called Vardar-Macedonia, with the capital city Skopje. East Thakria, with Adrianopel (Edirne) fell back to the Ottoman Empire. Romania received a large portion

of the Dobruja. Montenegro was able to slightly expand its territory, so that it bordered Serbia.

The first Balkan war had cost the Ottoman Empire huge swaths of land. Militarily, the second Balkan war brought Bulgaria into the same situation as Turkey during the first: attacks from all sides through the superiority of a coalition that surrounded it. Both states found the results deeply hurtful, and craved revenge. They could only hope for that with an alliance with Germany and Austria-Hungary. The two states therefore became potential allies of the Central powers. Therefore, the scene was set for the outbreak of the First World War, just a year later.

The only country whose status and borders were set neither at the Conference of London, which had ended the first Balkan War, or the Conference of Bucharest, which ended the Second, was Albania. These decisions were to be made at the Ambassador Conference in London. In December 1912, the powers had decided that Albania should remain an independent state. A commission was to fix the borders in the north to Montenegro and Serbia, and in the south to Greece. But since the Serbs did not agree to the border lines that the Commission had drawn, armed skirmishes followed. The Serbs occupied areas that the Commission had already granted to Albania. Only when Austria-Hungary exercised massive pressure did the Serbs leave these areas again in October 1913. The Commission once again went to work, but it was unable to finalize the borders before the beginning of the First World War.

The peace conference had made certain decisions regarding the border to Greece in that it granted Albania the cities of Koritsa (Korçë) and Argyrokastro (Gjirokastër) and their surroundings, although the majority of the population that lived there was Greek. The Commission could only lay down the exact border. The question was, would the Greeks vacate this area that they had captured and occupied. After long diplomatic negotiations, a solution was found: The Greeks would evacuate the area they had called North Piraeus, but therefore they would keep the Aegean islands they had occupied, even though the status of these islands had not been clear. Only three islands in the immediate proximity of the mouth of the Dardanelles would remain in Ottoman power. But neither the Greeks nor the Ottomans accepted this solution.

On February 28th, 1914, the Greeks living in *Nordepirus* (South Albania), under the command of the chief commander of the local army, Georgios Christakis-Zografos, declared themselves independent. Zografos created a government. Of course, the Albanians opposed this. The Ottomans refused to agree to give up the islands. An armed conflict between Greek and Turkey was inevitable, but then the murder in Sarajevo happened. The skirmishes were displaced by diplomatic maneuvers from Germany and Austria-Hungary, and both powers forced King Constantin, who was married to a sister of the Emperor, to reconcile with the Ottoman Empire. Even though there was a pro-German Balkan alliance on the horizon, the dispute about the islands showed to be a massive roadblock.

Bulgaria turned more and more to Germany, and Romania moved closer to the Entente. Serbia maintained good relations to Montenegro and Romania and acted expressly hostile to Austria-Hungary. The peace of Bucharest had in no way created peace in the Balkans, but instead encouraged the split into two camps. The clear contours of the constellation of the First World War loomed.

Above: Oscar Potiorek (20.11.1853 – 17.12.1933)

The Politics of the Balkan States on the Eve of the First World War

At the time, Romania had 7½ million citizens. A further 1 million Romanians lived in neighboring Russian Bessarabia. Three million lived in Hungarian Siebenbürgen and in the Austrian Bukovina. Especially the *"Irredenta"* in the k.u.k monarchy burdened bilateral relations, even though they were allied since Bismarck had initiated the four-way alliance in 1883. *Magyarisation-measures* in Hungary increased the antipathy, and the anti-Russian atmosphere due to the loss of Bessarabia in 1878, declined. When Austria refused to support the Romanian demands on Bulgaria in 1913, relations turned cold. The Romanian king was of German descent, and sympathized with the Central powers, but his population was anti-Hungary. Because the Romanian army had a good relationship with Germany, Berlin believed that Romania would not align itself with the Entente. Vienna, however, no longer believed in Romanian armed assistance. In the second Balkan war Romania had won Bulgarian territory on the Dobruja-Border, which of course, brought antipathy from Bulgaria.[14] Altogether, Romania had expanded its territory by five percent and its population by four percent during the Balkan wars.[15]

At the time Serbia, had a population of 4½ million. In the Balkan wars it had won the third-largest territorial and population growth, 82 percent and 55 percent respectively.[16] This and being the Russian-designated outpost of Slavic culture strengthened its self-confidence. Serbia had enforced its territorial claims in the south and the east via the Serbs who lived there, but in the north and the west, another 5½ million *"undelivered"* Serbs lived in the area of the Danube monarchy, primarily in Bosnia-Herzegovina. The Russian representative in Belgrade spoke of the *"promised land"* in Austro-Hungarian territory. A result of these ambitions was the murder of the Austrian heir to the throne.[17] Serbia's only ally was the small Montenegro, that had a population of about ½ million. It had increased its territory by about 62% and its population by 100%

during the Balkan wars.[18] It was clear that Serbia would join Montenegro on the side of the Entente in the forthcoming war.

Greece was one of the big winners of the Balkan wars. Its territory grew by 68% and its population by 67%.[19] The country had had 4½ million people at the time. To secure the spoils from the Balkan wars, Athens maintained an alliance with Serbia. The country was a British *client state* since 1862, and therefore a part of the British empire's *lifeline* through the Mediterranean to India. Its expansive, difficult-to defend coastlines made it seem wise to avoid any conflict with any other power in the Mediterranean.

Greece's foreign policy goal was to realize the *Megali Idea*, the reestablishment of the Byzantine empire. This brought it inevitably into conflict with Turkey, because large parts of Asia Minor and East Thrace were populated by Greeks. In addition, Athens and Constantinople were at loggerheads with each other about the confiscation of the Aegean islands. If Greece wanted to survive the coming conflict unscathed, a neutral stance was the *Conditio sine qua non*. ("a necessary condition")

Bulgaria had a population of 4½ million. In the Balkan wars its territory grew by 29% and its population expanded by 3%.[20] It lost a part of the Dobruja to Romania. In the south it had to give up Thrace and parts of Macedonia to Greece, and further parts of Macedonia to Serbia, so basically the entire loot from the first Balkan war. The old feud with Serbia was revived, and the conflict with Greece and Romania were new. Since Russia did not support Bulgaria, the forces which were in favor of an alliance with the central powers rose in power. On July 24th, so a day after Austria gave an ultimatum to Serbia, Bulgaria turned to Germany and asked to join the three-way alliance. Berlin stalled.[21]

Only when Russia's mobilization was clear and the war seemed unavoidable did Berlin give up its caution. Generalstabschef Helmut von Moltke tried to assuage the conflict between Romania and Bulgaria and bring them to the side of the Central powers. The two states were, together with Turkey, to attack Russia through Bessarabia. The German-Turkish military alliance was created on August 2nd 1914, but kept secret. Romania declared its neutrality on August 3rd, a day after Italy. At the negotiations with Romania and Bulgaria, it was clear that their differences were insurmountable. If it came to an alliance, Romania was not even prepared to allow Bulgarian troops to pass through its territory. On the other hand, Bulgaria remained restrained, because it did not trust Greece. Turkey declared its armed neutrality on August 3rd, and to secure this, mobilized troops. Altogether the states were in a waiting game. Bulgaria observed the developments in Austria-Hungary and Serbia closely. A fast, effective military attack against Serbia was expected.[22]

NOTES

(14) Reichsarchiv, *Der Weltkrieg 1914 bis 1918*, Volume IX, *Die Operationen des Jahres 1915* (Berlin: Mittler, 1933). P. 134f.
(15) Stavrianos, op.cit., p. 540.
(16) Ibid.
(17) Reichsarchive, *Weltkrieg*, IX, p. 135.
(18) Stavrianos, op.cit., p. 540.
(19) Ibid.
(20) Ibid.
(21) Reichsarchiv, *Weltkrieg*, IX, p. 137.
(22) Ibid, p. 137f.

From Left to Right: Three stages in the destruction of an observation balloon; s1. The fighter attacks the balloon which starts to burn as indicated by the smokes. 2. The observer takes to his parachute. 3. The burning balloon trails a cloud of smoke as it falls to earth. Balloons were valuable for reconnaissance and artillery spotting and were protected by anti-aircraft guns and often by a fighter patrol.

The Attack on Serbia: Start of the First World War in the Southeast

The First World War began on the Balkan peninsula on July 28th, 1914, when the Austro-Hungarian Emapire declared war on Serbia. Greece, Bulgaria, Romania, and Turkey declared their neutrality. The Austro-Hungarian general staff had drawn up plans for both war scenarios. Plan "B" was for an isolated war in the Balkans. Plan "R" assumed a two-front war against Russia and Serbia. Plan "B" looked more like a punitive expedition, for which only partial mobilization was required.[23] Plan "R" was, of course, a general mobilization, but most of the Austro-Hungarian forces would march against Russia.[24]

In order to better coordinate the offensive, the Army was divided into three squadrons. Squadron "A" was to only engage Russia in Galicia. The third squadron, so-called minimal group with 11 divisions, was the weakest and would only go to Serbia. The "B" Squadron would support the one or the other, depending on the situation, which required a high level of mobility. Chief of Staff Franz Conrad von Hötzendorff had recognized this already in 1908, and had ordered the railway office of the general staff to plan accordingly. The "B" squadron was to be able to react flexibly, and swing from one to the other theater as needed, even while in their forward march. But the railway office did not follow Conrad von Hötzendorff's orders. We can only speculate about the reasons for that.[25]

When on July 25th 1914 the partial mobilization against Serbia was ordered, the minimal group and parts of the "B" squadron, together 26½ infantry and 3 cavalry divisions, as deemed in Plan "B", began to roll towards the Balkans. The Serbian army was estimated at 12 divisions. The Austrian 5th and 6th army were to march into Bosnia, in the west of Serbia. The 5th Army consisted of the VIIII and the XII Corps, with 4½ infantry divisions, the 6th Army

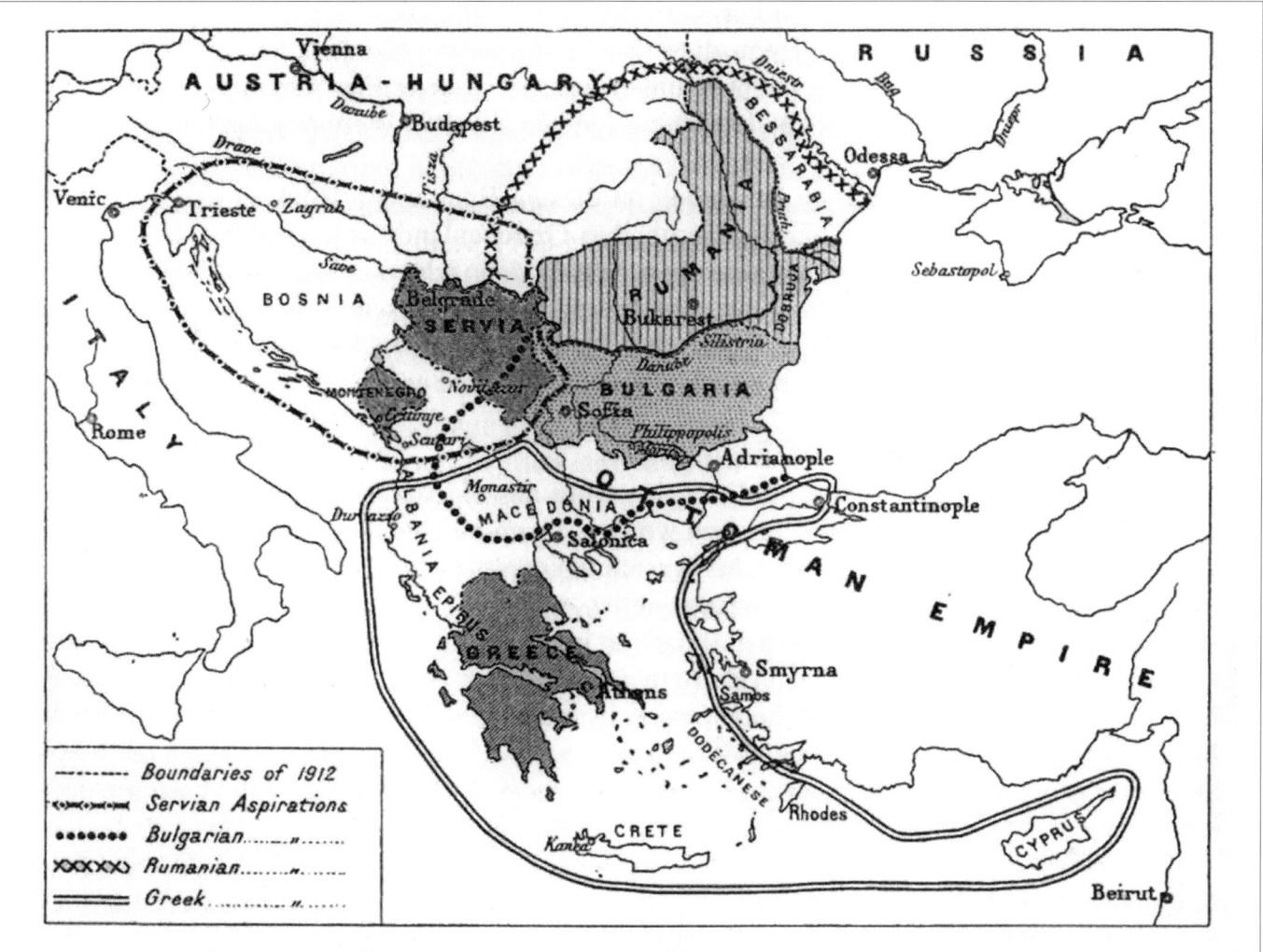

Above: The Ottoman empire and the territorial ambitions of the Balkan nations around 1900.

consisted of the XV and XVI Corps, with 6 infantry divisions. The 2nd Army, which belonged to the "B" squadron, was to stand at Syrmia, on the northern border to Serbia. It consisted of the IV and IX Corps with 6½ infantry and one cavalry division.[26] Since the railway network in Serbia was narrow-gauge, it was clear that the march would take a great deal of time, and the attack could only begin in the middle of August at the earliest. Vienna hoped that after the first successes Bulgaria would attack Serbia from the southeast.[27]

But when Russia ordered mobilization on July 30th, Austria followed a day later. Therefore, plan "R" kicked in. The plan stipulated that more than half of the currently planned troops in Serbia would shift to Russia. Only 11 divisions were to remain on the Balkan. The Austrian army leadership still hoped to inflict a significant hit on Serbia before this weakness occurred and would become noticeable.[28]

The plan that was developed for the attack on Serbia was not very realistic. It assumed that the Serbs would come forward to Sarajevo, to capture the Austrian province Bosnia-Herzegovina, where many Serbs lived. In addition, the Serbs would march forward in the flatlands between Drina and the Sava. These forces were to be surrounded and broken up in a sweeping attack. Even though this plan was criticized because it required a frontal attack over difficult terrain with rivers and steep, forested mountain ranges without roads, it remained operational.[29]

Berlin was appalled at Conrad von Hötzendorff's plans to attack Serbia, despite the looming conflict with Russia. Chief of Staff Moltke let Conrad von Hötzendorff know that the attack on Serbia was a purely Austrian affair, and that by no means should it lead to a weakening of the forces that were to march on Russia.[30] Emperor Wilhelm II went one step further, and at 16.40 sent a cable to Emperor Franz Josef: It is "*of great importance that Austria*

Above: The Viererbund (four allied countries of the Central Powers) on a color postcard.

deploy its primary forces in Russia, and not split them in a concurrent offensive against Serbia [...] In this massive fight, in which we are entering shoulder-to-shoulder, Serbia plays a very marginal role, that only requires minimum defensive measures."[31]

When Conrad von Hötzendorff asked the railway office if it could divert parts of the "B" squadron, which was on its way to the Balkans, to Galicia, the head of the office declined, saying it would lead to *"chaos"*. Only when the deployment of the "A" squadron was complete could they transport parts of the "B" squadron to Galicia. Conrad von Hötzendorff accepted this and so these units remained in Syrmia for the moment. Of course, they were missing during the fight against Russia in Galicia. Conrad von Hötzendorff hoped to defeat Serbia in about two months. In other words: Conrad von Hötzendorff hoped that in the short time in which the parts of the "B" squadron was in Syrmia, they could defeat Serbia. In principle, the timing of the evacuation was set, namely, as soon as the transportation of the "A" squadron was complete. Once these troops were pulled back, the minimal group of 11 divisions would be outnumbered by the Serbs. The only sensible solution would have been to follow German advice and quit the attack on Serbia. But there were also questions of prestige: *"A hammering by small Serbia [...] was surely more fatal than a failure against the very strong czarian empire; it would undermine the reputation of Austria-Hungary in the eyes of the whole world, but especially in the Balkans, where it was so necessary to show its population a success"*.[32] The attack on Serbia, as Conrad von Hötzendorff himself admitted, was a high-risk gamble.[33]

At that time, the Serbian army consisted of between 250,000 and 300,000 troops. It had modern equipment. The infantry was outfitted with 7 mm Mauser rifles. The field artillery had modern heavy guns with recoiling barrels. Every division had 36 cannons and 16 machine guns. More important, however, was the quality of the Serbian soldiers. Similar to the Turkish army, the Serbian army had great courage to do battle. The officers had gained a lot of experience during the Balkan wars.[34] Vienna underestimated the Serbian army like the British and French later did with the Turks: They were assumed to be a kind of aboriginal rag-tag troops.[35] In addition, the Serbs knew the territory well. Serbian army leadership had only developed plans for a defensive war. They expected an Austrian attack

Above: A cigarette company from Dresden marketed its tobacco with the name "*MACEDONIA*".

on Belgrade, so on the territory between the Kolubara river in the west and the Morava in the east. The northwest, where the Austrians wanted to attack, was only weakly protected. At first glance it seemed like the Austrian offensive plans had correctly identified the weak points of Serbian defensive planning. But as soon as they crossed the plain between Drina and Sava they would encounter the concentrated resistance of the Serbs, in the mountainous back country. But back to the chronological developments.

The supreme commander of the Austrian troops in the Balkan was Feldzugmeister Oskar Potiorek. Feldzugmeister is the second-highest rank of General officer, the step between Feldmarschallleutnant and Feldmarschall, for the Austrian Generals from the infantry and artillery. In 1911, the emperor named him to be the governor of Bosnia-Herzegovina. Therefore, Potiorek had a dual role: he was the civilian administrative head of the province, as well as the military commander of the troops that were stationed there. Since there was always some sort of pro-Serbian unrest in Bosnia-Herzegovina, Potiorek tended to answer this with violence. In his opinion, a military action against Serbia would end these skirmishes.[36]

Like the military leadership in Vienna, Potiorek expected a Serbian attack on Sarajevo, which he would then cut off. On August 4th, he received his first orders: he was to fend off Serbian intrusions into the region of the monarchy. The second army (part of the "B" squadron) was to support him, but on August 18th they were to begin demobilizing out of the region. When the Serbian attack failed to

Above: Eleftherios Venizelos (August 23rd, 1864–March 18th 1936). Venizelos was the Greek prime minister and one of the most dazzling personalities of his age. He led his country into the First World War and brought it significant territorial gains. In order to expand these further, he launched a war against Turkey from 1919 to 1922, which ended in a great defeat for the Greeks. After internal political conflict, he twice fled to exile in France, where he died in 1936.

materialize, Potiorek developed his own offensive plans: The Fifth army, three days earlier than previously planned, would cross the Drina in the north, and march toward Valjevo in the south. The Sixth army would advance to Užice as soon as it had completely arrived in the area, so on or about August 15th. Together, the two armies would crush the Serbian troops that were assumed to be in the area. Potiorek hoped that the presence of the Second Army, still stationed in Syrmia but unable to participate in this attack because it was getting set to leave the area, would scare the Serbs. On August

Right: A painting of an old bridge across the Istb.

CXCI. — ANCIEN PONT DE PIERRE A ISTIB.   Grav. et imp. par Gillot

12th, Potiorek informed Conrad von Hötzendorff and he agreed to the plan on August 14th, taking overall responsibility for the entire operation, probably to get a handle on Potiorek.[37]

The offensive of the Fifth army began on August 12th, under difficult conditions. The bridge-building equipment had not yet arrived, and the Serbian defenses proved rather efficient. They were perfectly camouflaged in the difficult terrain, and the artillery could not fight them, nor could the reconnaissance aircraft discover them. The Second Army launched a deceptive (fake) attack that was carried out so poorly that the Serbs quickly realized that they had nothing to fear from that side. The Serbian Woiwode (Field Marshall) R. Putnik ordered his troops, who were still near Belgrade, to the west. With these, he was far superior to the two divisions of the Fifth Austrian Army. When the Serbians attacked in full force in on the following days, the two Austrian divisions could not withstand the force. On August 16th one division capitulated. Before the Sixth Army could react, both divisions had to retreat to their positions of August 12th. The Fifth Army had lost 23,000 men, and the Serbs lost 16,000. The Sixth Army attacked Užice on August 20th, but Potiorek stopped them, and sent them to support the Fifth Army. But it arrived too late to turn the tide, and it, too, had to retreat. On August 25th not a single Austrian soldier remained on Serbian land.[38]

Numerous factors were responsible for the defeat. The most important were the leadership errors of Potiorek and the inexperience of an army which had not fought a war in quite some time. In particular, the soldiers of the Fifth army had to learn their craft on the battlefield. The artillery was still outfitted with Lafettenrücklaufgeschützen. Potiorek had stopped the modernization of the troops. At first he did not report to the Armeeoberkommando (AOK) (Army controlling command) in Vienna, so he could act without supervision. When Conrad von Hötzendorff took control of Potiorek on August 14th, it was too late and the first catastrophe was already happening. Potiorek should have been relieved of his duty, but instead he gained yet more power. On August 21st Potiorek regained his independence from AOK with the help of the Emperor's military chancellery. From now on he could do what he wanted in the Balkans.[39] For Conrad von Hötzendorff this was a *"crushing blow"*.[40]

The failure of this first Austrian offensive was fatal. Militarily, the Serbs had proven themselves equals, if not the stronger force. The political impression on the possible allies of the Central Powers was terrible, both Bulgaria as well as Turkey kept their distance. Sensibly, they turned to defense. But the political leadership in Vienna did not agree, and gave Potiorek more troops than would have been necessary for a defensive position, but too few to win. The AOK[41] could no longer influence the developments in the Balkan states, in order to pull back unneeded troops from there to send them to Galicia, where they would have been desperately

SALONIQUE. LA TOUR BLANCHE.

Above: The white tower in Saloniki.

needed. Potiorek had 13½ divisions, most of them newly created. Strengthened in this way, Potiorek planned a new offensive.

But the Serbs, too, were active. After the Second Army retreated from Syrmia, Russian military leaders persuaded the Serbs to attack Austria. On September 6th, Serbian troops crossed the Save near Mitrovica and Kupinovo. But near Mitrovica they encountered the 29th Austrian infantry division, which forced them into retreat within two days, under heavy losses. Further to the east, Near Kupinovo, the Serbs were able to get a foothold, but the Seventh Austrian division prevented further advances.

The Serbian attack forced Potiorek to go on the attack himself. He sent the Fifth Army over the Save, which brought significant losses and also the following attack had little success. If the Serbs eventually did retreat from Syrmia, that could be chalked up to a success of the Sixth Army. It approached from the Southwest, and encountered weak Serb forces. Serbian leadership recognized that this toehold could become dangerous, so it sent all of the troops retreating from Syrmia to the area west of Valjevo to Krupanj. It was there that the decisive battle would be fought. The retreat of the Serbian troops made it possible for the Austrian 29th division to cross the Save. In the meantime, attacks of the Serbian forces at Užice began, but they were fended off. Nevertheless, the battles elicited high losses for the Fifth and Sixth Armies, but the Serbs too suffered losses. A ceasefire came at the end of September, probably due to all-around exhaustion. Since the Serbs had embedded themselves near the lines of the Fifth Army, a trench war began.[42]

By the numbers, the Austrian Fifth Army was just as strong as its adversary, the Second Serb Army, but there were great qualitative differences. The Serbian troops were experienced warfighters, the Austrians were badly-trained and -equipped land posses. In addition, there were not enough officers, and the artillery suffered from a shortage of munitions. The result was that the attack of the Fifth Army stalled again on October 5th. The Sixth Army was under

Right: Colorized photograph of workers of Saloniki.

CLXXXVI. — Portefaix a Salonique.

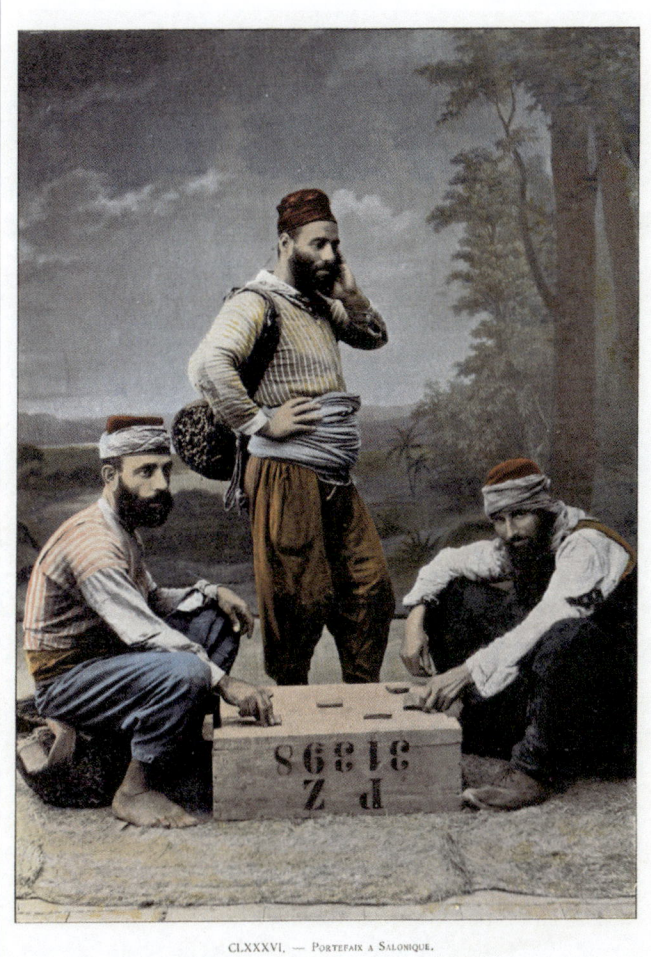

Below: Soldiers of the FFA1 at a "shindig". On the shoulder of the solder, who holds up the sign "ein prosit", one can see the number of the FFA 1. Since there is no activity insignia, we assume these are mechanics of the FFA1. (Feld-Flieger-Abteilung 1).

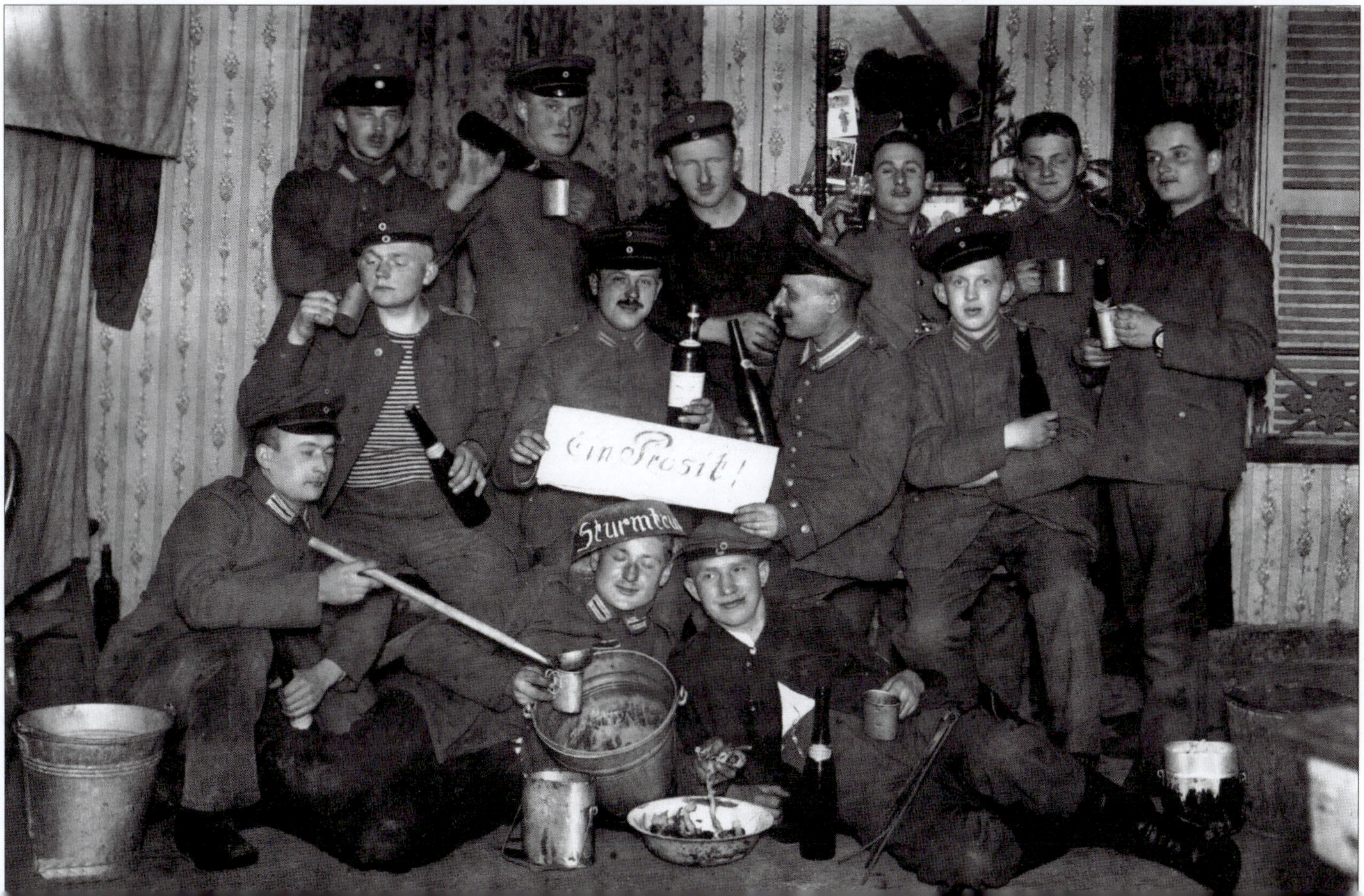

Left: Four soldiers, members of the alternative pilot unit 1, have themselves photographed in a studio clearly showing the insignia on their sleeves. This photo was probably taken as a souvenir prior to the departure to the Macedonian theater.

pressure from attacking Serbian forces. The Serbs who had entered Bosnia could be pushed back to behind the Drina by October 30th. All of this did not stop Potiorek from preparing a new offensive for the Fifth and Sixth Armies.[43]

On October 24th, the Fifth Army once again attacked the positions of the Second Serb Army. It was able to push the Serbs into retreat but could

Above: Hangar of FFA 1. Of the four aircraft that are seen, from the right, is Albatros C.V 1218/16, then an Albatros C VII without visible military number, on which the mechanics are working. The radiator on the leading edge of the upper wings was a field modification to replace the as-built "ear radiators". Behind that, just able to be recognized, the Albatros C.V 3327/16, and to the extreme left another two-seater aircraft.

Below: 21 men of the ground crew stand in front of a Rumpler C. IV aircraft.

Above: All of the mechanics of FFA1 are photographed in front of Rumpler C.IV 845?/16.

Below: Five mechanics prepare a training flight of an Albatros C.XII. There are no bombs loaded and the aircraft is chocked. Pilot and observer are already in their seats. The Rumpler C.IV and Albatros C.XII both used the 260 hp Mercedes D.IVa engine but the Rumpler offered much better climb and ceiling so was greatly preferred for long-range reconnaissance missions.

Above: All of the mechanics of FFA 1 are photographed in front of the Rumpler C IV 845?/16. Here in work clothes, therefore there is no insignia on the sleeves. It is also interesting here that the "hangar door" is actually a "hangar wall" that is folded down to open, and offers the aircraft a solid ground for the first three meters.

Below: In this group photo the officers and the soldiers of FFA 1. On three uniforms you can clearly see the insignia of the FFA 1 on the left sleeve, in addition there is a man with a sleeve insignia "63" on the right sleeve of the uniform coat. Presumably this is a group photo before leaving to the Balkans. Date and location are unknown.

Left: This photograph is a Greek "Venizelos fighter", a rebel against the Greek government.

not completely defeat it. The Sixth Army also successfully attacked at the beginning of November. It liberated Krupanj, and moved forward to Valjevo, which it then occupied. But as it attempted to cross the Kolubara river, it ran into trouble with its supplies. The troops were badly equipped, lacked munitions and artillery. In addition, the river ran through an area that flooded after the autumn rains. In the north, the Fifth Army had crossed the Kolubara, and reached the region south of Belgrade. But there too there were problems with the replenishment of supplies. Potiorek had the impression that the Serbs were on the retreat. But when winter came, with its massive snowfalls,

Above: Of all of the Allied pilots, the English ROYAL FLYING CORPS (RFC) was the strongest enemy. Here an image on a postcard of the time. The aircraft depicted is an Henri-Farman HF-20. Below are color images of RFC insignia.

the Sixth Army – unprepared and unequipped for winter weather – was no longer operational. But Potiorek was of the opinion that only a little effort was needed to crush the Serbian army. He gave the Sixth Army a few days off, and ordered the rail line to Valjevo be repaired and taken into service. He ordered the Fifth Army to take Belgrade. But the Serbs evacuated the city before this attack could begin. This convinced Potiorek that the Serbian Army was near collapse. The Fifth Army built a defensive position on the northeastern side of Belgrade. Thereby, there was a gap between the two armies.[44]

As the first supplies reached the army by rail, but had not yet been given to the troops, the Serb army attacked. They had been pushed back into the interior by the Austrian advance, but at the same time they had the advantage of the inner line, because they were closer to their supply depots. Their losses had been equalized through new recruits, and their artillery reserves were full. In addition, the weather improved. The attack was aimed at the weakened troops of the Sixth Army, which retreated. A relief attack by the Fifth Army in the north was not successful. The retreat of the completely exhausted Austrian troops looked like a flight. Valjavo was lost. Also at the Kolubara river there was no holding the line. Potiorek hoped to be able to hold Belgrade and Sabac as bridgeheads, but this too was impossible. The majority of the

Above: An Albatros C.I of FFA is being prepared for flight. On the fuselage side is the Hazet-style cooling radiator for the water-cooled Mercedes inline six engine. Radiators were mounted along both sides of the fuselage.

Austrian units were on the other side of the Save by December 15th, 1914. Belgrade was evacuated. They stood at the same point from which they had started the Autumn offensive.(45)

All three Austrian offensives had failed. It was impossible to defeat the Serbs. Altogether 450,000 men participated in the fighting in the theater, as well as 12,000 officers (including all temporary soldiers and garrison troops). There were 28,276 dead, 2,046 prisoners, 74,644 missing, 122,122 wounded, and 46,716 ill. So altogether there were losses of 275,000, including almost 7,600 officers. The losses were therefore about 60 percent. If we exclude the number of soldiers that were not actively participating in the fighting, we arrive at an even higher loss percentage.(46) The Serbs also dealt with heavy losses, of 130,000 men. As a result, they were weakened and not able to continue their attacks. Therefore, for the following months, a ceasefire held in this theater.

From a military perspective, this campaign was a disaster for Austria-Hungary, and politically the results of the Austrian defeat were catastrophic. The neutral Balkan states, which had sympathized with the central powers, held on to their neutrality. The land bridge that Turkey and Germany had wished for did not come to being. But the psychological effects were far worse: Tiny Serbia had beaten the great Austro-Hungarian Empire - the Shvaba (Swabians), as the Serbs disdainfully called the Austrians and the Germans.(47)

At first glance, this seems to make sense. But if we look a little closer, we discover that this campaign was not led by the Army command in Vienna, but rather by the local commander, Potiorek. He understood how to get out from under Conrad von Hötzendorff's control, with political support. He had, in a way, led a private war. It was his errors in judgement and weaknesses that had led to the catastrophe. When the political leadership blamed

Above: Here is FF Oblt. von Chappius of FFA 1 with his mechanics and an Albatros C.V/16, , the earlier version of the Albatros C.V. It has the two-side radiators and water header tank above the inline-8 cylinder Mercedes D.IV. of 220 hp.

him for the disaster in December 1914, he resigned. The new commander of the reduced Balkan armed forces was Archduke Eugen, and the leftovers of both armies were unified in a new Fifth Army, and placed under the Army's command.[48]

NOTES

(23) Reichsarchiv (ed.), *Der Weltkrieg 1914 bis 1918*, Band II, *Die Befreiung Ostpreußens* (Berlin: Mittler, 1925), p. 25. According to this, in the case «B», 2/5 of the army should be mobilized.

(24) 4/5 of the army was to mark in Galicia. Reichsarchiv, *Weltkrieg, II*, p. 25; Christian Ortner, "Die Feldzüge gegen Serbien in den Jahren 1914 und 1915", in: Jürgen Angelow (ed.), *Der Erste Weltkrieg auf dem Balkan* (Berlin: be.bra, 2011), p. 123f.

(25) Keegan claims, without documentation, that the railway office had created these plans. John Keegan, *The First World War* (New York: Vintage Books, 2000), p. 152.

(26) Österreichisches Bundesministerium für Heerwesen (ed.), Österreich-Ungarns letzter Krieg Band 1 *Das Kriegsjahr* 1914 (Wien: Militärwissenschaftliche Mitteilungen, 1930), p. 91.

(27) Reichsarchiv, *Weltkrieg*, IX, p. 139.

(28) *Ibidem*, p. 139f.

(29) Ortner, *op. cit.*, p. 125.

(30) Reichsarchiv, *Weltkrieg*, IX, p. 140.

(31) Reichsarchiv, *Weltkrieg*, II, p. 29. It is to be noted that this telegram is not mentioned in edition IX in 1933.

(32) Reichsarchiv, *Weltkrieg*, IX, p. 140.

(33) Reichsarchiv, *Weltkrieg*, II, p. 26f.

(34) Ortner, *op. cit.*, p. 125f.

(35) Keegan, *op. cit.*, p. 152.

(36) Ortner, *op. cit.*, p. 127.

(37) Reichsarchiv, *Weltkrieg*, IX, p. 140; Ortner, *op. cit.*, p. 127.

(38) Reichsarchiv, *Weltkrieg*, IX, p. 140; Ortner, *op. cit.*, p. 128.

(39) Ortner, *op. cit.*, p. 129f.

(40) Österreich-Ungarns letzter Krieg, I, p. 147.

(41) AOK: Armeeoberkommando

(42) Ortner, *op. cit.*, p. 131f.

(43) *Ibidem*, p. 132f.

(44) *Ibidem*, p. 134f.

(45) 32, *Ibidem*, p. 135.

(46) Österreichisches Bundesministerium für Heerwesen (ed.), Österreich-Ungarns letzter Krieg, Band 1, Das Kriegsjahr 1914 (Wien: Militärwissenschaftliche Mitteilungen, 1930), p. 759f. In the Reichsarchiv, op. cit., p. 146 230.000 troops losses is discussed.

(47) Keegan, *op. cit.*, p. 154; Österreich-Ungarns letzter Krieg, I, p. 151

(48) Ortner, *op. cit.*, p. 136; According to German sources, Reichsarchiv, op. cit., p. 146 Oskar Potiorek was released and retired on January 1st, 1915.

Above: Always a favorite motif of the German soldiers: these foreign, fascinating oriental landscapes. For the Germans at the time this was an alien culture, that was taken home in the form of photographic memories.

Facing Page: Map of the army movements of the invasion of Serbia.

Below: Sergeants and mechanics during a Christmas celebration of FFA 1, probably in 1915, before they were deployed to Xanthi. The first soldier sitting on the left has the number "1" embroidered on his left upper sleeve.

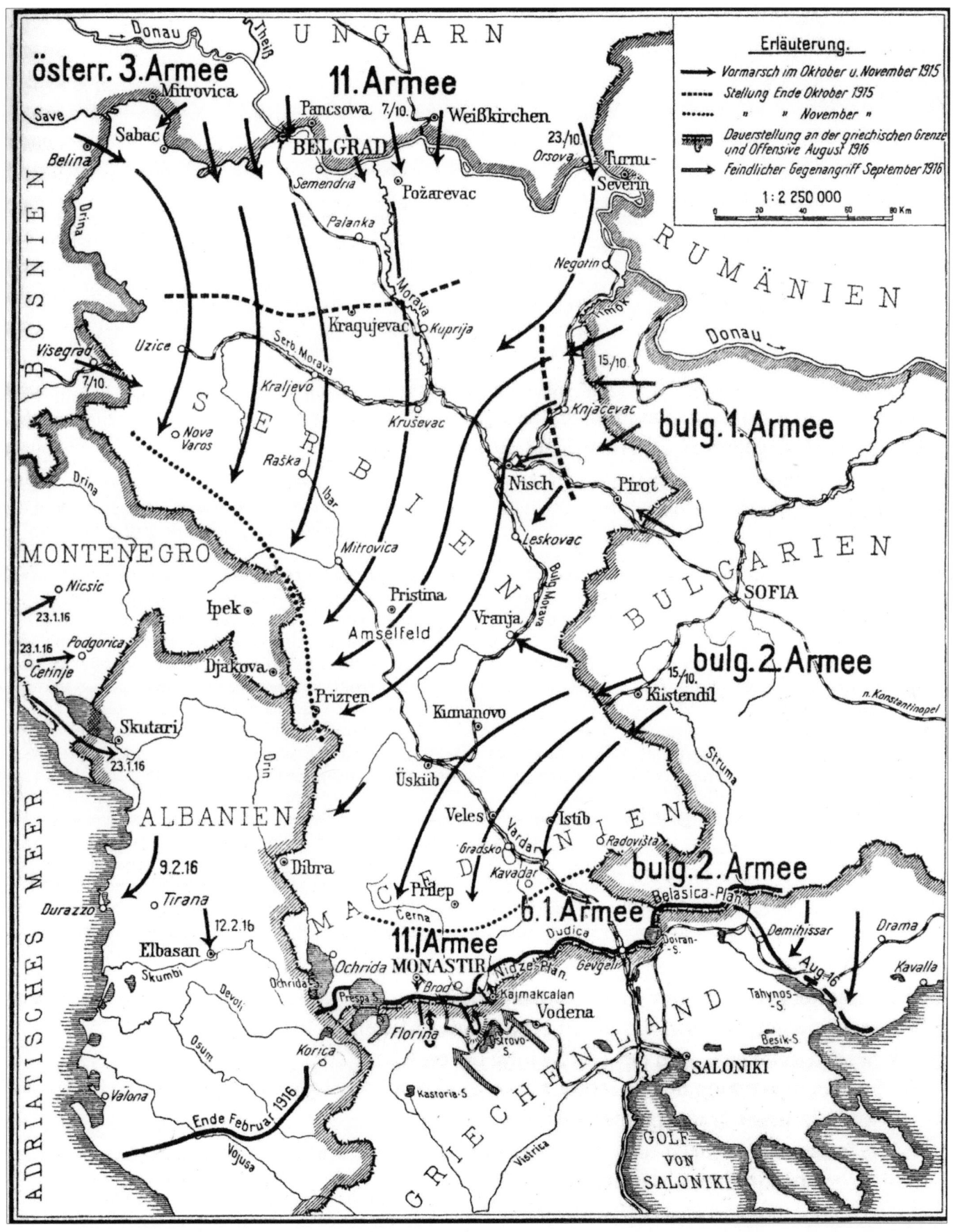

The Politics of the Other Balkan States Before the Beginning of the War

Following describes the politics of Bulgaria, Romania and Greece in the time before they entered the war. At the same time, we analyze the influence of the Central Powers and the Entente upon them, and the way in which this applies to the events of the war in the great theaters of the west and the east.

Bulgaria

Bulgaria had declared its neutrality ahead of the conflict between Austria-Hungary and Serbia, on July 28th, 1914. At the same time, Bulgaria said that it would maintain its friendly relations to Romania (49).[1] It would even renounce its claim on Dobruja. Romania, however, had to pledge to not oppose Bulgarian claims to Macedonia. After the world war broke out on July 30th, Foreign Minister Vasil Radoslavov reiterated Bulgaria's neutrality. On August 2nd 1914, he suggested to the Central Powers that Bulgaria join the triumvirate, under three conditions:

1. The triumvirate was to guarantee Bulgaria's territorial integrity.
2. The triumvirate must promise to support Bulgaria's territorial ambitions.
3. If Romania also joined the triumvirate, Bulgaria would make no claims on Romanian territory. Bulgaria's territorial goals lay in the west. Should Romania align itself with Russia, Bulgaria would once again lay claim to Dobruja and launch a campaign against Romania.

Berlin agreed and so the German ambassador Gustav Michahelles began negotiating with the Bulgarian government, but it was not completely clear to him if Bulgaria's goal was a benevolent neutrality, or if it was to enter the war. Chief of Staff Moltke would have liked most if Bulgaria had let loose on Serbia. (50)[2]

But there were several reasons that Bulgaria wanted to remain neutral. First of all, the country was in financial difficulties because of the Balkan wars, and second the army had not yet sufficiently recovered. Third, a large part of the population was *russophile*. But Prime Minister Radoslavov went further. He wanted to drive up the price of Bulgarian neutrality: He wanted to achieve territorial profit, without going to war. His aim was to play off the Central Powers against the Entente and vice versa. A bargaining began, like at an oriental bazaar.

This tactic was possible primarily because of Bulgaria's geographical position. For the Central Powers, Bulgaria was the land bridge to Turkey, and a Bulgarian attack on Serbia would make its defeat easier. For the Entente, Bulgaria played the role of locking bolt (hinge?) between Turkey and the Central Powers, which cut it (Turkey) off and isolated it. As a result, both sides wooed Bulgaria, offering up various territorial proposals to bring the country onto their side. This led to the aforementioned haggling (51).[3]

Without knowing it, Finance minister David Lloyd George and Naval Minister *Winston Churchill* as well as the Greek *Prime Minister Eleftherios Venizelos* supported this haggling with the idea of establishing a new Balkan League. While the two British politicians hoped that the League would go to war against the Central Powers, Venizelos wanted to hinder Bulgarian and Turkish acquisition of territory at the expense of Greece. The entire concept of the Balkan League was illusory, because all of the Balkan countries were much too divided and fragmented; but even British Foreign Minister Edward Grey found this to be a good idea because it would keep Turkey out of the war. So Grey instructed the British envoy in Sofia, *Henry Bax-Ironside*, to take up negotiations with the Bulgarian government.(52)[4]

It was clear to Bax-Ironside that Bulgaria was expecting to be offered territorial gains, if it was to be convinced to join its side. He suggested that they should convince the Greeks to give up Kavalla, and that the Serbs should give up portions of Macedonia. The Romanians would leave Dobruja to the Bulgarians. But neither Athens nor Belgrade were prepared to do this. Radoslavov understood that he could gamble for more, and let the British envoy know that the Central Powers had offered him more. If the Allies wanted to do business, they would have to offer Bulgaria all of Macedonia, up to the area that was undisputed. This, however, would have cost Serbia, and Belgrade was not prepared to give these areas up without a great compensation from Bulgaria.(53).[5]

On August 19th 1914, Radoslavov achieved a diplomatic victory when he signed a Bulgarian-Turkish defense pact. According to the document, the contract parties would support each other if they

Above: Again 12 mechanics in front of a Halberstadt C.V aircraft. This photo has some optical errors, but is still a very nice picture from the FFA1.

Right: Two pilots pose for the camera. The pilot on the left – Johannes Schaaf – wears the pilot insignia and the Turkish "Iron half-moon" medal.

were attacked. Therefore, the southeast flank of Bulgaria was secured. At the same time, Sofia sidled up to the Central Powers with this agreement. Berlin called promptly. Radoslavov was prepared to move closer to the Central Powers but the price was high: Bulgaria wanted Serbian Macedonia and the Greek part as well, if Greece entered the war on the side of the Entente. The central powers were immediately prepared to agree to the Serbian part, but regarding Greece, everything remained in limbo. Before anything could be finalized, Austria was defeated by Serbia on August 24th and Bulgaria distanced itself. (54)[6] The next round of negotiations with the Allies began.

This time, though, it was on Russia's initiative. Petersburg offered the Bulgarians the disputed parts of Macedonia and guaranteed this to the Bulgarian monarchy as long it showed a benevolent neutrality towards Serbia and would attack Turkey or Romania in the case that either of these attacked

34

Right: Messenger.
Facing Page, Above: Serbia.
Facing Page, Below: Zeppelin over Macedonia.

Russia. Considering the real situation, Bulgaria was unimpressed by this too-small offer. Britain once again took over leadership of the negotiations and once again offered Serbian and Greek areas: Bulgaria would return to the size it was in 1912 after the first Balkan war. France also supported this plan, but neither Athens nor Belgrade were prepared to agree to these concessions (55)[7] The negotiations reached a dead end.

Only when Turkey entered the war in October 1914 did any movement enter the discussions, because now the Entente could offer Bulgaria Thrace (Turkey) and the city of Adrianople (ed note: today called Edirne). On October 18th, Bax-Ironside offered Macedonia until the Vardar River, Thrace to the

Below: The pilots Johannes Schaaf goes goat hunting in the mountains with locals in Macedonia. For information: This pilot J. Schaaf was a flight instructor in Düsseldorf in October 1914. When the British attacked on October 8th, 1914, leading to the destruction of airship Z-IX, this flight instructor, along with his infantryman, who was armed with a carabiner, was ordered to chase the attackers.

Enos-Midia-Line, and financial support, if Bulgaria attacked Turkey. Radoslavov declined. (56)[8] The Austrians found out about this, and made their own suggestion, offering up Serbian Macedonia all the way to Niš. When Bulgaria did not react, Vienna promised the Bulgarians the entire area of Serbian Macedonia as well as the Pirot- and Niš-Region. In addition, it would support Bulgaria with its ambitions to win Kavalla and Saloniki as long as Bulgaria pledged to remain neutral. So, by the end of November, both the Entente and the Central Powers offered the country territorial gains under the condition it remain benevolently neutral. But for many weeks, things remained at a standstill.

On January 4th, 1915, the Entente guaranteed Bulgaria that following the war it would receive all of the areas in Serbia upon which it had historical and ethnographic claims. It would be permitted to occupy these lands with its troops. (57)[9] Bulgaria was to remain neutral, so do nothing, and therefore be rewarded with great territorial gains. Bulgaria had two offers between which to decide. A written one from the Entente with smaller territorial gains and a verbal one with greater profits but less security from the Central Powers. Radoslavov decided to wait a little longer.

In London an idea was born in January 1915, to send an expeditionary force to Saloniki in order to support Serbia. It was hoped that Greece would enter the war. Lloyd George and Churchill supported this quest. They were able to get France into the boat as well, which then offered to send a division also. Minister of War Herbert Kitchener was prepared to bring the 29th Division, which would later be used in Gallipoli. Later, Russian troops were to come as well. They offered Greece territorial gains if they participated. But Greece declined. Venizelos feared an attack from Bulgaria and Turkey. With that, the operation was suspended.(58)[10]

At the same time, another idea was developed, to convince Serbia to allow Bulgaria to take Macedonia up to the Vardar, if it remained neutral. If Bulgaria entered the war on the side of the Entente, it was to receive the area of Macedonia, including Kavalla, which it had won in the first Balkan war from Serbia and Greece, as well as the area up to the Enos-Midia Line. The Greeks would be compensated with territory from Turkey. When the British informed the Serbian Prime Minister Nikola Pašić of these plans, he declined and threatened to resign. Once again, things were at a standstill.

Then, Germany developed a new idea: Berlin offered Sofia the first installment of a Bulgarian bond, if the country mobilized, but this also ran into a dead end. In the following months both sides continued their efforts to attract Bulgaria. As the Entente offered larger and larger areas of land, Radoslavov gambled higher and higher, until he demanded the areas that Bulgaria had acquired in the Peace of San Stefano. This was accepted on May 29th under the condition that Bulgaria entered the war on the side of the Entente. Radoslavov delayed an answer until June 15th, when he explained that certain details had yet to be worked out. The bargaining continued uninterrupted. At the beginning of August, it was clear that the Entente would pay almost any price to pull Bulgaria onto its side. (59)[11]

On September 1st, Serbian prime minister Pašić made a large concession. Bulgaria would be permitted to take Macedonia with the exception of the areas of Prilip and Monastir. Therefore, the Allies had to guarantee Serbia the areas of Croatia and Slovenia, and Bulgaria was required to immediately attack Turkey. In addition, Serbia was to be recognized as an equal treaty partner, which would permit it to participate in the peace conference. The Allies agreed to the political demands, but denied the territorial claims, as well as the immediate entry into war. On September 14th, the Entente made a new offer to the Bulgarians: Bulgaria would receive the area that had fallen to it in 1912, if it attacked Turkey. They were convinced that they had won over Bulgaria.(60)[12]

But the Allies did not know that since July, Bulgaria had moved closer to the Central Powers. Czar Ferdinand found the Entente's offer vague, whereas the offers from the Central Powers were more concrete – they guaranteed the acquisition of all of Macedonia. In addition, following Russia's defeat in the battle of Lemberg and the British and the French standstill in the offensive on the Gallipoli peninsula, the tide seemed to be turning. Czar Ferdinand had the impression that the Central Powers would win. Negotiations for a ceasefire between Bulgaria and Tukey began. Turkey was prepared to cede a limited amount of territory to Bulgaria. But when Radoslavov tried to poker here, he found the Turks, who had in the meantime stopped the Allied attack in Gallipoli, unreceptive. They had decided they no longer had to rely on Bulgaria. (61)[13]

In the end, they reached a compromise: Turkey agreed that Bulgaria should receive the entire western Bank of the Maritsa, as well as a 2-kilometer wide band on the eastern bank. This would give Bulgaria control over the rail line to the Aegean port city of Alexandroupolis (Dedeağaç). On September 6th 1914 Turkey and Bulgaria signed a military alliance. On the same day, Bulgaria agreed to an alliance with

Germany, in which it was promised the entire area of Vardar-Macedonia. If either Greece or Romania attacked Bulgaria, Bulgaria would be justified to annex the areas it had lost to these powers in the second Balkan war. In an Appendix, Germany promised to attack Serbia within 30 days. Bulgaria would attack Serbia five days later as well. (62) [14]

In August, Germany and Austria-Hungary lent the Bulgarians 200 million gold-francs. The following begin of negotiations with the Central Powers led to a treaty on August 16th: As long as Bulgaria remained neutral, it received nothing. If it joined the conflict on the side of the Central Powers, it would get all of Macedonia, Northeast Serbia and the territory it lost to Greece and Romania during the Second Balkan War. The treaty was signed on September 6th and entered into force a day later. In this military alliance, Bulgaria committed to attack Serbia five days after the Central Powers began their offensive. If Romania or Greece engaged militarily, Bulgaria would also get the Dobruja and the region around Kavalla (63).[15]

Since the treaty was kept under wraps, the Allies continued their efforts. On September 22nd 1915, Bulgaria prepared to attack. On October 4th, Radoslavov rejected the Allies' final offer: The Entente could not offer Bulgaria as much territory as it needed. The Allied ambassadors understood that the chips had fallen, to the advantage of the Central Powers, and demanded their passports on October 5th. On the 11th, Bulgaria attacked Serbia.

Romania

Since 1883, Romania had belonged to the triumvirate. In 1912, negotiations began to continue this alliance, and concluded in 1913 with a two-year extension. So it could have been expected that Romania would enter the war on the side of the Central Powers. But like Italy, Romania also had some unfinished business with Austria, and declared its neutrality as the war broke out. Romania wanted to acquire the Austro-Hungarian border region of Transylvania, which was mainly inhabited by ethnic Romanians. These Romanians hated being second-class citizens, and were subject to *Magzarization*. But the country had further plans to take over also other regions where Romanians lived. In addition, the goal was to establish borders that were better defensible. In June 1914, as the prospect of war rose on the horizon, Russian *Foreign Minister Sergei Sasonov* had the impression that Romania would join the stronger side, from which it could achieve the greatest profit for itself (64).[16]

Romanian Prime Minister *Ion Br tianu* personally directed the country's foreign policy. Foreign

Above: Pilot J. Schaaf protects himself against the icy cold of Macedonia with a warm leather suit.

Minister Emil Porumbaru hat little say in the matter. The Romanian parliament had decided to hold no foreign policy debates after the war broke out. Therefore, Br tianu had *carte blanche*. Similar to his Bulgarian and Greek colleagues, the Romanian premier was convinced that the only way for the country to reach its national goals was by entering the war. But for Br tianu this was the *ultimo ratio*. He was determined to get the most out of the negotiations for his country. He was supported in this quest by Russian military leadership.

For the Russians, a Romania that was benevolently neutral was extremely valuable, because it would hinder an attack on Russia from the Black Sea to the Carpathian Mountains. Br tianu had contact to the Russians as early as August 3rd 1914, when the Romanian crown council decided on neutrality. On October 1st 1914, Russia acknowledged Romania's right to annex the Austro-Hungarian territories – Transylvania and the Burkovina - in which Romanians lived, by occupying these territories at the right moment. Since these areas were under Austro-Hungarian

Above: This photo cannot be of the Feldfliegerabteilung 1 because it was taken after March 1st 1918. This can be determined by looking at the iron cross on the fuselage next to the third person on the right. Presumably the photo shows a battle squadron. These had only pilots, few officers, many sergeants (as pilots) and crews (as MG-marksmen). On the photo no officer can be recognized (not even the three pilots in the middle).

Above: General Sarrail on a postcard of the era.

control, it was an attempt to draw Romania into the war. But Brătianu was careful and stayed neutral. After General August von Mackensen broke through the Russian front near Gorlice in Galicia pushing the Russian forces into a retreat in May 1915, the Russians began to try to entice Romania to join the conflict. They had to promise Romania large territorial gains. London and Paris agreed, after all, the Romanian army had 500,000 men.

Negotiations began in the summer of 1915. But the Romanians not only demanded Transylvania and the Romanian-inhabited southern part of the Bukovina, but also the northern part, in which primarily Ukrainians lived, as well as Hungarian territory on the Tisza River and the Serbian Banat. Sasonov found these demand to be exaggerated and wanted to reject them. But the western allies supported the Romanian demands: They had to draw Romania into the war, cost what it may. As the Allies suffered defeats in Galicia (Przemysl) and Poland (Warsaw), Sasonov relented. On July 21st, Sasonov accepted the Romanian demands and sent them a contract for signature.

But Brătianu did not even think of entering the war. He found unconvincing, comfortable excuses to deny his signature. In reality, he was of the opinion that the moment to enter the war had not yet arrived, since the Russian forces were retreating in the east, the Dardannelles battle was close to being lost, and in France the allied offensive was mired. When the Central Powers tried to bait him with Bessarabia, he rejected these also. He knew that 90% of the Romanians sympathized with the Entente. The Austrian envoy in Bucharest thought attempts to pull Romania onto his side were fruitless, since the Entente could always offer more. At the same time, Brătianu played the Allies off against each other. This was noticed and brought forth some anger, but Brătianu did not care.

In the fall of 1915, he let the Entente know that Romania would enter the war if the Russians sent a strong army to Romania. With this move, Russia, which had been in a desperate situation in Serbia, hoped to be able to save Serbia after Romania entered the war. But Brătianu's price was too high: He demanded the allied army in Saloniki and a Russian army of 500,000 men take part in this offensive. The Allies were simply in no position to provide this kind of support. So the negotiations dragged on through the entire Spring of 1916 until the successful *Brussilov-Offensive* in June 1916. Romania's entry into the conflict that followed will be dealt with in a later chapter.

Greece

Greece and Serbia had agreed on a defensive alliance on June 1st 1913, which was supposed to last for 10 years. It was ratified on June 21st. The treaty stipulated mutual support in the case of an attack from another country. At the beginning of a conflict, Greece would transfer 90,000 men to the region around Saloniki and its fleet into the Aegean Sea. Serbian would amass 150,000 men on the border to Bulgaria. These numbers were not permitted to be lower without the agreement of the other's general staff. There was an alliance obligation should Greece or Serbia be attacked by Bulgaria. The alliance was, therefore, primarily one against Bulgaria.(65)[17] At the last moment and against the protest of the Greek military, the passage that mentioned the obligation in the case of a Bulgarian attack was watered down with the addition of *"and a third power"*. The text read in its German translation: *"…in the case of war between the allied states and a third power, the circumstances of which are fulfilled in the Greek-Serbian Alliance treaty…"*(66)[18] The meaning of this wording was in no way clear, since the *"circumstances"* defined only a war with Bulgaria. Therefore, it was not surprising that the Greeks reacted the same way as the Serbs had in a similar situation shortly prior, when Greece had asked Serbia to help after a possible attack from Turkey, and Serbia had declined. The Serbian government was of the opinion that a possible Turkish attack on Greece was no *Casus Foederis*.

When the Serbs asked Greece on July 25th 1914 for support in the case of an Austrian attack (67),[19] the ministerial council met under the leadership of King Constantine. Constantine and Venizelos agreed that the country should maintain a course of benevolent neutrality to the Entente, and the treaty with Serbia was reserved only for a Bulgarian attack. (68)[20] On August 2nd Venizelos declined to help in the case of an Austrian attack. The country would retain its stance of benevolent neutrality. A Bulgarian offensive, however, would be rebuffed. Venizelos was not even prepared to place a larger number of troops in the region of Saloniki, because that could have provoked the Bulgarians. (69)[21]

King Constantine behaved similarly coolly, when on July 27th Wilhelm II demanded the country align itself with the Central Powers. (70).[22] Reichs Chancellor Bethmann-Hollweg formulated this order on July 30th, 1914, and passed it on to Wilhelm II, who had it forwarded to Athens. In it, he said: *"I consider it a matter of course that the memory of your father, who died at the hands of a murderer, will prevent you and Greece from acting against me and the triumvirate, and taking the side of the*

A glass window in the Imperial War Museum in London. The name "SALONIKA" is honored as a heroic fight.

Serbian assassins. But even just from the standpoint of sheer usefulness for the Greek interest, it seems that your country's and your dynasty's proper place is on the side of the triumvirate. Even Serbia, which cannot be protected from its fate through Greek support, will realize, that force majeure will determine Greece's position. [...] Should you position yourself on the opposing side, against my optimistic expectations, Greece will be subject an immediate attack from Italy, Bulgaria and Turkey, and even our personal relationship will forever suffer under that decision." This text was transmitted only verbally.(71) [23]

Constantine's answer on August 2 was clear: *"It never even occurred to us to assist the Serbs. But it also seems to me that it is not possible to consort with their enemies and attack them, since they are our allies. It seems to me that it is in the interest of Greece to remain absolutely neutral, and maintain a status quo in the Balkan region, as created in the Treaty of Bucharest. If we would let this fail, then Bulgaria will expand by annexing the part of Macedonia it won from Serbia [...] Therefore, so that this does not happen, I have no guarantees. These considerations force us into neutrality, and also to do everything in our power, together with Romania, to prevent Bulgaria from getting involved..."* (72)[24]

In a typical aside, Wilhelm II considered this answer "crap" (*"Blech"*) On August 4th he let Constantine know, via the Greek aide Nikolaos Theotokis, that he had sealed an alliance with Bulgaria and Turkey to fight against Russia, *"and will treat Greece as an enemy, if it does not immediately join [...] If Greece does not immediately come along, it will lose its place as a power in the Balkans and its wishes will never be supported, but rather it will be treated as an enemy."* (73)"[25] Theotokis immediately informed Constantine about Wilhelm's outburst. (74).[26]

"The answer from King Constantine was tactful but final. His personal sympathies and his political views, he said, were on the side of the Emperor. But unfortunately! – what the emperor asked of him was completely out of the question. Greece could not imagine pitting itself against the Entente under any circumstances: The Mediterranean Sea was in the hands of the French and British fleets which could destroy the Greek royal and merchant marines, take over the Greek islands and make Greece disappear from the map. Since this was the case, he explained, the only solution for Greece was neutrality. He concluded addressing the Emperor's threat with a threat of his own, which, despite all obscuration, was sharp, and with an assurance that the friends amongst my neighbors (meaning, Bulgaria and Turkey) will not be bothered (attacked) as long as they do not touch our interests in the Balkans."

At this point we can determine that when the war broke out, Venizelos and the king were of the same opinion. First, if the Central Powers attacked Serbia, the Greek-Serbian military alliance was null and void. Second, that neutrality was the only option for Greece. But if there was the impression from the outside that Athens spoke with one voice, this was misleading, because there were significant internal leadership differences between King Constantine and Eleftherios Venizelos.

King Constantine was born in 1868. His father, King Georg I was the second son of Christian IX of Denmark (who ruled from 1863-1906) and his mother was the Grand Duchess Olga Konstantinowa Romanowa, a Russian. The Danish dynasty was not particularly enamored of the Germans since the Prussian-Danish war of 1864, when the Danes lost Schleswig and Holstein. Constantine's three brothers were married to princesses from France, England and Russia. In addition, Constantine was the nephew of the English king and the Russian czar. His father succeeded Otto von Wittelsbach as the Greek king. Therefore, he knew well that he should never enter a conflict with Great Britain, the protector of his country, or he would risk losing the throne.

Constantine had attended the Prussian military academy and was married to Sophie of Prussia, the sister of Wilhelm II. in 1913, Wilhelm named Constantine as a Prussian field marshal. Through this and his marriage, he was considered a Germanophile in allied circles. In Greece, Sophie was often compared to Queen Friederike, the wife of King Paul, who was also herself a granddaughter of Wilhelm II. The comparison is weak, though. Friederike was a domineering woman who practiced her own politics, lorded over her husband and was actively involved in Greek politics. Sophie had no such ambitions, and the marriage to Constantine deteriorated in later years, they fought constantly. In the end, she had no influence on Constantine's policies. To imagine that Constantine was a devotee of Germany because of her is absurd. An American contemporary wrote: *"If Sofia suggested something, Constantine was against it; a queen who was pro-German meant that the king would lean towards the Allies."* The American war correspondent Hibben described it as follows: *"It is one of the sad aspects of the war, that otherwise intelligent people fall under this evil influence, if they accept such a childish intellectual world."*

As a trained soldier, Constantine anticipated that the war would last a long time, and probably due to the numerical superiority of the Entente, would

CXC. — Abreuvoir sur le Vardar. Grav. et imp. par Gillot

likely end in a *Remis*. So he was of the opinion that the best course of action for a small country like Greece would be to remain outside of the fray, and to stay neutral. In addition, his country needed peace, so as to integrate the newly acquired regions in the north. (75)[27] If he threw himself on the side of the Central Powers, his country would be undefended and at the mercy of a maritime attack by the Entente. In addition, there was the threat of a blockade, that would have meant thousands would suffer starvation. He also knew that Turkey and Bulgaria had some unfinished business with Greece, and would take the opportunity of the Great War to reclaim its lost territories. He was prepared to defend himself in an armed conflict. Most Greek politicians and military leadership agreed and therefore wanted Greece to remain neutral. Only prime minister Venizelos was of a different opinion. Since he understood little of the military, he thought that the war would be short, and the Entente would win. If Greece cast itself on the side of the Entente, it would be possible to fulfill the country's foreign policy dream of the *Megali Idea*.(76).[28] On August 18th, 1914 he stated to the ministerial council: *"He was convinced that the war would end in three weeks with the complete defeat of the Central Powers."*

Born on August 23rd 1964 on the island of Crete, Venizelos entered politics and became popular through the various uprisings on the island. The *Revolution of Goudi*, architected by the Greek military in 1909, brought him to Athens. In 1910 he founded the Liberal Party, which won the majority in the election of the same year. Venizelos became prime minister of the country. Abbot write about his early work: *"In the course of his early career, M. Venizelos was a complete organizer of administrative departments and a keen manipulator of independence movements. He made a name for himself as a rebel against the authorities and came to Athens with that attitude. Internal and external obstacles, however, made subversive undertakings impossible. With the quick adaptability of his nature, he turned and became a guardian of established institutions: The enemy of the revolution and a sharp agent for reforms. Supported by the crown, he raised his voice as a "revisionist" over the sea of voices which demanded a "representation of the people"*. (77) [29]

He introduced deep reforms. The rotten political system healed. Greeks began to look to the future with optimism. Venizelos was an admired leader of the nation.: *"To complete this task, he required a better-endowed brain than that of the others who were connected with this effort. His initiative was untiring, and his decisions were quick. As opposed to many of his countrymen, he was not satisfied with ideas that came without work, and his polished thought process did not serve as an alternative to his deeds. In addition to these talents, he had an eloquence that was irresistible for the Greek people, even though these skills do not always go hand-in-hand with high intelligence, but when it is so, it is worth more than all the talents of profound politicians and experienced lecturers, put together. He instinctively understood the character of every man he met, and dealt with them accordingly. This, together with an obliging smile and clear openness, gave his personality a charm that only those who had experienced it could value."(78)* [30] Abbot, therefore, had a very positive picture of Venizelos.

Cosmetatos came to a very different and differentiated conclusion: Venizelos was *"a clever politician, an agile diplomat, an excellent speaker – he had all the required talents that would be expected from a distinguished statesman. But in addition to these high qualities came a restless and subversive spirit. He was adventurous, an excellent actor, and could play all the roles perfectly. Nothing could stop even his smallest ambition or mood. He could have formed the renaissance of Hellas, but instead he wasted the resources of his intelligence by satisfying his passions through civil war."*

Until the end of the Balkan War, Venizelos had every right to the appreciation of his country. But immediately after the Bucharest Treaty, a shadow fell over the shining image of Greek unity. Venizelos' fame was surely brilliant, but he had no monopoly on it. He had to share it with Crown Prince Constantine, who became king in 1913 and who was the one who actually had commanded the armies that defeated the Turks and Bulgarians. Venizelos couldn't bear to share this fame. His autocratic temperament was hurt. He wanted to be the only one in Greece who gave orders. When the First World War came, a clash between the two men became inevitable. [...]

Venizelos, through his revolution against his king, revolted against the almost unanimous will of the Greek people, who were concerned about preventing the war – a stance that only Constantine could implement." In a letter to General Korakas dated November 7th 1916, Venizelos described his actual thinking: *"I was never in the habit of basing my calculations on purely logical and historical foundations, but used the principle of psychological changes, general concepts, however incomprehensible they may be. The law of violence and domination is stronger than all written and unwritten laws, real or hypothetical."* (79)

"[31] This is Machiavellianism in its purest form, appropriate for a power-greedy politician of a great nation, but not for one of a small country, where such a policy ultimately leads to a catastrophe.

As shrewd and cunning as Venizelos was in domestic policy, he was naive in foreign policy. He bet mainly on the British, and fully relied on British politicians' vague promises. He failed to understand that even the British Prime Minister's verbal promises were not binding. He transferred his ideas of the powers of a Greek prime minister, who actually had the say in Greek politics and whose word counted, to the British cabinet system, and did not understand that decisions were made there by majority vote. In addition, he apparently counted on the philhellenism of the British. This is the only way to explain the fact that he relied on the promises of British politicians, and did not nail them down with a written contract. Ultimately, he was the only leading politician in the Balkans who had not contractually secured his territorial ambitions at the end of the war. At the Paris Peace Conference, he appeared empty-handed.

In terms of foreign policy, Venizelos counted on the Mediterranean Sea powers. Only with them by his side could he realize the old dream of the *Megali Idea*. The successes, achieved together in the Balkan Wars, made Venizelos and Konstantin appear to be the ideal team to lead Greece, but in reality the two did not like each other. Their characters were too different. Constantine the straightforward, realistic, sober soldier and Venizelos the cunning politician who dreams of realizing the *Megali Idea*. At some point this had to lead to a collision, and then Venizelos would be unscrupulous enough to denounce Constantine's policy of neutrality as pro-German, and to counter it with his old methods of conspiracy and insurrection.

At the cabinet meeting on August 18th, 1914, Minister of War M. Streit reported on the informal remark by the Russian envoy that Greece should send 150,000 men to Serbia to join the Serbs in fighting against the Austrians. Venizelos was enthusiastic: He wanted to accept this invitation and subordinate all Greek armed forces to the Entente. M. Streit pointed out to Venizelos that this was not an invitation but a private remark. In addition, it was important to learn more details about a possible

Above: FLS-Abt. 34: Danube ferries near Cemendria. The invasion of Serbia by Austrian and German soldiers begins on October 7th, 1915. Info: FLS = Feld-Luftschiffer-Abteilung 34. These field-airship-units were the prime sources of aerial photography for the army units until 1918.

deployment location of the troops. After all, it was critical to find out how this would benefit Greece if the Entente were to win. M. Streit asked to continue the discussion in the afternoon. Venizelos replied that there was no time to lose because the war would be over in three weeks. Thereupon M. Streit announced his resignation.(80)[32]

On August 19th 1914 Venizelos offered the Entente the forces of the Greek army and navy. He could deploy 250,000 men and would be ready to open all ports in the country. King Constantine approved Venizelos' offer, but the Greek troops were not to be deployed so far away that, in case of doubt, they could not defend Greece against Bulgaria. London was very pleased with the offer. The English king thanked Constantine. The Russians and French reacted similarly, the latter believing that the Greek troops should be used to keep Turkey neutral. (81)[33] Since the Allies suspected that Venizelos wanted to steer them into war with Turkey for selfish motives, the offer was ultimately rejected. After all, they wanted to keep Turkey neutral at the time. (82)[34]

Churchill, however, was impressed by the offer. He was convinced that Turkey would eventually enter the war, sooner or later, and it was important to be prepared for this case. As soon as Turkey joined the conflict, the powers had to occupy the Gallipoli Peninsula with an amphibian surprise attack as soon as possible, and enter the Marmara Sea with a fleet of ships. The Gallipoli Peninsula would open the path to Constantinople. If that city fell, it would be a fatal blow for Turkey. (83)[35]

Churchill believed that *"the Greek army and navy were substantial factors and a link between the Greek armies and navy, and the British Mediterranean squadron was a quick and effective way of resolving the difficulties in the Dardanelles. The Gallipoli Peninsula was currently poorly occupied by Turkish troops and it was well-known that the Greek General Staff had well-developed plans and was prepared for an attack. (84)* [36]

Of course Churchill understood that Venizelos' offer was aimed at acquiring Constantinople as part of the *Megali Idea* policy. Since he was convinced that after the arrival of the two German ships, *Goeben* and *Breslau*, Turkey would sooner or later enter the war on the German side, he thought the offer worth considering.

At this point I would like to begin with the photographic material, so as to prepare the reader for the south-east front war theater. We ask to excuse that the images are not always analog to the text. It was important for us to give the reader an idea of the happenings at that time and the overall situation, and not to present the areal history of the war.

But Foreign Secretary Gray was opposed: *"After very concerned deliberations, Sir Edward Gray convinced the cabinet to reject Monsiour Venizelos' proposal because he doubtlessly had legitimate reasons to fear that an alliance with Greece would result in immediate war with Turkey, and possibly Bulgaria. [...] He was particularly concerned about not promoting any Greek venture against Constantinople that could in some way have offended Russia."* (85) "[37] Churchill had to accept this, but that did not prevent him from continuing to work on the creation of the Balkan Federation.

Constantine wrote to the British *Admiralty* that Turkey would not be attacked, but Greece would defend itself if it did attack. This attitude was exactly the same as that of the Entente at the time. Venizelos thought this was wrong. For him, Turkey was Greece's archenemy. In his eyes, an attack with so many strong allies was a once-in-a-lifetime opportunity. He did not understand that the Entente wanted to keep Turkey out of the war and why Greece should stay out of it as well. He definitely wanted to maintain the Greek offer. Should the Allies ever launch an attack on Turkey, Greece would still be ready to provide its troops. Apparently Konstantin was infected by Venizelos' enthusiasm, because he passed on plans for the conquest of the Dardanelles, which had been drawn up by the Greek General Staff, to the British Admiralty via the British Commander-in-Chief of the Greek Navy, Admiral Mark Kerr. Later, too, he sent the British information that the Greek secret service had collected.(86)[38]

At the beginning of November 1914, the envoys of France, England and Russia supported a Serbian request for help to Venizelos. Venizelos continued to decline: *"Venizelos had no problem showing the Allies that it would be crazy for the Greeks to do this. What was required of Greece was that they send their armies 400 kilometers beyond their borders, regardless of their means of communication, which for hundreds of kilometers - in Macedonia - were exposed to a potential Bulgarian attack. In Athens it was known that the Bulgarians, had in the meantime aligned themselves with the Central Powers, but Paris and London obstinately refused to acknowledge this fact. M. Venizelos and the General Staff therefore made the Greek promise dependent on the active cooperation of Bulgaria and Romania, but this could not be guaranteed by the Entente."*(87) [39] He, Venizelos, was only prepared to support Serbia if Romania guaranteed Bulgaria's neutral stance, i.e. if in doubt, attack Bulgaria if Bulgaria attacked Greece. But Romania refused to give such a guarantee. (88)[40] Therefore, this topic was off the table until early January 1915.

But nevertheless, Greece supported Serbia by delivering artillery munitions. During the Austrian offensive, King Constantine agreed to leave Serbia almost all of the Greek artillery's reserves, since it was suffering a shortage. About 20,000 grenades were delivered to Serbia, which contributed significantly to the success of the Serbs' defense against the Austrians. France promised to replace the ammunition, which also occurred, except for the fact that the shells were of the wrong caliber. Complaints were of little use, as France was suffering from a shortage of ammunition at this point as well. So the Greek artillery was without enough ammunition for the next few years. This was also one of the reasons why Greece refused to enter the war when Germany and Bulgaria attacked Serbia in autumn 1915. (89) [41]

On January 8th, 1915, Lloyd George sent a letter to Venizelos, in which he demanded that Greece support Serbia. England would give Greece a financial leg up, and send an Army corps to support the Greek troops. Venizelos answered as he had before: He would be prepared to support Serbia if and when Romania stopped Bulgaria from attacking Greece. (90).[42]

On January 7th, 1915, the British ambassador in Athens informed the *Foreign Office* that Venizelos had told him, that he would be prepared to participate in an offensive against Turkey, under the condition that the Allies guaranteed him certain territorial gains in Asia Minor. At the same time, he would have indicated that he rejected supporting Serbia, because that would lead Greece into a war with Austria. *Grey* enthusiastically accepted the offer. (91)[43] On January 11th, Grey's private secretary asked the *First Sealord Admiral John Fisher* what the chances of occupying Constantinople were. Fisher answered that the chances were very good, if Greece supported a British attack with its army, and conquered the Gallipoli Peninsula. Grey let the Russians know that a Greek intervention on the Gallipoli Peninsula would be very welcome. On January 13th, 1915, discussions between Foreign Minister Grey and his Russian colleague began. Grey

Above: Austrian propaganda card after the completion of the Serbian campaign.

was prepared to promise Greece a strip of land on the coast of west Anatolia and Cyprus as well if it attacked immediately. The Russian had no objection to Greece getting Smyrna and its hinterland, but he made it clear that *"if the Greek army were to operate strategically in the vicinity of the strait and Gallipoli, Greece should be told that Russia would not agree to support future expansion of Greek sovereignty in this area."*(92) [44] The Russians also made it clear that they rejected a Greek expansion towards the strait and Constantinople. Fisher on the other side was also still not convinced of the purely naval operation. He agreed under the condition that Greece provide ground troops and occupy the peninsula.

But *Premier Asquith* continued to want Greece to support Serbia, because without it Serbia would otherwise be defeated in the forthcoming Austrian offensive. Gray agreed. On January 20th, Asquith signaled his readiness to come to the aid of the Serbs with troops. Gray did nothing at first, but Lloyd George addressed the Greek ambassador in London: They would be ready to send 40,000 British troops to Saloniki if Greece supported Serbia. He promised even more troops if the Greeks asked for them. Venizelos indicated that he wanted to know beforehand what territorial gains Greece would derive from it. Lloyd George offered Venizelos parts of Western Anatolia and Cyprus on January 24th, if Greece would support Serbia. This was the first time that the Greeks were offered territorial gains in Asia Minor. Venizelos was excited about the offer: The realization of the *Megali Idea* seemed to be advancing. (93)[45]

On January 23rd, Foreign Minister Gray demanded Venizelos provide military support to the Serbs. If Greece supported Serbia, it could count on great territorial gains on the coast of Asia Minor. In order to keep Bulgaria neutral, one must offer it territorial gains in Macedonia. This would primarily be at Serbia's expense. Venizelos thought the proposal was absurd. Bulgaria will never allow itself to be impressed by hypothetical territorial gains and remain neutral. It was impossible for Greece to help Serbia as long as its flank was threatened by Bulgaria. But then he began to think about it, and came up with an idea: Kavalla should be left to the Bulgarians, provided it entered the war on the side of the Allies. (94).[46]

On January 24th, Venizelos presented Constantine with a memorandum, in which he developed on these ideas: One had to help Serbia, otherwise the

Austrians and the Germans would march through the country to Saloniki. The Central Powers would grant Bulgaria Serbian Macedonia. Ceding Kavalla would be compensated by large land gains in Asia Minor. In addition they'd get the the *Dorjan* (95)[47]-*Gevgeli Area* of Serbia. If Bulgaria accepted the offer, the Balkan League would rise again – and it could assert itself against the Central Powers and help the Allies to a victory. If Bulgaria remained intransigent, they would have to work with Romania. (96)[48]

On January 27th, he informed Grey of his conditions: Bulgaria should also support Serbia. If this could not be achieved, the Romania must enter the war on the side of the Allies. If Bulgaria did not declare itself benevolently neutral, then the Allies had to send to Greece enough armed forces to scare it off. The Allies must support Greece financially, as well as with war material. When Venizelos found out that the British support would be meagre, he reduced his territorial demands, but these were still too large to be seriously considered. Kitchener explained that he would be prepared to send an army of 500,000 men to Serbia, once the *Dardanelle Operation* ended successfully. At the moment, however, nothing could happen because the Balkans were covered in snow.

Constantin, for whom a surrender of Greek territory was a non-starter, rejected the plan. At most, he could imagine working with Romania. When the Greeks asked, the Romanians evaded. So they returned to the original plan of *"buying"* Bulgaria by ceding Kavalla. In a second memorandum, dated January 30th, Venizelos tried to make the plan palatable to the king again: 125,000 km² in Asia Minor would be gained for the surrender of 2,000 km² around Kavalla. In addition, the loss of 30,000 souls in the Kavalla region will be compensated by the acquisition of more than 800,000 new Greek residents in Asia Minor. Many of the Greeks in the Kavalla area would emigrate to the newly-won areas. Destiny would offer the Greeks the fulfillment of their wildest national dreams.(97)[49] The historian Smith judged Venizelos' breathtaking optimism as follows: *"The Ionian vision blinded Venizelos to military dangers of his suggested actions, just as the ruinous effect of the Kavalla suggestion on national unity."* (98)[50]

In contrast to Venizelos, who succumbed to dreams, Greek military leadership was skeptical. Already on January 27th 1915 the General Staff warned that Greece would not have the soldiers or the financial means to maintain a permanent occupation or an efficient administration of the distant new areas. There was talk of an occupation army, that would have had to comprise of at least 100,000 men. The new territories would have to be constantly defended against Turkish attacks, and there was also still the very real danger of a Bulgarian invasion. Greece was about to launch itself into a very dangerous colonial situation that was beyond the country's resources.(99)[51]

On February 2nd, the General Staff officer *Ioannis Metaxas* spoke up: The Austrian attack on Serbia was imminent. So much time had passed before the Greek army was mobilized and marched to Serbia that the Serbian army had meanwhile been defeated and a clash with the Austrians threatened Greece. Bulgaria would attack the Serbian and Greek armies from the flank and stab them in the back, cut off supplies and thus destroy both armies. Even if Romania intervened, it would not be able to stop the troops of the Central Powers. The landing of a British corps in Saloniki was completely inadequate. Four corps would be needed. Not even if the Bulgarians entered the war on the side of the Allies, would the combined armies of the Balkans be able to defeat the Central Powers. Should Greece attack a country allied with Turkey, the Turks would take action against the Greek minority in Turkey. (100).[52]

Venizelos was not impressed and found arguments against it: The rapid development of the administration in the Macedonian area had shown what Greece was capable of. The Turks in Asia Minor would become loyal Greek subjects after the disappearance of the Ottoman Empire. The Greek inhabitants of Asia Minor would do their part to ensure that peace and order would prevail, so that the Greek army could quickly return to defending the European fronts. The Entente would meanwhile keep Bulgaria in check. Should the Bulgarians attack anyway, Serbia would defeat them. (101)[53] One can only marvel at such loss of reality.

About the same time, it became known that Austria-Hungary and Germany had given Bulgaria a loan of 150 Goldfrancs, and a contract between Bulgaria and Turkey had been signed which guaranteed unhindered trade from Germany to Turkey. Thus, the Turks could finally receive war material on a large scale. It was then clear that Bulgaria had fled to the camp of the Central Powers. This brought Venizelos back into reality, and he let the Allies know that Greece would not enter the war because of the threat posed by Bulgaria. In addition, Romania had once again emphasized its neutrality. (102).[54]

On February 15th, Foreign minister Grey once again demanded Greece support Serbia. The Entente was prepared to send one French and one English division to Macedonia for security. Venizelos

Above: FLS-Abt. 34 Captive balloon near the Danube in 1916. This balloon could be filled with gas bottles. During the Serbian campaign the artillery observer could make out the covered batteries, and fight them. This better overview had a disadvantage, though, in that it was easily visible. The waterproofing agent was also yellow. Balloon observers also received parachutes, because they were a preferred target in the air.

declined, adding that these forces would be inadequate. Romania was again not in the mood to engage. (103)[55]

In the meantime, the British had begun to fire at the forts on the Dardanelles, and the unified French and British fleets were preparing to force through the straits. The attack on the Dardanelles, with the goal of reaching Constantinople excited the Greek public. Venizelos believed that the time had come for Greek intervention and decided to put an army corps and the entire Greek fleet at the disposal of the Allies. On February 28th, he announced his decision to the King, who voiced concerns and recommended that he get in touch with the Allies first. On March 1st, Venizelos informed the British envoy of his decision, claiming the King agreed. He then instructed the General Staff to draw up an operational plan and dictated a memorandum in which he proposed to the King that a corps should take part in the Dardanelles operation. The other four corps should secure the Bulgarian border. The British government replied in the affirmative, and was delighted with the subordination of the Greek fleet. (104) [56]

But neither the King nor the army leadership agreed with Venizelos' plans. The General Staff had commissioned a study of the attack on the Dardanelles. It was of the opinion that after the bombardment of the Dardanelles forts the previous November, the chances of success were slim. At that time, the Turks had been warned, and had prepared. In addition, an attack by the Bulgarians was rather likely. The transfer of a corps to the Dardanelles would have considerably increased the risk of such an attack. (105).[57] On March 3rd the British military attaché wrote: *"The unanimous opinion of the Greek General Staff was that the naval attack should be supported by land operations."* Churchill was delighted with that idea. When Foreign Minister Gray reacted very cautiously, Churchill was upset: They needed the Greeks. If one did not support Venizelos, one will soon have a new pro-German prime minister.(106) .[58]

When the Russians discovered that the Greeks planned to take part in the expedition against Constantinople, they reacted angrily, and made this clear to the Greeks. (107) [59] The Russians had long dreamt of taking Constantinople and the straits under their own control, and were under no circumstances prepared to retreat from that now. The Greek ambitions of making Constantinople the capital of a resurrected byzantine empire – as propagated in the *Megali Idea* - were well-known in Petersburg. Once the Greek army reached Constantinople, Russian diplomacy would never be able to get them to leave. (108)[60]

On March 2nd, Venizelos presented his memorandum to the King. His eloquence resulted in the King changing his mind and approving the plan. But when Venizelos left the palace, he ran into the incumbent chief of staff, Metaxas, who declared his resignation in protest on the grounds that a minister was concluding military agreements with foreign states without first consulting his own military. (109)[61] In addition, he said, the fighting would not end when Constantinople was conquered because the Turks would continue the war further inland. (110)[62]

In order to build support for himself, Venizelos convened the Crown Council, to which all former prime ministers belonged. These were Stefanos Dragoumis, Georgios Theotokis, Kyriakos Mavromichalis and Dimitrios Rallis. At the first meeting on March 3rd, 1915, Venizelos presented the facts. He admitted that there were differences between himself and the military. Should the council decide on a course of neutrality, he would step down. At the second session of the Crown Council on March 5th, he announced that not three divisions would be sent to Gallipoli, but rather just one, and it would immediately be replaced by the summoning of a reserve division. After defusing the situation, he told the members that he would not be available to lead a course of neutrality. The politicians were impressed by this declaration, and backed Venizelos. Constantine said that he would abdicate rather than join this suicidal undertaking. Again, no decision was made. At the next meeting on March 6th, the King declared that he had decided that Greece should remain neutral for the time being. Venizelos thereupon announced his resignation. *Dimitrios Gounaris* became the new prime minister. The new government was sworn in on March 10th. (111) [63]

Later it would be claimed that Constantine fired Prime Minister Venizelos, and therefore violated the constitution. But in fact, he did not dismiss him at all, which he would have had the right to do under the constitution, but rather Venizelos resigned because the King was not prepared to give his consent to entering the war. This was in accordance with Article 32 of the constitution, according to which the King alone had the right to declare war or make peace. So Constantine acted in accordance with the constitution, (112) [64]

Gounaris, who was born in 1866, studied law in Germany, England and France. In 1902, he was elected to parliament, representing Achaia on the Peloponnesis and in 1908, for a short time, he was finance minister in the cabinet of Georgios Theotokis. He was a socially-minded conservative.

King Constantine dissolved parliament and called for new elections. Actually, according to the Constitution, these should have taken place in a short period of time, but since the residents of the territories gained in the Balkan Wars were also permitted to take part in these elections for the first time, the new elections were delayed until June 13th, because the electoral lists in the new territories had to be compiled first. In addition, the King was ill for a long time. (113).[65]

As soon as he gained power, Gounaris declared in a government statement that he would pursue a policy of neutrality, recognizing the obligations arising from the alliance with Serbia, but at the same time indicating that he was by no means prepared to jeopardize the territorial integrity of Greece. The diplomatic representatives of the Entente in Athens were somewhat annoyed that he had not mentioned the Allies. Foreign Minister Georgios Christakis-Zografos, who had previously been head of the provisional government of Northern Epirus, instructed the Greek envoys in London, Paris and St. Petersburg to inform the respective governments that the new Greek government would hold on to the pro-Allied policy. The only difference to the Venizelos government is that they did not want to enter the war immediately. Venizelos was still of the opinion that the country should take part in the Allies' Dardanelles campaign, but Gounaris hesitated. For him, the risk was too high. (114).[66]

On March 12th, Gounaris informed the Allies under which conditions Greece would be prepared to join the Entente. On April 12th, the representatives of the Entente signaled that it would be prepared to agree to territorial gains in the Aydin Province (Vilayet), south of Smyrna, as long as Athens cooperated in the fight against Turkey. Gounaris wanted to know what exactly this cooperation would entail, and how large the territorial gains in Aydin Province would be. The British envoy responded that he had no exact instructions for the first part of the question, and for the second part he referred to foreign Minister Grey's message in January, in which called for *"very significant concessions on the Asia Minor coast"*. When Gounaris demanded more precise information, the British diplomat said that that meant Smyrna and its back country. His Russian and French colleagues agreed, although they had received no instructions from their ministers. But the Allies were not prepared to close a deal with Greece in which the military obligations and the territorial gains were fixed in writing. The reason for this hesitation was Italy, which also had ambitions for Turkish land across from its Dodekanes islands. When Gounaris wanted to know whether the diplomats had passed on Venizelos' interpretation of the territorial gains to their respective governments, and what replies they had received, he learned that neither the French nor the Russian had received any information from Venizelos. The Briton stated that he had only been informed informally and had not passed this on to London. The result was that Venizelos' grandiose Asia Minor land gain ambitions shrank to Smyrna and its hinterland. There had been no talk about the territorial integrity of Greece, although the Entente intended to give Kavalla to Bulgaria in order to draw Bulgaria to war on its side. The Allies hoped that Venizelos would once again come to power in the elections, because he had agreed to the planned cession of Kavalla. (115)[67] One gets the impression that the Entente was not playing a completely honest game with Venizelos either.

On April 14th, the Greek government submitted its ideas about entering the war on the side of the Allies to the Entente: If the Allies gave a formal guarantee of the territorial integrity of Greece, including the islands and northern Epirus during the war and for some time afterwards, the country would be ready to enter the war against Turkey with all its military forces. The aim of this war must be the dissolution of the Ottoman Empire. If it persisted, the position of the Greeks in Smyrna would not be secure. The Allies would have to contractually establish territorial compensation and the financial and material support for Greece. Only under these conditions would Greece be ready to enter the war. The military details would be negotiated between the staffs. As long as Bulgaria did not join the war on the side of the Allies, Greek troops could only be deployed in European Turkey. If the position of the Bulgarians remained unclear, Allied troops would be deployed to Macedonia and Western Thrace, in order to repel attacks by the Bulgarians and Turks. (116).[68]

The reaction of the Allies was stunningly arrogant: The suggestions were simply ignored. In Paris, the Greek envoy was told that since his government saw the Dardanelles operation differently from the Allies, an agreement was impossible. The Greek government indicated that it was still interested in cooperation and ready to submit new proposals, but would like to know beforehand what would be acceptable for the Allies. Foreign Minister Théophile Delcassé turned the Greek envoy away: he could not even tell him unofficially which proposals would be acceptable. The Greeks then offered up their navy to participate in the war against Turkey. The army should be prepared to defend against Bulgaria. Delcassé accepted the offer on the requirement that the Greeks did not attach any conditions to it.

Before entering the war, however, Athens wanted a guarantee of Greek territorial integrity during the war and at the peace conference, financial and material assistance, and an assurance that territorial gains in Asia Minor would be about the size of what Venizelos had described. Delcassé was not even prepared to listen to this offer. Greece should enter the Entente, without making conditions, and rely on the good will of the alliance. King Constantine was prepared to do without the financial and material help, but not the guarantee of territorial integrity. *"The main thing [...] is that the Entente powers give us an earnest promise that they will respect our territorial integrity and that it will induce others to respect it also, and that it will not be harmed by them in a future peace accord. [...] Greece has the right to be amazed that the friendly powers accept her as an ally, but refuse to make a clear statement to her,"* (117) [69]

The reason for the refusal of a guarantee was still the intention of the French to buy the Bulgarian entry into the war on the Entente side by ceding Kavalla. Constantine regretted this and was unwilling to rely on the goodwill of the Allies. He announced that Greece would pursue an extremely benevolent neutrality towards the Entente. After that, Constantine fell seriously ill, so that he could not take part in the political life of his country for a few weeks. (118) [70]

The Greeks' mistrust was justified. In the end, the supposedly neutral Bulgaria allowed German war material to pass into Turkey. Constantine had repeatedly warned the Allies that Bulgaria had allied itself with Germany and that sooner or later it would join the war on the side of the Central Powers. But in Paris and London no one wanted to listen. This information from Berlin was viewed as a German maneuver to bluff the Entente. (119) [71]

At the end of May, both Serbia and Greece received word from Paris that the Entente had offered parts of their sovereign lands to Bulgaria. Serbia declared that it was by no means prepared to give any territory to other states; the constitution forbade this. When the information arrived in Athens, the government could not believe that the Entente had offered Bulgaria large parts of Greek Macedonia, namely Kavalla and its backcountry. Foreign minister Zografos wrote: *"It is completely contrary to the principles of justice and freedom proclaimed by the Entente Powers - it seems to us absolutely inadmissible to plunder a neutral state, especially one whose friendly neutrality has so helped the Allies to deal with their territories, in order to buy help from a people who have so far done everything to help the enemies of the Entente. With what right and for what reasons can they mutilate our country? The opinions once expressed by M. Venizelos and since filed by their author do not give sufficient grounds for plunder. The whole thing is an unthinkable outrage: it shows that our fears were justified and that our demand for a guarantee was absolutely indispensable."* The scandal was so great that Paris and London tried to downplay it. The ceding of Kavalla had only been planned if Greece had voluntarily consented. Greece had no reason to be afraid. But Foreign Minister Zografos considered his efforts to have failed and declared his resignation. The Gounari government rejected further talks about entering the war. (120) [72]

Venizelos emerged the victor of the elections of June 2nd 1915. The new parliament was to be constituted on August 17th. But before this happened, on August 3rd, the scandal reached its high point, when the British envoy formally offered the Bulgarian government Kavalla and an undefined part of its back country as well as the Macedonian part of Serbia, and assured it that Great Britain would put pressure on the affected states. Territorial gains will be made dependent on their acceptance of this cession. The Serbian Prime Minister Paši complained to the Greek envoy in Niš about it. He noted bitterly that since one was dependent on England, one had to think about it. The British envoy Francis Eliot informed the Greek government shortly thereafter and demanded the cession of Kavalla on behalf of the Entente. Gounaris protested vehemently on August 15th: First England had promised to respect the territorial integrity of Greece and now it was disregarding this and wanted to force the Greek government to accept a decision made by the Entente as its own. He will not cede the recently-won region of Kavalla. London was not deterred by the Greek protests: the Entente hoped that Greece would accept its proposal after all; as soon as Bulgaria accepted the offer, the compensatory territorial gains of Greece in Asia Minor would be precisely defined.(121) [73]

Venizelos' election victory was not as large as it had been in 1910, when his party had won 146 of 182 seats in Parliament. It now controlled 185 of 314 parliament seats, which was still a powerful majority. On August 30th, 1915, Venizelos once again took control of government.

On September 1st, under pressure from the Entente, the Serbian Prime Minister Paši made the concession that promised Bulgaria large parts of Macedonia if Bulgaria remained neutral. He informed Venizelos, who accused his Serbian colleague of questioning the balance of power in the Balkans and of violating the foundations of the

Above: FLS-Abt. 34: In the course of time, filling the gas became more and more difficult, especially in the southeast. The last gas-filling station was near Vienna (Fischamend). So the balloon would be filled with gas but not emptied again. The "Division-sausage" had to be camouflaged.

Serbian-Greek alliance. A little later, the Serbian General Staff contact for the Greek General Staff informed the Greek government that Serbia was not in a position to deploy the contractually agreed 150,000 men on the Bulgarian border, at most 2 or 3 newly established divisions on the northern part the border with Bulgaria. Eight regiments could be deployed further south near Greece. When the Greek General Staff heard of this, it told the government that under such circumstances entering the war would lead to the destruction of Greece.(122)[74]

On September 6th, Athens learned of the conclusion of the Bulgarian-Turkish alliance, and on September 21st of Bulgaria's mobilization the day before. The latter was justified by saying that it was a precautionary measure, because a German-Austrian army had attacked Serbia and the combat zone was approaching the Bulgarian border. Venizelos responded in a characteristic way: *"The news put Venizelos in a fever of excitement. [...] Venizelos was a statesman with broad thoughts, hated dry facts and firmly believed in his own lucky star. Time factors were insignificant to him and considerations of uncertainties meant little to him; he displayed a sovereign disdain for the cautious attitudes of professional soldiers and other unimaginative individuals. At no time did these qualities seem more evident than on this September 21st."* (123)[75]

On the afternoon of September 22nd, Venizelos visited Constantin in Tatoi. The topic of the audience was the mobilization. Venizelos had prepared the mobilization order and demanded the King sign it. At the same time, he recommended entering the war on the side of the Entente, against Bulgaria. Constantine agreed to the mobilization for security reasons, but refused to enter the war. Constantine's main argument was that of the General Staff, that Serbia could not field 150,000 men against Bulgaria. Greece itself could only mobilize 160,000 men, while the Bulgarian army numbered 300,000 men. Greece would be defeated soundly. In addition, Greece had far too little functional artillery and almost no ammunition. Venizelos responded with military arguments that showed that he understood little about it. When these failed, he argued politically: *"Your Majesty, I said, I was unable to convince you. I am sorry, but because I am representing the people at this moment, it is my duty to tell you that you have no right to disagree with me this time. In the last elections, the people agreed to my policies and gave me their trust. [...] Therefore, you cannot deviate from this policy at this moment - unless you repeal the Constitution; in that case you must express this clearly, suspend the Constitution by decree and take responsibility."* Constantine replied that he felt bound by the will of the people with regard to domestic affairs, but if this was a foreign affair involving the nation, *"it is my view that as long as I judge a matter to be right or wrong I have to insist that it be done or not done because I feel responsible before God."* Venizelos exploded and accused Constantine of behaving like a King by the grace of God, an absolute ruler. Under these circumstances, all that remained for him to do is tender his resignation. Constantine appealed for his sense of responsibility in the face of the Bulgarian threat, and Venizelos dropped his threat to resign. (124)[76]

Since Venizelos knew that the Serbs would be in no position to send the contractually agreed 150,000 solders to the southern front against Bulgaria, he suggested asking the British and the French if they shouldn't send the 150.000 men instead. Constantine agreed, but thought they should not be colonial troops. (125).[77] Another version of the story says that the King declined to discuss the latter, before he had made up his own mind about it (126) [78]

There was only agreement on the topic of immediate mobilization. Venizelos still wanted an immediate entry into the war against Bulgaria, which Constantine continued to refuse. Furthermore, according to his own statements, it was clear that the King was not prepared to deviate from his course of neutrality even if the French and the English sent troops. Venizelos hoped, however, one way or another, to persuade him to enter the war. (127) [79] In addition, he soon wrote a long letter to the King, in which he recommended the landing of the Allies in Saloniki. Constantine was irritated and feared that Venizelos might act rashly, and do things that he had not approved. Therefore, he sent his court-marshal Alexandros Merkatis to Venizelos. But before he arrived, Venizelos had already acted. (128)[80]

Shortly after his return to Athens at 7pm, Venizelos had called together the diplomatic representatives of the Entente. He informed them about the mobilization. *"For further action, I would have to know whether the Allied powers would be interested in standing up the 150,000 soldiers that Serbia was obliged by treaty to provide for joint action against Bulgaria. They promised to telegraph, sending immediately a particularly urgent telegram, and said they would let me know their answer."* This happened around 8 p.m.(129)[81]

At around 8.15 p.m. Merkati met with Venizelos. He informed him that the King had meanwhile changed his mind and demanded Venizelos stop his request to the Allies. Venizelos had to admit that the request had already been made. He had asked the Entente envoys to land 150,000 soldiers in Saloniki as soon as possible, *"in order to fulfill the Serbian obligations towards Greece in the event of war against Bulgaria. Monsieur Venizelos, who seemed very embarrassed, quickly calmed his astonished visitor. His overtures, he said, were of a strictly personal character and had nothing to do with the state!"* (130)[82]

On the next day, Venizelos repeated this statement to the King in order to excuse and justify himself. The King refused to accept this excuse, demanded that Prime Minister Venizelos abide by the Constitution (Article 99) (131)[83] and terminate the matter by calling it off. Venizelos promised to do so, but did nothing. Instead, he got in touch with the French ambassador, Guillemin, who met with the King two days later and, against his better judgment, assured him that the matter had been settled and that the action by Venizelos would be considered irrelevant. But this was just a maneuver to calm the King down. (132).[84]

According to a statement by Foreign Minister Théophile Delcassé, the French government considered Venizelos' request to be official. On September 23rd, he replied to Venizelos as follows: *"The French government would like to support Greece so that it may fulfill its obligations of the treaty with Serbia, and is ready to do its part to provide the requested troops."* (133) [85] The English Foreign Minister Gray, however, was skeptical. On the same day he informed the envoy Elliot that it would take a long time to send so many troops to

Above: An English propaganda card: "For right and liberty". An incoherent slogan considering the invasion and the occupation of Greece was a breach of international law.

Saloniki: *"A relatively small contingent could of course be sent to Saloniki in time to give Bulgaria the impression that we are proving our intention of supporting Serbia and Greece."* (134)[86]

The positive response from the Entente, committing 150,000 troops to Saloniki, arrived on September 24th, 1915. Venizelos informed the King. He forced Venizelos, citing Article 99 of the Constitution, to provisionally reject the offer: As long as Bulgaria does not attack Serbia and thus raise the question of Greek assistance, no Allied troops should be sent because this would constitute a violation of Greek neutrality. Constantine and the military believed that the best way to serve Greece's interests was to remain neutral until the Allied guarantees were set in stone or the Bulgarians attacked. (135)[87] On September 24th, Venizelos informed the ambassadors of the King's decision. (136)[88] However, according to the report of the British envoy of September 25th, Venizelos added that he hoped that preparations for the dispatch of the troops would continue. *"When the final request to land them is made, it may be necessary for Greece to formally protest in order to save King Constantine's face with the Germans."* The British government made clear that it did not like these diplomatic tricks but permitted the preparations for the landing to continue. (137)[89]

In London meanwhile, Minister of War Kitchener suggested to the so-called Dardanelle-Committee (a committee of the cabinet) that the positions in Suvla Bay on the Dardanelles Front should be vacated. The 10th and 11th divisions located there were to be transferred to Saloniki. There, according to the General Staff's plans, they were to support the Greek army in securing the flank on the Serbian border. On September 25th the Commander-in-Chief on the Gallipoli front, General Ian Hamilton, was informed of this, and that the French would withdraw a division from the front at Cape Helles. After some back and forth on the Gallipoli front, it was decided that the French division and the two British divisions should be embarked as soon as possible. Hamilton and some staff officers were sent to Saloniki to prepare for the October 1st landing. (138)[90]

On September 27th, Venizelos asked the French envoy to send as many allied troops to Saloniki as possible, as quickly as possible. They were to let him know 24 hours before they were due to arrive so that Greece could formally protest. Thereupon, in addition to the French division, the 10th British division was also set to march. (139) [91] On September 29th the French consul, accompanied by the captain of a warship and two officers from the Dardanelles appeared before the Commander in Chief of the 3rd Corps, Konstantinos Moschopoulos. The explained to him that they – as agreed between Venizelos and the French envoy – were to prepare the landing of the French troops. On September 20th, Hamilton and his staff officers arrived. They said that they would occupy a part of the city and the harbor, and place it in a defensible state. (140)[92] Moschopoulos said that he had not received such orders from the government or the King. Without these, he saw himself obliged to reject the occupation of Greek territory. After all, Greece was neutral. He suggested postponing the scheduled landing until 4 p.m. The delegation agreed and left him. Moschopoulos then informed Athens. Venizelos realized that the Allies had not kept to the agreement to warn him 24 hours before landing. He spoke to Moschopoulos on the phone.

But something unexpected happened. It was well-known that the Entente had made an attractive offer to Bulgaria on September 14th, in order to pull it over to the side of the Entente. Since Foreign Minister Gray knew nothing of the rapprochement between Bulgaria and the Central Powers, he made very friendly comments in the House of Commons about the country on September 28: *"There is no hostility to Bulgaria in this country and there is traditionally a warm feeling of sympathy."* Then followed a remark about the British Balkan policy, which was aimed at satisfying the legitimate territorial claims of all Balkan states. But he also pointed out that if Bulgaria sided with the Central Powers and attacked other Balkan states, Britain would come to their aid. Venizelos only heard of the first friendly remarks and completely misunderstood them. He had the impression that the Allies were sending troops to Greece so that Bulgaria could enforce its territorial claims. (141)[93]

Venizelos turned to the French envoy and complained that he had been deceived and that his trust in the Allies had been abused. He feared that the landed troops would be used to put Greece under pressure to make territorial concessions to Bulgaria. He would reject a landing and it would have to be postponed if it did indeed take place. Otherwise he would be forced to stop the landing by force. The French envoy telegraphed to the Dardanelles headquarters in Moudros on the island of Limnos. Further troop transports should be postponed and those already on the way should be recalled. Therefore the landing operation was interrupted for a while.(142) [94]

On October 1st, Venizelos appeared in the palace. In mock indignation he informed the King that an English admiral had arrived in Saloniki who, together with the French consul, was preparing the landing of Entente troops. *"I will protest with great force,"* cried M. Venizelos, trembling with anger. I will protest against this unspeakable breach of our territory by the Entente!" " Certainly," replied the King," a very stern protest must be made!" And M. Venizelos went quickly to his ministry to formulate this protest. (143)[95]

Venizelos drew up a long memorandum in which he described the reasons for his protest (144) [96] The decisive passage said: *"There is a grave misunderstanding emerging between Greece and the Entente Powers over the issue of sending international troops through Saloniki to Serbia. When I asked for 150,000 men to be made available to complete the Serbian contingents in the event of a joint fight with Serbia, I asked for assistance not for Greece, but for Serbia, in order to get rid of the objection raised by our allies, because the treaty lapsed due to Serbia's inability to meet this obligation. By accepting the principle of such a consignment, the powers helped Serbia and their own cause in the East above all. I had also made it very clear that as long as Greece was neutral, a landing of international troops in Saloniki could not meet with our official approval. Our neutrality forced us to protest for the sake of form, and after that things would go on as in Mudros,"* (145)[97]

If one reads the text closely, and also between the lines, one gets the impression that the text had the function of establishing an alibi. King Constantine was supposed to be deceived by the *"protest"* about Venizelos' actual policy, at which the Allied landing was aimed. The memorandum was more or less a substitute for the 24-hour report. Venizelos was by no means as angry as he seemed. The reference to Mudros is revealing: Mudros is the capital and port of the island of Limnos. In connection with the Dardanelles operation, the Allies occupied the island, Greece formally protested, but let the Allies have their way. At the same time, this *"protest"* was the proof that Venizelos had never thought of withdrawing the invitation to the Entente, as he had promised the King.

On October 2nd, 1915, the French ambassador wrote to Venizelos that Allied troops would land in Saloniki shortly. These troops were allegedly only there to help the ally, Serbia. It was hoped that the Greek government would do nothing against it. London and Paris also let Venizelos know that all

Above: FLS-Abt. 34: View of Mount Olympus. The Feldluftschiffer-Abt. 34 was responsible for the analysis of the aerial photos of the Vardar and Saloniki front. In Spring 1917 a book featuring photos from this unit was published.

the offers they had made to Bulgaria were now null and void.

This was the alibi that Venizelos wanted and he answered that the troops should be sent at once. (146)[98] He wrote: "*As Greece remains neutral in this European war, the royal government is unable to give its approval for your planned actions because it would disrupt Greek neutrality. This is especially the case because these are actions by two powers currently participating in the war. Therefore, the Royal Government is obliged to protest against the passage of foreign troops through Greek territory.*" The commander of the Greek troops in Saloniki, Moschopoulos, was instructed on October 2nd to do nothing to prevent the landing. On October 3rd, 1915, the first troops landed in Saloniki. (147).[99]

The King finally grasped that Venizelos had, in collaboration with the Entente, tricked him into bringing Greece into the war. At the same time, Venizelos had indirectly violated Articles 1 and 3 of the *Hague Land Warfare Regulations*, which forbade such actions, and also directly, because they prohibited neutral states to tolerate military actions by third states on their own territory. It was also clear that Venizelos had broken the Constitution by disregarding Article 99. From that moment on, his relationship to Constantine was strained. Venizelos knew that the King could now dismiss him at any time. But this dismissal should be done according to his rules, so that he could portray himself as a victim of royal arbitrariness.(148)[100]

On October 4th, Venizelos delivered an absolutely bellicose speech to Parliament. He now proclaimed the view that Greece was contractually obliged to declare war on Bulgaria and Turkey. When a member of parliament asked him what would happen if the country aided Serbia and encountered German and Austrian troops, did one have to fight them too? Venizelos replied that it was a question of national honor. This appeal bore fruit. The passionate debate lasted into the early hours of the morning. With 152 in favor and 102 against, the vote produced a clear result: Greece would join the war on the side of the Entente. (149).[101]

King Constantine and the General Staff, on the other hand, wanted Greece to remain neutral. The king asked Venizelos for a meeting and explained to him that the General Staff considered entering the war "absolute madness". The Allies had only landed 13,000 men, and not 150,000 as they had promised. The Allies wanted Greece to enter the war in order to save troops. Venizelos reminded the King that he had won the election and that meant that the majority of the Greeks approved of his politics of intervention. (150)[102] When Venizelos remained stubborn, Constantine demanded he change his policy or tender his resignation. Venizelos saw no alternative, resigned, and Constantine dismissed him from office according to Article 31 of the Constitution. (151)[103] His successor was Alexandros Zaïmis.

On October 6th, Russian envoy Prince Demidoff judged Venizelos' behavior in a telegram to Petersburg as follows: *"The hasty statements made by Venizelos in the Chamber provoked a break between himself and the King. Unfortunately, Venizelos allowed himself to be carried away and declared his policy of intervention, contrary to our advice and without waiting to complete the mobilization and the development of the serious events that were taking place. In his haste it was actually stupid, especially since he knew the king's efforts not to be drawn into this war."* (152)[104]

But Venizelos had achieved his goal: *"From that day on, just as he had wished, he was seen in the eyes of the Allies as a great and noble victim of their cause, who fell because of his desire to fight the Germans. And King Constantine, who sensibly did not want to enter the war without rifles and cannons, or turn his country into fodder for the Germans, was therefore seen as an agent of Kaiser Wilhelm! The journalists brought the full force of their extravagant dialectic as a means of serving this infamous campaign. [...] Finally, starting in October 1915, the Entente press kept declaring that the Allies were going to Saloniki at the invitation of M. Venizelos. And that was the truth. For a long time, M. Venizelos denied this with furious energy, and it went so far that various gentlemen of the Entente countries, who had a more generous attitude, defended him in good faith against their own countries."* (153)[105]

But also in France, the decision to land in Saloniki was not made on the basis of sober considerations. Up until the last moment it had been believed that Bulgaria could be drawn into the war on the side of the Allies, and the warnings from Athens that Bulgaria had already forged an alliance with Germany had been ignored. The French government knew that the invitation came from Venizelos and that he had acted without the King's approval. It was further known that the Russian government rejected the enterprise. Delcassé knew that the commander-in-chief of the French troops, Joffre, considered the Saloniki expedition to be "absolute madness" in light of the personnel and material situation, and therefore rejected it. It was also known that the British were hesitant. (154)[106]

The decision came out of frustration at the failure of the French policy on Bulgaria. *"The Saloniki expedition was decided upon in an instant without any study or preparation. No review of sea and land transportation, health care and topographical difficulties in Macedonia was completed prior to the decision. Nobody knew what options there were to send troops or war material and how long it could take. Confidence was placed in what was seen to be the boldness of the idea and relied primarily on the Greek army. M. Briand said in his secret committee on June 16th, 1916 that the powers were hoping that this action would allow M. Venizelos to include Greece. As unlikely as it may seem, Paris made no inquiries about whether the Greek army was in good enough condition to enter the war within one day."* The French troops that landed in Salonika did not even have maps of Macedonia. (155)[107]

So therefore, sending the troops had become useless. The Allies sent troops so that Greece would be forced to keep the pact with Serbia. But the new government did not wish to give up Greek neutrality and enter the war. The balance of power shows the absurd situation: So far 200,000 war-weary Serbs, 150,000 Greeks and 20,000 Allied soldiers stood opposite 200,000 German and Austrian troops, augmented by 300,000 Bulgarians and an as yet unknown larger number of Turkish troops. With Greece's renewed neutrality declaration, the entire operation was pointless. In addition, the allied forces would arrive far too late at the Serbian front to actually accomplish anything. The landing on Greek territory also contained the danger that Bulgaria would attack Greece, and the allied support would not even last long enough to fend off such an attack. A responsible allied government would have stopped the undertaking. Perhaps Britain would have had the courage to make such a decision, but in the meantime the initiative had moved to France, and Paris just carried on. (156)[108]

The arrival of the 20,000 allied soldiers in Saloniki brought Greece indirectly into the war, and that is

Above: FLS-Abt. 34: The "iron gate" near Üsküb: "WILHELM II, GERMAN KAISER, KING OF PRUSSIA, ORDERED HIS SOLDIERS TO BUILD THIS ROAD, 1916" These narrows in the Danube between Serbia and Romania near the Mraconia monastery are no longer visible today due to a dam.

why the further description of the developments inside the country will follow in a later chapter. Here is now the history of the assistance expedition to Serbia.

The Allied Expedition to Serbia October – December 1915

Political Maneuvering in the Background

In July, General Maurice Sarrail[109] was named the new commander-in-chief of the French troops in Gallipoli. Sarrail had been the commander of the 3rd Army in 1914, where he held up the German attack on Verdun and turned into a national hero. He was a soldier who was not afraid of openly giving his opinion. He commented on politics and also criticized his superiors. He had been a staunch Republican, anti-clerical, and a Dreyfus supporter. He was close to the radical socialists and they politically supported him. After the bloody and costly Argonne offensive, he had openly criticized the French commander-in-chief Joseph Joffre for the high losses. Joffre was outraged and relieved Sarrail of his command without informing the Ministry of War.

The cabinet was angry because it feared run-ins with the socialists, who had already begun to maintain that Joffre had booted out Sarrail for personal reasons. Due to his political connections, the cabinet offered him the post of commander-in-chief of the French forces in Gallipoli. But Sarrail declined, because this would have been a demotion. In addition, he did not want to have to report to Hamilton. He said he'd only be ready to take over command of an independent army in the middle east. In order to assuage him and to get him out of the city, the French government said it would be prepared to send four divisions to the Dardanelles Front, which, together with two divisions which were already there, would form the *Armée d'Orient*. Originally, Sarrail would operate from the Asian side of the Dardanelles, but Joffre and Kitchener wanted to keep an eye on him, and it was decided that he would be sent to Saloniki. So Sarrail could credit political pressure at home for his appointment to command of the *Armée d'Orient*.[110]

On September 28th the French high command informed him that he should not go to the Gallipoli front but to Saloniki. On October 7th, the day of the German-Austrian attack on Serbia, he embarked in Toulon. The French intended to send four divisions there, but when the offensive in the Champagne-Artois began on September 25th, the relocation of these troops was postponed.[111]

Above: A German U-boat in Greek waters. These few German U-boats were a constant danger for the Allies. As a propaganda argument, the allegedly secret U-boat bases in Greek waters were used as an excuse to march into this neutral country.

Below: Two Macedonians in traditional garb, smoking.

Above: Overview map of the southeastern war theater from the series *"Schlachten des Weltkrieges, Volume 11"*, 1925. The thick black line is the front.

On October 5th, minister of war Kitchener and Naval minister Arthur Balfour met with their French peers Alexandre Millerand and Victor Augagneur in Calais. Millerand promised to send three more divisions to Saloniki, and Balfour offered to transport them with the Royal Navy. Shortly thereafter, when Kitchener learned of Venizelos' resignation, he stopped the transfer of the 10th Division and ordered its commander General Bryan Mahon to return to Moudros. The Yeomanry Regiment from Egypt, which had been slated to move to Saloniki, was also held back. He felt a landing without Venizelos' support seemed dangerous. But on the very next day *the War Committee* in London decided to go ahead with the troop transports.[112]

On October 7th, Prime Minister René Viviani und Naval Minister Augagneur arrived in London, where they held consultations with Asquith, Grey, Balfour and Kitchener. The French suggested sending 400,000 men to Saloniki (67,000 French and 333,000 British). They were of the opinion that the operation should be carried through, no matter what, even if Greece did not agree. They did not achieve a result. On October 8th, Kitchener went to France to speak to Secretary of War Millerand and Commander-in-Chief Joffre. Kitchener gave in to pressure from the two and promised to send 90,000 men to Salonika.

The French would send 60,000 men. But only 30-40,000 men could be sent immediately. The rest would follow in early November.[113]

Joffre opposed the Saloniki operation but he but bowed to political pressure. On October 9th, he made the following suggestion to Kitchener regarding the Saloniki operation: Allied forces should hold Saloniki as a base for British, French and Serbian troops. They were supposed to secure the railway line to Skopje and thus ensure the connection to the Serbian army. In addition, they were to cover the right flank of the Serbian army against the Bulgarians. 150,000 men would be enough for that. The main thing was to block the Germans' path to the Mediterranean near Saloniki.[114]

The British general staff did not think much of these suggestions. Even if they sent 150,000 men to Saloniki, they would arrive at the beginning of December at the earliest. It was doubtful if these few troops would be enough to block the German munitions transports to Turkey. These troops would be operational in mid-December. It would be better to use the forces earmarked for Saloniki in Gallipoli. Kitchener submitted this assessment to the War Committee. There they recognized the problem: For military reasons, the Saloniki operation was rather pointless. But if Serbia was abandoned, the French,

Above: The crew of FF Oblt. von Chappius and BO Oblt. Trenkmann in their Albatros B.III aircraft. Here the observer sits in the front and also works the radio. The antenna is let out with the crank on the side of the fuselage. The pilot also has a nose cover in addition to his googles, to protect against the cold.

who chose to pursue the Saloniki operation, would feel the same way. Letting the Serbs down would certainly have a negative impact, but for Kitchener, abandoning Gallipoli was *"the most catastrophic event in the history of the Empire."*

The majority of the cabinet was in favor of continuing the Dardannelle operation, only Lloyd George and a small minority demanded that reinforcements be sent to Saloniki. A compromise was agreed: reinforcements would be sent to the eastern Mediterranean, so primarily to Egypt, where they could be quickly sent to Saloniki or to the Dardanelle front.[115]

But not only was British leadership split on this question, also the French had reservations. George Clemenceau rejected a Balkan operation even if Greece participated on the side of the Entente. The French press approved an intervention in favor of the Serbs. On October 13th the French parliament decided to send a sufficiently large force to Saloniki, upon which Foreign Minister Théofile Delcassé announced his resignation.

General Sarrail arrived in Saloniki on October 12th. As he himself admitted, he had no idea about Greece. *"I arrived in Greece on October 12th, 1915, without any information, without official instructions, without knowledge of the country, not knowing anything about the people, and nothing about the events that had taken place since the beginning of the war. What was my Ariadne (Ariadne-DFaden ???) to help me find my way in this chaotic environment? I could only rely on my experiences and my philhellenism. Based solely on my upbringing and my memories of the classics,*

Above: Pilot Oblt. Herbert von Chappius, born on June 22nd 1887 in Stettin, died on January 12th, 1916 in Kavala-Orfano (near Saloniki), FFA 1.

Above: Portrait of observer Oblt. Georg Trenkmann, born on November 25th, 1890 in Berlin, died on January 12th, 1916 in Kavala-Orfano (near Salonika), FFA 1. Both aviators were shot down on the way back from their first enemy flight, from Xanthi to Saloniki, probably by British fighters from Thasos.

my unequivocal republican ideas and my absolute conviction, I had come to the conclusion that we must come to an immediate result in the war instead of hypnotizing ourselves about the post-war period."

The Orthodox priests, the churches and the religious population irritated him. He did not know that the Orthodox clergy was the bearer of the national idea and he persecuted the followers, desecrated churches and confiscated them for secular purposes. *"The most important thing for the general was to make Greece a republic. The word "kingdom" offended his innermost convictions. Then, his politics became a strange mixture of inconsistency. In the background, Sarrail conceived a threat from Constantine. He used this fictional story whenever he had to explain his inactivity to the public."*[116]

On October 12th, General Sarrail arrived in Saloniki and immediately transferred an infantry regiment and an artillery unit to the Vardar Valley to secure the railway line from the Greek border to Strumica. The British General Mahon was not subordinate to him and had orders to remain at the landing site, although Britain declared war on Bulgaria on October 14th. British troops sent to the eastern Mediterranean were not necessarily destined for Saloniki. To resolve their differences, Joffre and Millerand came to London on October 17th. Millerand and Lloyd George believed that the Dardanelles operation had failed and that Allied efforts should be transferred to Saloniki. But the British refused. The next day the news reached London that the Bulgarians had blocked the railway line at Skopje. Sarrail and his troops went north to the Vardar Valley. In contrast to the French, Kitchener was of the opinion that any aid to Serbia would come too late. Joffre threatened to resign if the British were not ready to send reinforcements to Saloniki.[117]

Above: Postcard from the city of Xanthi. In the Spring of 1916 not only the FFA was stationed here, but also the Naval Aviation Station Xanthi. At this time, this was Bulgarian territory; today it is Greece.

On October 29th, the government of Viviani resigned. Aristide Briand became the new prime minister and General Joseph Galieni became minister of War. Both men supported the continuation of the Saloniki operation. Joffre understood that his position would also become weaker if he was not able to bring the British into the boat when it came to the question of Saloniki. So on October 29th he came to London.[118]

During the meeting at 10 Downing Street, Joffre insisted on supporting the Serbs. If the attack by the Central Powers could be halted for a while, Greece could still join the war on the side of the Allies. The British were supposed to secure the port of Saloniki and the Vardar Valley railway line to Krivolak. He assumed that the port of Saloniki could handle 2,000 tons a day with the pier that was newly built by the French. The British Admiralty estimated the loading capacity at a maximum of 500 tons. Joffre threatened to resign as commander in chief if the British response did not meet his expectations. Even the climate in the alliance would suffer.[119]

The British General Staff vehemently opposed the Serbian enterprise. Still, the War Committee decided to make a concession. The cabinet also agreed and Joffre received the following written answer: They were ready to secure the railway line to Krivolak, the French were to take over the security to Veles. But if contact with the Serbian army was not established and maintained, the Allied forces would withdraw. But now the expedition to Serbia got sucked into the Dardanelles operation.

On October 14th, the Commander-in-Chief of the Gallipoli Front, General Hamilton, was replaced by General Charles Monro. On October 31st, Monro informed London that he was in favor of the evacuation of the Dardanelles Front. But just at that moment, Commodore Roger Keyes of the Royal Navy presented a plan for how a victory could still be possible. Kitchener was very impressed and when he visited Moudros he saw that Monro and the local Navy Commander in Chief, Admiral John de Robeck, were also impressed. He therefore suggested that the government leave the elite divisions intended for Saloniki (27th and 28th divisions) in Egypt and instead send three worn-out divisions from the Gallipoli front to Saloniki. But now the General Staff and the Admiralty protested, who disliked the idea of the new attack, Paris expressed political concerns. At a conference in Paris on October 17th, the project was buried, in order to maintain peace in the alliance.[120]

906 SALONIQUE — Aviatik descendu par nos canons dans les Jardins de la Tour Blanche. — L.L.
Visé Paris n° 2851

Above: A crashed German Albatros B.II is photographed for propaganda and presented at the White Tower in Saloniki. It is not clear if it is the aircraft (January 12th, 1916) of Oberleutnant Herbert von Chappius and Georg Trenkmann.

Le Taube captif à Salonique le 11 Février 1916

Above: The mechanics of the FFA 1, in their service dress are photographed in front of an aircraft. The sewed-on "1" on the left sleeves are clearly visible.

Right: A German field artillery station with minimal camouflage in a corn field. In the southeast, corn was the primary staple, not potatoes.

Facing Page, Below: An Albatros B-type (not a "Taube" as stated in the postcard) that made an emergency landing on February 11th, 1916 is shown off as a trophy in Salonika.

The Allied Expedition to Serbia in October–December 1915

On October 17th, Kitchener visited Saloniki and found the situation there extremely worrying: the three French divisions and the 10th British division were across the Greek border in southern Serbia. The latter division had received approval to do so shortly prior to that. Between these troops and Saloniki lay two Greek army corps. The Greek military authorities had not allowed the Allied troops to set up defensive positions near Salonika. Sarrail expressed concern that if the Greek forces took a hostile position against the Allied forces, it would be impossible for them to re-embark.[121]

Indeed, there had been another change in Greek domestic politics. As mentioned, the previous head of the National Bank, Alexandros Zaïmis, had been Prime Minister of Greece since October 6th. Zaïmis was an atypical Greek politician: *"A capable, patriotic and proven administrator, sympathetic to the Entente and - unlike Venizelos - not a politician. He has no ambitions to satisfy and no need to maintain a political organization at the expense of the general public."*[122]

He held no big speeches, gave no interviews, and remained silent. He was neither a Venizelist nor an anti-Venizelist, he belonged to no party. Also on the subject of foreign policy, he did not ascribe to either of the two views. He tried to steer a sensible, pragmatic course. Venizelos tolerated him, even if he occasionally criticized him.[123]

Zaimis had announced that Greece would steer a course of neutrality that was benevolent towards the Entente. The Greek mobilization was purely a precautionary measure and was directed exclusively against Bulgaria. The prime minister was encouraged in this course by Lieutenant Colonel Metaxas, who presented him with a corresponding memorandum on October 8th. It contained an analysis of the situation in which it was determined that an intervention by the Greek army in Serbia would lead to a catastrophe, since the available forces were insufficient to stop the Austro-German army. Greece should remain neutral until an appropriate time for entering the war arrives. On October 10th, he informed the Serbs that the Greek-Serbian treaty could not be fulfilled because compliance would lead to a clash with the Central Powers. However, he was ready to do anything for Serbia, that was also in the interests of Greece.[124]

While the British tried to persuade Athens to enter the war, the French ambassador in Athens was of the opinion that pressure should be exerted. One should cut off the grain supply to Greece, leading to a famine after two weeks. But the British envoy opposed such measures. Such methods would only turn the population against the Entente and damage British trade with Greece.[125]

The British government still hoped it could convince Greece to enter the war. On October 16th, foreign minister Grey offered Greece Western Thrace and the island of Cyprus, if it joined Serbia in fighting Bulgaria.[126] Zaïmis, however, stuck with his course of neutrality, that would be favorable to the Entente, even though this angered the Entente, especially Great Britain. But on November 4th, Germany announced itself. Berlin demanded that Greece should refuse to allow the Allied troops in Serbia to return to Greek territory. In return, Berlin would ensure that the Bulgarians did not attack Greece. Zaïmis again sought advice from the General Staff, which said: Greece should remain neutral and keep its army operational. The German request would not have caused any particular excitement, but when the Entente-friendly Zaïmis lost a vote of confidence in parliament on November 3rd and had to resign, the situation became worrying because the new Prime Minister Stefanos Skouloudis was known for his neutralism, *"not to promote German interests, but to protect those of his own country"*. After being sworn in on November 7th, he first tried to form an all-party government; but when that did not succeed, parliament was dissolved on November 11th and new elections were scheduled for December 19th, 1915.[127]

In the meantime, a note from Austria-Hungary arrived, calling for Allied troops to be disarmed if they retreated to Greece. Skouloudis replied on November 8th that the *Hague Land Warfare Regulations* would be applied, that is, disarmament would be carried out. But if the allies resisted it, Greece could not do anything. On the other hand, he would not accept any border violation by the Bulgarians. On November 9th he made a similar statement to the French envoy. On November 10th he repeated this: It was the duty of Greece to disarm the Allied troops after they cross the Greek border.[128]

The allies understood this as a threat and applied pressure. They demanded that the Greek army leave Macedonia. The allied fleet would occupy the Cycladic island of Milos and be ready to steam to Piraeus to put the Greek government under pressure. At the same time, all ships with cargo on board that were destined for Greece were detained in the ports of Egypt and Malta. An Anglo-French naval demonstration was prepared, consisting of three French and two English battleships, as well as other vehicles in the sea off Piraeus and Salonika. On November 12th, the British envoy warned that if

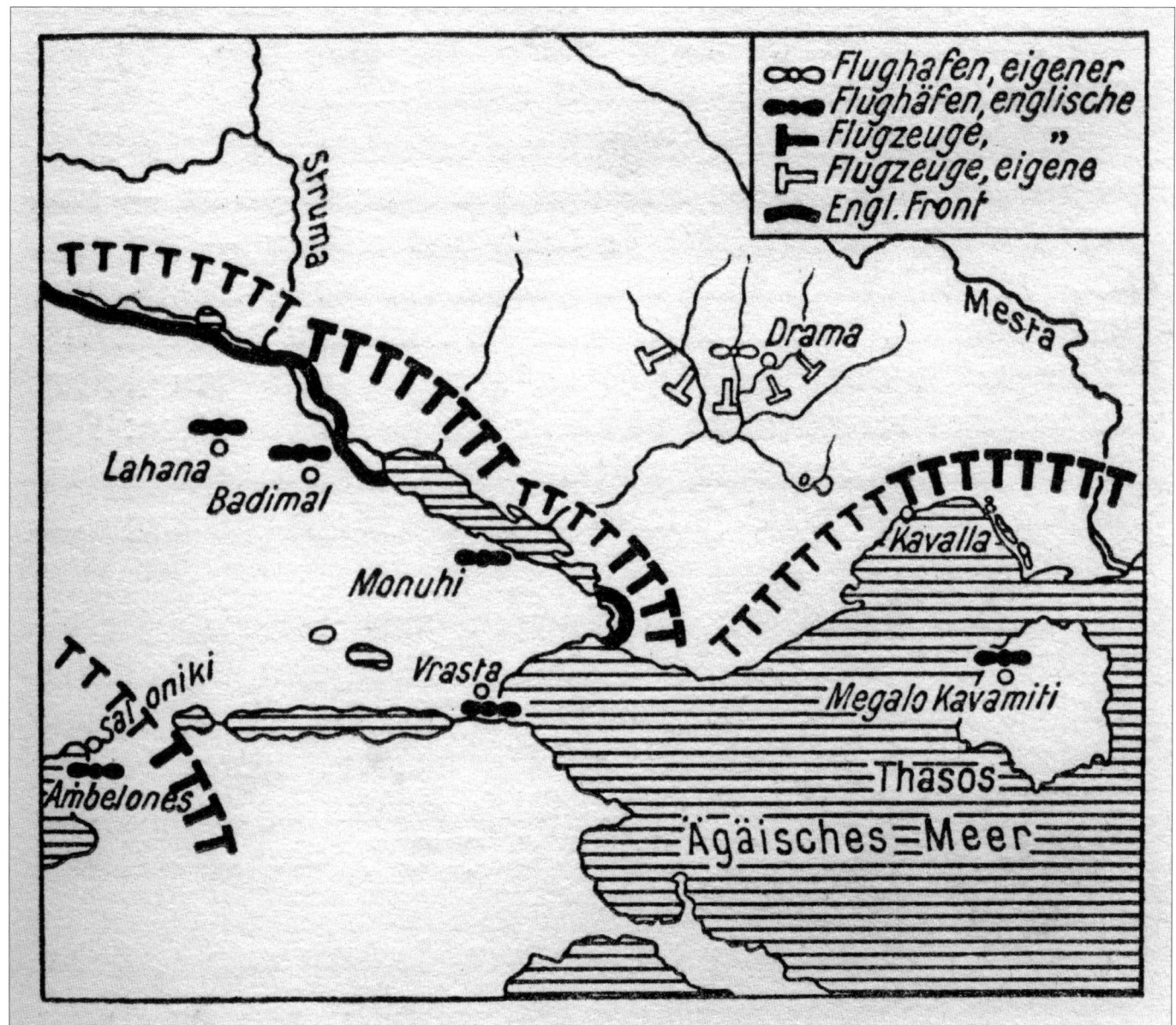

Above: This map shows the material inferiority of the Central Powers regarding the number of aircraft. The places that are only rarely mentioned in the text are shown here on the map. (Map from the book Haupt-Heydemarck, Georg Wilhelm: *Feldflieger über Mazedonien*)

a single soldier was touched in order to disarm him, there would be grave consequences.[129]

On November 18th, the French Finance Minister Denys Cochin appeared in Athens. During the conversation with King Constantine, the latter gave him his word of honor as commander-in-chief of the Greek armed forces that the Greek army would never attack the Allied troops retreating from Serbia. He was ready to support the Entente in every way, but he would not join the war. He was ready to protect the flank of the Allied troops with the Greek troops. *"The Greek king made it clear to the French minister that he was taking an Entente-friendly position. In his opinion, however, Greek interests made it necessary to remain neutral, at least at this point in time."* Cochin traveled to Saloniki the following day to inform Sarrail.[130]

On November 19th, the Greek government replied: The Greek army in Macedonia was there to protect the country against Bulgaria. A disarmament of the Allied troops in accordance with the *Hague Land Warfare Regulations* was not planned, but the Allies did not trust these assurances from the Greeks and occupied Milos on November 21st.[131]

In the meantime, minister of War Kitchener, coming from Saloniki where he had met with Sarrail, arrived in Athens on November 20th. During the meeting, held from soldier to soldier, there was full agreement: *"The British minister of war explained that he had never supported the Serbian adventure. It only came about because of the persistence of the French. In his opinion, the war would be decided in France and not in Bulgaria. The effort that was being made on smaller fronts like Macedonia was useless and a waste."* These statements gave Constantine hope that the *Allied War Council* would soon stop the Balkan enterprise, especially since the Commander-in-Chief of the Gallipoli Front, General Monro, had the day before also called for this front to be closed.[132]

The mobilization of the Greek army targeted only Bulgaria. Should the Allied troops in Saloniki be strengthened in such a way that they alone could repel a Bulgarian attack, the Greek army could be demobilized or at least withdrawn from Saloniki. At the moment the Greek army was protecting the Allied troops by its mere presence in Macedonia. If demobilization was demanded immediately, it would translate into a clear invitation for Bulgaria to enter Greece.[133]

King Constantine gave Kitchener his word that he would do nothing against the Allied troops, or disarm or intern them. At the bottom of his heart he was anti-German. He only rejected an attack on Bulgaria because he did not want to destroy his country. He was in an impossible situation. If the Bulgarians were to attack the retreating Allied troops on Greek territory, it would be a violation of Greece's borders. But the Germans had given him to understand that they would interpret any intervention against Bulgaria as an act of war. .[134]

Kitchener believed him. He told the British Naval Attaché, *"He considered King Constantine to be an honest, sincere soldier, inclined towards the Allies, with a clear understanding of the war situation in the East. When Kitchener left Athens all matters seemed amicably settled."*[135]

Further statements by Constantine to Kitchener, mentioned in the official German history of the First World War, in which he said he would do nothing against troops of the Central Powers advancing into Greek territory but rather only against Bulgarian troops, does not seem entirely credible. They must have been made, but hardly to Kitchener. It was a signal to the German side. Constantine's later politics supported this.[136]

During a later discussion between Kitchener and Premier Skouloudis, Kitchener said, among other things, that London was wondering why the Greek government had invited the Allies to Saloniki, but did not themselves join the operation. He had agreed to the operation because he believed Greece would take part. Skouloudis replied: *"If Venizelos invited you as you say, and which I believe, although he vehemently denies it, then he did it without involving the official bodies. Neither the crown, nor the parliament, nor the cabinet were aware of this step. So I would like to reiterate that Greece never invited you."*[137]

On November 23rd, upon the instigation of Sarrail, the Entente envoys demanded an official declaration from Skouloudis that there would be no attempt at disarmament. They would not want to force Greece into war, but firmly reject any restriction on the freedom of movement of the Allied military forces on land or at sea. Greece must provide the railroad network and the port of Saloniki for the removal of the troops.[138] On November 26th, the French and British ambassadors repeated and refined these demands: *"Withdrawal of the Greek troops from Saloniki, control of the railway lines to the border, permission to build defensive lines around the city and across the Chalkidiki peninsula. Furthermore, the Allied Navy's right to control merchant ships, and should enemy submarine bases be found in Greek waters, to destroy them."*

In order to enforce acceptance of these demands, the commercial and economic blockade was stepped up. On November 28th, Athens accepted these demands under protest, but suggested that the details of the implementation should be done with due regard for Greece's neutrality, for which a liaison officer would be sent to Sarrail. The Allies accepted this, and so Lieutenant Colonel Konstantinos Pallis went to Sarrail's headquarters.[139]

Sensible solutions emerged from the talks between Sarrail, Pallis and Moschopoulos: the 11th Greek Division would remain in Salonika. The railroad should be operated jointly. The Greeks would allow the construction of fortifications, but would officially protest against them. Should the Bulgarian army attack the Allied troops on Greek territory, the Greek army would stay out of the fighting. The Allies agreed and the Greeks moved the other divisions of the 3rd Corps to the east. The area between the border and Saloniki was, for all practical purposes, under Allied control. Milos was evacuated and the naval maneuvers called off. They had come to agreement. On December 10th, Berlin said that if the Greeks granted the Allies such privileges, the Germans should be granted the same rights.[140]

But another question arose. Should the intervention in Serbia continue or not? London was nervous, especially in light of military setbacks.

Above: Hand-drawn postcard of P. Hiemann from Macedonia: "On Lake Doiran".

The Serbs were battered and poured into Albania en masse. General Sarrail had to slowly pull out of southern Serbia. The British expedition in Mesopotamia went badly. The situation on the Dardanelles was bleak if the Keyes plan did not succeed, but that required fresh troops and these could only be the divisions intended for Saloniki. Given this situation, the British Cabinet was in favor of evacuating Saloniki as soon as possible.

On his return to London at the end of November 1915, Kitchener reported his experience in Greece to the General Staff so negatively that *"the reluctance of the British General Staff to operate in this war zone led to the unequivocal recommendation that the Saloniki mission be terminated. The General Staff ordered that we withdraw from the Dardanelles and Saloniki and pool our forces in defense of Egypt."*[141]

On December 1st, he informed the cabinet that it was the official stance of the French to continue the operation and that the British should stay there also. In his opinion, one had to evacuate Salonika if one wanted to keep Egypt. The majority of the cabinet favored the withdrawal from Saloniki. They decided that the Prime Minister, the Minister of the Navy and the Minister of War and their advisors should discuss the situation with their French counterparts in Calais on December 4th.[142]

There had been political changes in France that had an impact on developments in Saloniki. In October 1915 Aristide Briand had become the new Prime Minister and Foreign Minister. He had been in favor of the Saloniki Project from the start. In addition, his coalition partners were the Radical Socialists, of which Sarrail was a well-known member. If one were to withdraw from Saloniki, Sarrail would lose his post because Joffre would no longer give him any further command. In order to include Joffre in the decision-making, Briand made him commander in chief of all French armed forces, making him Sarrail's superior. So Joffre's appointment also made him a proponent of the Saloniki operation.[143]

On December 1st, the British made it clear to the French that they were in favor of the withdrawal. Briand replied that they should stay in Saloniki and increase the number of troops to 300,000. Should the British be of a different opinion, this would lead to bilateral tensions. The British realized that if they did not give in, the Briand government could fall.[144]

At the conference in Calais on December 4th, Aristide Briand went to great lengths to convince the British of his point of view. He even wanted to increase the Allied forces in Saloniki to 350,000 to 400,000 men. He was convinced that glory would be earned in Serbia.[145] But the British remained steadfast. Ultimately, Prime Minister Herbert Asquith expressed his view: The troops could not be left in Salonika because there was a threat of catastrophe. They had to prepare to evacuate. The French agreed.[146] But when Briand informed his cabinet, a majority was in favor of continuing the Saloniki operation.[147]

Two days later, Chief of Staff Archibald Murray met with his French, Russian, Italian and Serbian colleagues. All were against evacuation. When they found out about this in London, they feared major quarrels within the alliance and instructed Gray and Kitchener to go to Paris and negotiate a compromise. When the men arrived on December 9th, they encountered an atmosphere of deep suspicion. They refrained from discussing the question of evacuation, but limited themselves to talking about how the British troops could best be deployed on the ground. On December 13th, Kitchener was able to report to the *War Committee* that tensions had been cleared up and that preparations were being made to build defense positions at Saloniki. So the French had prevailed. In addition, the decision to leave the Gallipoli peninsula had been made on December 8th. This made it clear that they had committed to stay in Saloniki indefinitely. This was a decision that was not supported by British politicians, the army or the navy. Falls correctly writes: *"A stranger episode would be hard to find in the annals of military history."*[148]

On December 14th, one of Sarrail's staff officers announced the view of his chief to some war correspondents: *"You can consider this as a conclusion, [...] the Allies will not evacuate Saloniki until a European peace is signed."*[149]

As mentioned earlier, the Greek parliament had been dissolved on November 12th and new elections were scheduled for December 19th, 1915. Venizelos called for a boycott; he claimed that the elections would be rigged and that a large part of his voters had been called into the military and would not be allowed to vote. The result was that the election campaign was fought between two conservatives, Dimitrios Gounaris and Dimitrios Rallis. While around 730,000 voters voted in the June 1915 elections, there were now only 230,000.[150] On the one hand, Venizelos' call for a boycott worked, on the other hand, those conscripted into the army were not permitted to vote. But the election result did not mean that the country was now on a pro-German course or, as Venizelos and the Allied propaganda claimed, that people had, by abstaining, voted for Greece to enter the war under the leadership of Venizelos. Constantine and the conservatives continued to be neutralists, benevolent to the Allies.

At the beginning of December, Constantine gave the American AP Correspondent Hibben an interview. He welcomed the fact that he could speak with a representative of a neutral country. *What is happening to Greece, could happen to any neutral country. "The fundamental reason for the threatening attitude of the Entente towards Greece today and the dire situation of my country is the Entente's unfounded assumption that Greece will betray the Entente to Germany at the first opportunity."*

Greece had already offered three times to join the war on the side of Entente. Constantine gave the facts. *"Despite all this goodwill, the Entente is now demanding - in the form of an ultimatum - that Greek troops be withdrawn from Saloniki, that is, from all of Macedonia. In this way we leave our population defenseless against Bulgarian comitadjis (guerrillas) or war atrocities. Why does the Entente treat me as if I were a Negro king of a Central African tribe?"*[151]

The Greek populace had had enough of war. When Venizelos maintained that he, Constantine, had agreed to the Allied landing in Saloniki, this was false. The entire Allied Balkan operation had been erroneously planned, and would fail due to too few troops. He declined to pull Greek troops out of Macedonia. Greece must remain a sovereign state, and be able to protect itself. When Hibben asked what the king would do if the Entente exercised further pressure, he answered: *"Then we will demobilize our armies and wait for things to run their course. What else should we do?"*[152]

At the end of December, the French chief of staff Noël Édouard sought out the king after a visit to Sarrail. Similar to Kitchener, the two soldiers understood each other immediately. Castelnau understood that Constantine was prepared to do everything in his power to assuage the Entente, other than enter the war. Castelnau did not visit Venizelos, which prompted his supporters to show a great sympathy in order to impress him, which in turn strengthened the diplomats' resolve.[153]

After this overview of the *political* developments between October and December 1915, allow us to return to the *military* events during this time.

The Austrian-German Campaign Against Serbia in the Fall of 1915

After the unsuccessful campaign of Feldzeugmeister Potiorek, plans for a new campaign against Serbia had been developed over and over again, but none of them were actually finalized. From July to September 1915, Lieutenant Colonel Richard Hentsch was assigned to the General Staff of the Austro-Hungarian Army by the German Supreme Army Command, in order to explore possibilities for a renewed attack across the Sava and Danube in agreement with the High Command of the Austro-Hungarian Balkan Armed Forces. He made some appropriate proposals.[154]

Three of the four Serbian armies were in the north of the country, that is, on the Sava-Danube front. The main thrust would be directed against these. But there would also be attacks from Austrian Bosnia and Bulgaria, in order to surround the enemy and to direct it southwards. When Bulgaria signed the military alliance with the Central Powers on September 7th, the situation was clear. Serbia would be attacked from three sides at the same time.[155]

On September 16th, the German General Field Marshal August von Mackensen took over the command of the Army Group named after him, which consisted of three armies: the 11th German, the 3rd Austro-Hungarian and the Bulgarian Army. The Army Group was to smash the Serbian army and open and secure the connection via Belgrade and Sofia to Constantinople. The 11th German Army consisted of 130 battalions and 136 batteries, including 38 heavy and seven heaviest. The 3rd Austro-Hungarian Army had 74 battalions and 124 batteries, including 31 heavy and five heaviest. The 1st Bulgarian Army, reporting to Mackensen, which would attack only five days after the Central Powers began their attack, consisted of 88 battalions and had about 48 batteries.[156] In northern Serbia, including the defenders of Belgrade, there were about 163 Serbian battalions with 437 guns.[157]

The campaign against Serbia was led predominantly by German and Bulgarian troops, because the Austrian leadership had only two and a half-strong divisions available, while the German side had 11 divisions.[158] The offensive began on October 6th with the crossing of the Danube near Belgrade, and the Save near Smederevo. Thanks to the careful reconnaissance work of Lieutenant Colonel Hentsch, this went off relatively smoothly despite the enemy ready to defend itself. Belgrade fell on October 9th, but the Serbian army continued to offer fierce resistance and the mountainous country south of Belgrade also slowed the advance. It was impossible to encircle the enemy. It was not until the end of October that the combined forces of the Central Powers succeeded in pushing the Serbs back on Kragujevac. On October 14th, Bulgaria declared war on Serbia. The 1st Army was subordinated to Army Group Mackensen. The independent 2nd Army received the order to interrupt the railway line from Saloniki-Skopje-Niš in order to cut off Serbia from the Allied supplies. Despite all its efforts, 1st Army made very slow progress against the Serbian defenders. The 2nd Army, however, encountered far less resistance and reached the Niš - Saloniki railway line near Vranje, north of Skopje, on October 16th, and cut it off. Further south, two divisions of the 2nd Army occupied the railway line at Kumanovo and in the Vardar Valley near Veles on October 20th. Even further south, Macedonian legionaries advanced against Strumica. Thus, towards the end of October 1915, a 100 km long section of the line was in Bulgarian hands, and thus the only useful supply connection between the bulk of the Serbian army and the Entente Army in Saloniki was interrupted.[159] On November 1st, Kragujevac, about 150 km south of Belgrade, was captured by the combined forces of the Central Powers and the Bulgarians, but the Serbs managed to evade the enclosure by moving to the southwest.

This withdrawal movement into an area that could no longer be replenished made it possible for the German units, which were also suffering from lack of supplies, to push the remnants of the Serbian army further south-east to Prizren. From there, about 140,000 Serb soldiers, including 50,000 untrained recruits, but still with 25,000 horses, 7,000 draft oxen and 95 artillery pieces, fought their way to the Adriatic coast near Durazzo (Durrës) and Skutari (Shkodra) in Albania. Here they were picked up by Italian and later by French ships and brought to Corfu in disregard of Greek neutrality. There, after a period of recovery, they were trained again and combined to form newly formed associations. These were later moved to the Saloniki front.[160] At the end of December 1915, after a seven-week campaign, all of Serbia was occupied.

Above: Albatros C.I on a mission.s

After the capture of Niš on November 5th, when the railroad from Germany via Bulgaria to Turkey was opened, and on the same day the first tow trains with ammunition for Turkey started, the goal of the Serbian campaign had basically already been achieved. Now it was time to plan the next steps. For this purpose, the two commanders in chief of the armies of the Central Powers Erich von Falkenhayn and Franz Conrad von Hötzendorff met in Pless. Colonel Peter Gantschev from the Bulgarian General Staff also took part in the talks. Conrad von Hötzendorff suggested that after the end of the Serbia campaign they should turn together against the Saloniki army *"until the enemy alliance was completely discredited in all Balkan states."* This was the only way to prevent Romania and Greece from joining the Entente. Falkenhayn had agreed to an attack on the Salonika army.[161]

Falkenhayn was very well-informed about the Saloniki army through the German consul general in Saloniki and the military attaché at the German embassy. In addition, Falkenhayn maintained contact with the Greek government through the Foreign Ministry and had a lively exchange of views. They supported King Constantine, who wanted to keep his country out of the war. It was recognized

Facing Page, Above Left: Kagohl 4, squadron 20. Observer Lt. Peres. Portraits of some members of Kampfstaffel 20. (Slg Jörg Kempf)

Facing Page, Above Right: Lt.d.R. Georg Schlenker, born on October 1st, 1890 in Weissnig (Kreis Torgau), died on February 24th, 1947 in the NKVD special internment camp Buchenwald. (Slg Jörg Kempf)

Facing Page, Below: Damaged Rumpler C.I .541/16 (?) of Oblt. Fritz Heising. On the very right the marking of the Kasta 2 [3], Kampfstaffel 23, is visible. Oblt. Heising did not go to the Balkans, he stayed with Kagohl 4, squadron 23. The same aircraft is pictured in the book by Jörg Mückler, *Deutsche Bomber im Esten Weltkrieg*.

that an armed conflict with Bulgaria could arise at any time. The Entente controlled the sea and was able to stop the supply of food to Greece. Since the Greek supplies only lasted for a month, the government could be blackmailed. If Greece could be kept neutral, a great deal would have been achieved. It would have been desirable if Greece, as a neutral state, had expelled the Entente troops from its territory.[162]

When the offensive against the Serbs ended at the end of November, it was revealed from Athens that the Greek government had announced on November 9th that it would demilitarize and intern Serb forces who came across the border. It even indicated that it would have to proceed in a similar way with the Allied troops if they did not embark again immediately. In addition, trains going north from Gevgeli were not carrying any troops, but only supplies. Falkenhayn got the impression that it might be possible to drive the Saloniki army back to Greece. On November 22nd, he sent a request to the commander-in-chief of the Bulgarian army, General Nikola Jekow. On November 24th it was learned that the British were exerting massive pressure on the Greek government and that Kitchener was in Athens. Konstantin explained to him that *"if the Franco-English troops did not 'at least leave' when they retreated into Greek territory, but continued the fight, he would be forced to let his troops evade to both sides and open the borders to the pursuing troops. But it was also clear that Bulgarian troops alone would in no way be tolerated on Greek soil."*[163] This statement was a signal to the German side that Constantine would not resist an invasion of troops of the Central Powers.

It was now clear that a unilateral Bulgarian attack on the Saloniki army, across Greek borders, would lead to Greece's entry into the war. To avoid this, German troops would have to go into pursuit. But before any decisions could be made, the Saloniki Army initiated its withdrawal from Serbia. Troops of the 2nd Bulgarian Army pushed in and drove the Allied troops back to the Greek border. On December 13th, the last soldier in the Saloniki Army left Serbia. The Bulgarians respected the Greek border and stopped pursuing them 2km from the border.[164]

A Bulgarian general staff officer said: *"For us Bulgarians the war is over. We have everything we wanted. Why should we follow the enemy beyond the Greek border if we cannot keep any of the territories conquered? It can only be an advantage for Germany, too, if we do not force the enemy to embark again in Saloniki. We thereby tie up a large Entente army and force the enemy to keep sending reinforcements, and we would certainly repel any attack into Macedonia."*[165]

Disputes between Falkenhayn and Conrad von Hötzendorff, as well as transport problems for any German troops that might be arriving, meant that a possible attack was postponed, and then finally abandoned at the end of December – with reasons similar to those the Bulgarian staff officer had stated.[166]

Facing Page, Above: Kagohl 4, Staffel 20: An Albatros C.III of Kasta 20 in flight on the western front. Behind the iron cross one can recognize the individual marking of the pilot or the observer. This half squadron was transferred to Romania/Macedonia, to KG 1. (Slg Jörg Kempf)

Facing Page, Below Left: Kagohl 4, Staffel 20: Lt. Gotthard Rossteuscher, born on September 23rd, 1891 in Charlottenburg, died on December 16th, 1962 in Baden-Baden. (Slg Jörg Kempf)

Facing Page, Below Right: Kagohl 4 Staffel 20: Staff leader Oblt. Franz Kaestner. (Slg Jörg Kempf)

Above: This photo was taken after December 1917. On December 4th, 1917, Alfred Keller (named: Bomb-Keller) was awarded the Pour le Merite and is photographed standing on the right with the award around his neck. To be exact, the photo must have been taken before December 12th, 1917, since on this day Rudolf Kleine (leader of Kagohl III) fell [the noticeable back of the head about 8.5cm from the left in this photo]. (Slg Jörg Kempf)

Left: Kagohl 4, Staffel 20: Before leaving on a hunt with a car. The lettering on the car says "Kgohl 4, Staffel 20". In front next to the driver is Lt. Fritz Kempf. The officer with the gun is Oblt. Kaestner, and Vzfw. Werner Voss stands to the right of the car. At this time the Staffel 20 was in Carignan, France, on the Chiers river. (Slg Jörg Kempf)

Albatros C III xx6/16 at Kagohl 4, Staffel 20. Crash of Lt. Otto Hunzinger in Carignan. The painted "20" in front of the iron cross points to membership in this unit. The individual marking on the rear fuselage is painted around it. (Slg Jörg Kempf)

Kagohl 4, Staffel 20: Lt. Friedrich Löhr in Carignan. (Slg Jörg Kempf)

Kagohl 4, Staffel 20: On this photo he placed a flower on the jacket of his uniform. No other life dates of Lt. Otto Hunzinger are known.

The Allied Advance into Serbia

The Allied expedition to Saloniki was intended on the one hand to induce Greece to enter the war and on the other hand to save Serbia. Both were doomed from the start. There was no military plan for conducting the operation. There was not even any agreement on what the Allied forces would do militarily after the landing. They had not even agreed on a unified high command. Integral staff talks between the staffs of the Allied, Serbian and Greek armies had not taken place. From the beginning there were political differences. From the beginning, French politics was influenced by domestic political motives. On the British side, the military tended to predominate; the general staff rejected the principle of the whole enterprise.[167]

The French *Armée d'Orient*, commanded by General Sarrail, consisted of the 156th, 57th and 122nd Divisions, which were disembarked at the port of Saloniki between October 12th and November 5th, 1915. The British were initially only represented by the 10th Division, which was very battered. Their first units arrived in Saloniki on October 5th. General Mahon arrived on the afternoon of October 8th. On October 17th, all units of the 10th Division had arrived in Saloniki. Then there was an interruption of the transports until November 5th, when the first units of the 22nd Division arrived.[168]

On October 15th, at the urgent Serbian request, Sarrail sent a regiment and artillery of the 156th Division by train to the Serbian train station Strumica (halfway between the border and Veles) and on October 16th another regiment to Gevgeli. The French thus controlled the Vardar (Axios) Gorge. At that time the railway and a rather dilapidated road, which was only passable in parts, ran through this gorge. There were no road bridges over the river until Veles and the railway bridges were not passable for road vehicles. Here and there was a ferry, but they couldn't transport larger vehicles. The area above the gorge is mountainous and barely accessible by road.[169]

On October 17th, the Serbian unit that had previously secured this route was relocated to Veles because the Bulgarians were approaching. Sarrail replaced it with a unit of his own. On October 20th, General Maurice Bailloud and his troops took over the security and on October 21st rebuffed the first Bulgarian troops at Rabrovo. In the next few days, Sarrail pushed further troops to secure the railway line north to Krivolak. Due to the small number of troops and the difficult terrain that was also difficult to secure, Sarrail could not secure the railway line further north. The Bulgarian 2nd Army conquered Kumanovo, Skopje and Veles on October 22nd, thereby breaking the connection between the French units and the Serbian army. The French army secured both sides of the railway line from about Gradsko to the south.[170]

All along, Sarrail had been urging his British colleague Mahon, who did not report to him, to help secure the railway line in Serbia. But Mahon had strict orders to stay in Saloniki and not to cross the Serbian border. The British government clung to the fiction that it only went to Greece to enable the Greeks to support the Serbs. It wasn't until October 22nd that the British government allowed Mahon to move his troops within Greece. Mahon wanted to move his troops along the railway line from Saloniki to Constantinople on Lake Dojran, that is, east of the French positions in Serbia. But now the Greeks were causing problems: on October 25th, they forbade British troops to settle so close to the border. On October 26th, Kitchener informed Mahon that under no circumstances should he abandon the French. Should it be necessary to cross the border, Mahon would get permission to do so. Although there were again disputes with the Greeks about the use of the railway to Lake Dojran, Mahon avoided this point of conflict by transporting the troops of the 10th Division to Gevgeli and marching east from there. In London the *War Office* ordered that the 26th Division, which was still in France, should be transported to Saloniki.[171]

On October 31st, Sarrail received a message from the Serbian Supreme Commander, Voivode Radomir Putnik, that he intended to retake Skopje with two divisions. He asked for French support from the south. On the same day, Sarrail learned that the War Office had agreed on October 29th that the 10th Division would take over the railroad security as far as Krivolak. That was an exaggeration given the small strength of the British, but it still freed up the 156th French division. Sarrail got the impression that he could help the Serbs substantially and therefore attacked this division in the direction of Strumica in Bulgaria. He postponed the attack in the direction of Skopje until the 122nd Division landed in Saloniki on November 1st. The large attack with the 122nd and 57th Divisions against Veles lasted from November 3rd to 13th. The French managed to cross the Crna, despite

Above: Kagohl 4, Staffel 20: in front of their mess in Le Chatelet, Summer 1916. On the door are also the names of those who lived there: Lt. Wilms and Feldwebel Werner Voss. From the right it is possible to recognize Lt. Hauff, Lt. Fritz Kempf and Lt. Frommherz. Sitting in the middle, the squadron leader, Oblt. Franz Kaestner. Romania's threat to enter the war demanded German action: air units were transferred. (Slg Jörg Kempf)

Right: Kagohl 4, Staffel 20: Lt. Hauff. (Slg Jörg Kempf)

Bulgarian counter-attacks. The 156th Division had meanwhile reached the Kostorino border crossing. But in the meantime the French government had learned what a bad situation the Serbian army was in and that the offensive against Skopje was not making progress. On November 6th, the Serbian leadership informed Sarrail, *"If the connection with Saloniki is not established in the shortest possible time, the Serbian army can no longer be counted on. Joint operation by our and French troops is*

Above: Kagohl 4, Staffel 20: Vzfw. Werner Voss arrived from the FEA 7 to Kagohl 4, Staffel 20 in 1917. He also went to the Balkans with Lt. Fritz Kempf.

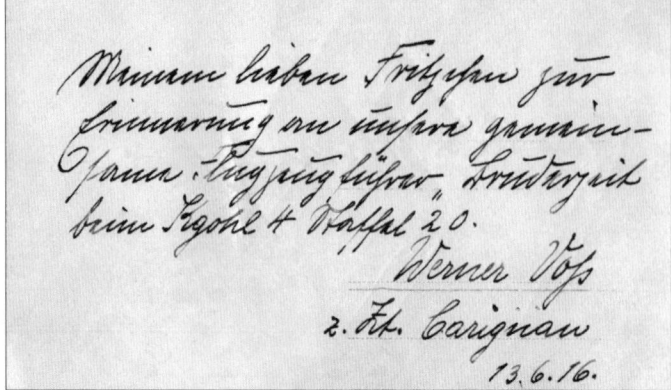

Above: "To my dear Fritzchen, as a remembrance of our "brotherly" time together at Kgohl 4, Staffel 20. Werner Voss, at the moment in Carignam, June 13th, 1916." (Slg Jörg Kempf)

therefore necessary." Sarrail had already stopped the offensive on November 12th, following warnings from Paris. He wanted to wait for the Serbs to arrive. In addition, since no one knew how the Greek government would react, Paris ordered Sarrail's withdrawal on November 14th. Sarrail delayed the withdrawal a little longer, but when he learned on the 22nd of the final failure of the Serbian offensive against Skopje, he initiated the withdrawal because there would be no further contact with the Serbs.[172]

The Allied retreat had barely begun when the German military attaché in Athens handed King Constantine a telegram from Falkenhayn dated November 29th, 1915. In it, Falkenhayn stated that if Greece did not succeed in disarming the retreating Allied troops or enforce their immediate embarkation, it could happen that, depending on the development of the fighting, the Germans and Bulgarians could be forced to cross the Greek border.

After consultations with Skouloudis, Constantine replied: Greece does not agree to a violation of its territory. However, if such a move was not hostile to Greece, no armed action would be taken against it, provided the Bulgarians guaranteed they would not raise any territorial claims against Greece. Neither the Bulgarian King nor his son were allowed to enter Salonika. The Germans would have to lead the command.[173]

Falkenhayn was ready to guarantee the territorial integrity of Greece. Occupied territories would be returned after the end of the campaign and damage caused would be paid for. Greece should withdraw its army from Macedonia so that there could be no accidental clashes. Constantine refused the latter, because the Allies had not been granted this either. Falkenhayn then suggested that Greece should prevent Allied landings at Kavalla and Katerini. The King also refused to do this on the grounds that the country was neutral.[174]

The withdrawal itself went differently. In the last days of November, the 10th Division got caught in the storm that led to massive frostbite losses in Gallipoli, and suffered terribly. Afterwards, Bulgarian troops had to be rebuffed, which also caused great problems for the French. Their retreat was accompanied by constant Bulgarian attacks. On December 12th, the British and French units returned to Greek soil. The Bulgarians did not follow suit, but respected the Greek border.[175]

The French lost 143 officers and 4,822 men, the British 33 officers and 1,176 men; in the latter there were 99 dead, 386 wounded and 724 missing. The total losses of the Bulgarians are said to have been 6,277.[176]

The whole operation had been a French one. The British had been more or less assistant troops,

Above: Kagohl 4, Staffel 20: Before departing Le Chatelet to Rasgrad in Romania: "from left: Löhr, Pfeffer, Gutknecht, Hauff, Peres, Schlenker, Rossteuscher, Frommherz, Voss, Hunzinger, Oblt. Dörstling. Seated: Squadron leader Oblt. Kaestner." (Slg Jörg Kempf)

nothing more. The British General Staff and the British government had rejected the whole operation as pointless, but had been forced to participate for political reasons. In fact, the whole operation, as performed, was completely pointless. The troops arrived in Saloniki far too late to do anything. The three French divisions were only able to protect the railway line against the attacks of the 2nd Bulgarian Army in the south as far as Gradsko for a while, further north this was impossible, because the Serbian army could not withstand the onslaught of this Bulgarian army. Whether the presence of the three French divisions saved the Serbian army from being surrounded by the 2nd Army, as the case assumes, can justifiably be doubted, because the French troops did not even advance that far north.

In order to make this operation a success, it would have been necessary to move considerably more troops to Saloniki much earlier, that is to say, at least the 250,000 men mentioned in the beginning. These troops would have had to have been fresh and should not have been units that had experienced the Gallipoli Battle, or that consisted of freshly formed reserve units inexperienced in combat. The argument that the Serbia undertaking indirectly contributed to the success of the Champagne offensive because the French troops intended for the Salonika front remained in Champagne, is an attempt to justify something that cannot be justified. Whether an earlier landing would have prevented the Bulgarians from entering the war is only speculation. What is certain is that the Saloniki enterprise was an enterprise forced by French politics that never stood a chance.

But who was actually held publicly responsible for the defeat in the Allies? The American war correspondent Bidden makes this crystal clear: *"General Sarrail refused to take responsibility. He did everything he could with the troops entrusted to him under terrible transport and living conditions. Of course, it was not unimportant to the governments of France and England to determine*

Above: "The whale DD of Oblt. Kaestner" was a Roland C.II (Slg Jörg Kempf)

who was responsible for the failure. The Serbs could hardly be blamed after all they had suffered – even if London and Paris had a strong tendency to blame the Serbs for their own defeat. The only people who could safely be blamed to take full responsibility were the Greeks. Not Venizelos, of course, because he had tried to bring the army into line with the Entente, and it was hoped he could still do it. So it was King Constantine on whom the Entente press

Above: Vizefeldwebel Johannes Keller, who came to the Balkans via Flugpark 11.

Above: On the loading train station in Autumn 1916 before the transfer from Le Chatelet to Rasgrad in Bulgaria. The loading is complete. For information: In Autumn 1914 the average speed of a military train was 15 km/h. This doubled to 30km/h by the end of the war in 1918.

Right: Lt. Fritz Kempf in the sleeping car in the southeast region. This sleeping car had one cabin for two officers. Mj. Siegert introduced this system with sleeping quarters, depots and workshops, who made his bomb teams very flexible in this way. Several trains could quickly transfer a whole unit and bring them to deployments.

dumped all the blame. *The Allied soldiers on Greek soil were of the same opinion. King Constantine's wife is the sister of the German emperor. Therefore, King Constantine must have been inclined to the Germans and he was forced to betray the Allied troops in Serbia to Germany.*[177]

Indeed, French supporters of the failed operation feared that the opposition might attack them and raise public criticism. So it was necessary to build a smoke screen that would hide the ignorance and mistakes of political leadership. Both groups had to be diverted from the facts and find a scapegoat to blame. *"They began to put a thick veil of relentless censorship over anything that could destroy the public's confidence in their abilities. With the help of Venizelos, they cleverly spread the following picture in the media, which proclaimed the official truth: If the Serbs were to withdraw and the Germans and Bulgarians would advance, then it would be Greece's fault. They stressed that Greece was Serbia's ally and since it did not come to the aid of its ally Serbia, it was betrayal of the allies, and Serbia had to go. Needless to say, the critics of Greece and its King never considered providing evidence to support their allegations. […] This was the origin of the smear campaign that was supposed to support one of the boldest falsehoods of the world war."*[178]

One can only marvel at this *"logic"*, but for the Allied public, it was the most convenient.

The Allied Occupation of Macedonia January–August 1916

Macedonia Under French Occupation

Normally, the following period in the English and Greek historiography would be described as *The Entrenched Camp of Salonika*.[179] In reality, this and other parts of Greece experienced a kind of occupation.

Sarrail enjoyed inserting himself into politics. "During his entire life, Maurice Sarrail had an insatiable hunger for political intrigues."[180] His future place of activity, Greece, or rather Saloniki, offered the right breeding ground for this obsession, as we will see later. In his political views and judgments, Sarrail did not differentiate. His maxim was: whoever is not for me is against me; Or to put it another way: whoever is not for France is an enemy, and this also applied to neutralists. In his eyes, the neutral King Constantine was a friend of Germany who had to be fought. Furthermore, Sarrail was a fanatical Republican who had wanted the King's removal from the moment he arrived in Greece. *"Acting as proconsul on the part of Greek territory that he had brought under his control a few months ago, he is neglecting ancient Greece. He went to Athens in April 1916 to make a courtesy visit to King Constantine. It made him feel "employed" in the Salonika region"*.[181]

The occupation began in Corfu. This island was given to Greece by Great Britain as part of an international contract in 1864, and was declared forever neutral. France was also a signatory. Corfu therefore had an internationally-accepted neutral status, just like Belgium. But now some French military ships appeared off the island, and an infantry division occupied it on January 1st, 1916. But the French government explained that it was not an occupation. The country justified this with the claim that Corfu was a German submarine base, probably because Wilhelm II had a summer residence (Achilleion) there.

Admiral du Fournet[182] wrote: *"The intelligence service had portrayed Achilleion as a perfectly organized submarine base, with a mooring that channeled the fuel to the sea. The occupation of Corfu allowed all of these legends to disappear. The quay was made for the palace's boats; The oil supplies were simply a little gasoline for automobiles. The pipeline was just the main sewer."*[183]

The whole thing was an untenable claim. In fact, the occupation was against international law. It was justified to the Greek government as having humanitarian motives. Then, beginning on January 14th, the last Serbian troops were brought to the island, which also became the seat of the Serbian government in exile.[184]

An Entente declaration spoke of a "pure duty of humanity". The "conquest" of the Achilleion as a submarine base was portrayed in the French press as a great feat, although the only people there were the castle administrator and a gardener.[185] A total of 133,000 Serbs were brought to Corfu, of which 3,681 died of exhaustion, typhoid and dysentery. When the survivors recovered, six divisions were formed. France supplied the weapons, the artillery and the animals to be transported. In mid-March the Serbian units were ready for transport.[186]

Another occupation of Greek territory had preceded this. On December 28th, 1915, French sailors from the cruiser *Jeanne d'Arc* occupied the island of Kastellorizo. The French consul had incited a group of French-friendly residents of the island into a revolt. The official justification for the occupation, however, was to prevent possible Turkish reprisals. When the Greek Evzones wanted to land, the French blocked them. The Turkish artillery then promptly shelled the island because of the French occupation.[187] The real reason for the occupation of the island was that the French wanted to establish a naval base on Kastellorizo to fight German U-boats.[188]

The fear of German submarines in Greek waters reached proportions that bordered on hysteria. Mainly in the French press, Greece was accused of *"treason"*: Greece was allegedly a hornet's nest of German submarines. The Greek coast was full of hiding places for German submarines. The submarines were said to have used the Greek ports without the slightest hindrance. In December 1915 it was reported that – in Saloniki of all places – a supply depot for the submarines had been discovered. Allegedly, the Allies found countless oil depots on the coast, and these would be tracked down again and again by the vigilance of the Allies. *"Some newspapers published Jules Verne-style inventions: one swore that there was a workshop on Athens' popular Phalerum beach. This diabolical,*

This Page: Three photos of Gotha G.II 204/16. Quote from J. Mückler's book *Deutsche Bomber im Ersten Weltkrieg*: Staffel 20 from Kampgeschwader 4 took over eight of the ten G.II built before they were transferred to the Romanian war theater. Without other noticeable changes the Mercedes D.IVa engine turned a Gotha G.II to a Gotha G.III in the fall of 1916, whose construction followed immediately after the G.II. (Slg Jörk Kempf, KG1) Below are factory photos of the Gotha G.II 204/16, probably before it was delivered to the Staffel 20 of Kagohl 4.

Above: Oblt. Oswald Boelcke, who carried the Pour le Merite award; at this time not permitted to fly by the Kaiser, finds himself on a tour of the Balkans. On August 7th 1916, he visits FFA 28. Boelcke in the middle, left of him Oblt. Walter Thiemann, right of him Lt. Sterns. As the first war hero, Boelcke loved to have himself photographed.

Above: Oswald Boelcke fell on October 28th, 1916 after shooting down 40 aircraft, when his aircraft collided in the air with his squadron colleague Erwin Boehme.

underground and underwater facility allowed submarines to approach submerged in full daylight. They could connect themselves to a pipeline 500 meters offshore and collect large quantities of oil. *Queen Sophie* (emphasis added) often came to Phalerum for "teatime" to watch this."[189]

Every day the press brought new "revelations" of this kind. Gullible readers were appalled and angry at the Greeks and their King who committed such crimes. In the eyes of readers in Paris, the Greeks were worse than the (German) "Huns". There was not a shred of truth in all of the reports: they came from the poisonous kitchen of the French secret service under the leadership of the Naval Attaché de Roquefeuil[190] and the Venizelist press, according to a reply by Admiral Dartige to an article in the *Revue des deux Mondes* of March 1917, which commented negatively on Greece: *"The revelations of the Venizelist press about the supply of German submarines in Greece are a tangle of absurd legends.*

Despite the most repeated checks, not a single one of these "facts" was true. A number of mischievous villains, however, made a living from trading in fake news and abused the naivety or negligence of our secret service. A large number of people who wanted to be informed, in their ignorance, repeated this nonsense. We have cited a few examples; they were numbered in the hundreds. Tanks, towers that were mistaken for submarines and places for refueling that were absolutely inaccessible are just a few examples, etc. etc."[191]

It can be assumed that the fear of the U-boats also influenced the planning of French policy in Greece, since at the beginning of January 1916 the Quai d'Orsay was rife with talk about occupying Piraeus, in order to bring Athens under its own control. The neutral King should be chased out and Venizelos brought to power by force, if needed.[192] But so far, these were just considerations.

In order to secure his position in Macedonia against possible attacks from the east, be it from the Greeks or the Bulgarians, Sarrail had the railway bridge of Demir Hissar (Sidirokastro) over the

Above: Razgrad [Rasgrad] in the northern part of Bulgaria. This was the final station of the transfer of the flying units from the west for the beginning of the war with Romania, from August 1916. (Slg Jörg Kempf, KG1)

Struma (Strymonas) as well as the bridges of the same railway line south of Lake Dojran on a branch line, blown up.[193] In doing so, he ultimately signaled to the enemy that he was not planning an attack to the east. The Greek *Fort Roupel* under construction on Struma-strait was about five kilometers north of the bridge, and through the demolition, it was cut off from supplies from the west and east of Macedonia. So it was at the mercy of Bulgaria. Sarrail paid no attention to proposals from either the Greek or French officers to take possession of the fort.[194]

On January 22nd, 1916, the British troop transport *Norseman* was torpedoed by a German submarine at Cape Karabournou. The captain succeeded in running the ship aground so that neither the people nor the 500 mules were harmed. But the attack convinced the French that some of the Greek officers in the Karabournou fortress had had contact with the Germans, and he decided to put an end to it.[195]

At the end of January, Sarrail had the *Karabournou Fort* at the entrance to the port of Saloniki and the battery at the mouth of the Vardar, occupied by a French battalion. French warships took over artillery security. Sarrail asked General Mahon, who was still not under his command, to occupy the eastern outskirts of Saloniki with two battalions in order to stop any Greek protest marches to the fort. Fortunately for Sarrail, the fort's crew did not fight back, and there was no unrest in Salonika. Mahon criticized Sarrail's behavior towards London: the bridges being blown up and the occupation of the fort would drive even the Ententefriendly Greek officers into the arms of the friends of the Central Powers. Something should be done to appease the Greeks.[196]

Briand and Joffre warned Sarrail to treat the Greeks a bit more kindly but that did not help. In January and February, more and more French naval units landed on certain islands and arrested foreigners and Greeks who had been blacklisted by

Above: "On a hot summer day". The men sit in the sun in front of their sleeping car. They had acquired a fez already, a fitting headdress to the oriental landscapes. (Slg Jörg Kempf, KG1)

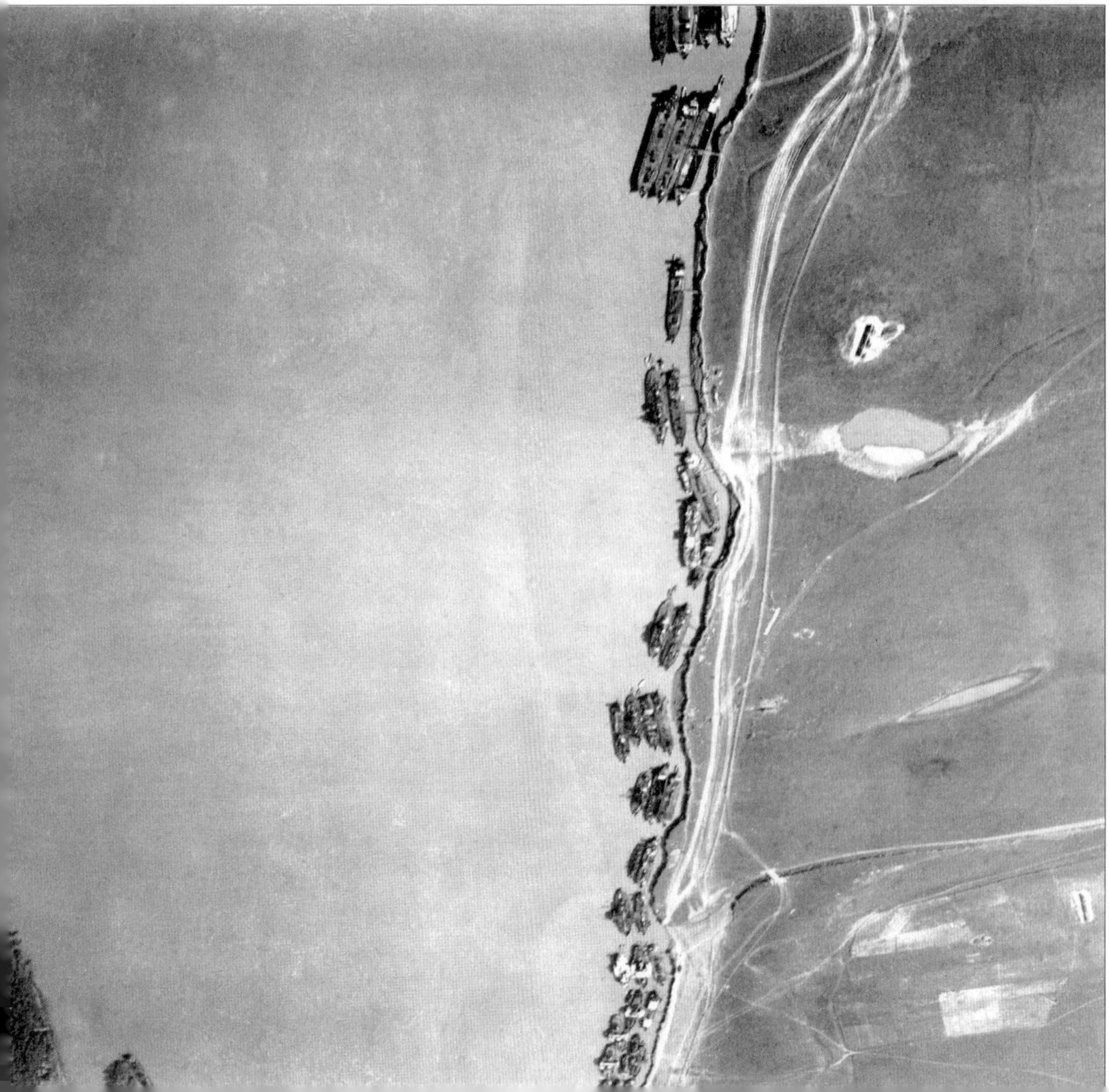

Left: Austria-Hungarian Danube flotilla near Belene (Bulgaria, near the Romanian border). (Slg Jörg Kempf, KG1)

Above: Aerial photo Razgrad in Bulgaria: one can recognize three Gotha large aircraft with their own personnel. Left of the train one can see nine tents. The "living cars" of the train are parked on two tracks. The front row – from left top – is made up of about nine covered cargo wagons and about 25 wagons for living quarters. (Slg Jörg Kempf, KG1)

the French secret service as friends or agents of the Central Powers. Petrol tanks belonging to private merchants were blown up in the ports because their owners were suspected of selling the fuel to German submarines.[197] Obviously, the Sarrail officers had no idea that the German submarines were powered by diesel engines. Such measures naturally turned the population against the Entente.

But it got even worse. At the end of December, at French instigation, imports of grain, flour, minerals, coal and the like were severely restricted. The reason given was that the stockpiling of supplies in neutral states bordering on those of the enemy should be prevented. However, it would be ensured that necessary imports could pass unhindered. But exactly this did not happen, and so *"normal"* goods came into the country in greatly reduced quantities. The result was that they became scarce and the corresponding stores raised their prices. This of course primarily affected the little people, who were outraged. The Allies turned the Greeks against them more than the Central Powers' propaganda could ever have done. When the Prime Minister and King protested violently, this turned the Allies against them and they reacted harshly. This, in turn, was grist on the mill of the Venizelists, who began to stir up against the King and the royalists. A national split loomed on the horizon.[198]

The British and Russians watched developments with concern and asked Paris to hold Sarrail back. Briand therefore admonished Sarrail and the French ambassadors on various occasions, to hold themselves back more, and not take any measures which would anger the Greeks. In this context, Sarrail and Constantine met on February 22nd. Briand had given Sarrail clear instructions. He should tell the king that there was no intention of humiliating him or his country. Certain measures were militarily necessary. They respected Greece's neutrality, but welcomed the cooperation. The conversation itself was harmonious and Sarrail relented on a few small points.[199]

Above: The officer corps of KG 1, Winter 1916/1917 is lined up in front of the restaurant car of the sleeping train of the French "Compagnie Internationale des Wagon-Lits". This is the oldest European company, which, since 1872, built and operated sleeping cars, dining cars, and luxury trains. Mj. Siegert used this so that he could transfer entire units with confiscated vehicles on railways. Number 1 in the photo is Hptm. Hermann Kastner, Number 2 is Lt. Kempf – these photos are from him – and Number 3 is Lt. Wolff. (Slg Jörg Kempf, KG1)

Skouloudis used this temporary relief to address the bond issue in February. The International Finance Commission had approved a loan of 150 million gold francs for Greece, which had not yet been disbursed. Greece urgently needed this money, because the government was on the verge of running out of money due to the mobilization. London and Paris were ready to pay it out if Greece made concessions. The British envoy wanted the Greek police to report to the Allies and his French colleague supported the demobilization of the Greek army. When Venizelos found out about this, he suggested that Greece enter the war. Demobilization was not an option for him. The Allies postponed the problem for some time.[200]

In the meantime, opposition to the Saloniki operation in the British General Staff increased massively. Chief of Staff William Robertson was preparing for the Battle of the Somme and wanted to withdraw all British troops from Saloniki and as many as possible from Egypt. On February 12th, he submitted a memorandum to the *War Committee* with his ideas: Britain must gain the diplomatic upper hand in the Entente. Peace should be made

Above: Glance into the restaurant car. On the wall is an overview map of Greece. (Slg Jörg Kempf, KG1)

with Turkey, with Russia forgoing Constantinople. If the latter was not possible, Bulgaria should be broken out of the Central Powers' front by offering it all of Macedonia from the Turkish border to Saloniki in the west and Veles in the north if Bulgaria becomes neutral again. The area could immediately pass into Bulgaria's possession. If Bulgaria accepted this offer, the supply of Turkey from Germany would be blocked. The 200,000 men could be withdrawn from Saloniki. The only state that would not benefit from this is Greece, but it did not support Serbia either. Although the *War Committee* did not agree with the ideas of a separate peace, it fully agreed with his ideas regarding Saloniki. There should be no offensive in the Balkans.[201]

France then tried in vain to get Italy to send troops to Saloniki. Afterwards, the Saloniki question was more or less forgotten until June because of the German offensive against Verdun. At a meeting on March 14th, it was decided that the Serbian troops should be transferred from Corfu to Saloniki.[202]

On April 5th, 1916, the French and British ambassadors visited Prime Minister Skouloudis and informed him that the Serbian troops were operational. They were to be brought to Patras by ship and transported to Thessaloniki by train via Athens and Larissa. The diplomats hoped that the Greek government would agree. Skouloudis protested violently. The transport of 100,000 men across Greece meant that civil rail traffic and thus the country's economic life would practically be shut down for two months. In addition, such a transport violated the neutrality of Greece. He, Skouloudis, must emphatically reject the project: *"No, gentlemen, we will not allow such a thing. I will officially declare this to you." "Our government,"* replied the French minister, *"has not instructed us to ask your permission, but to inform you of the decision. "[…] Skouloudis said:" I declare that it is the decision of my government to not to allow the overland passage and further inform you that I see myself compelled to blow up the railroad "*. With increasing anger, he continued: *"You have left nothing for our country - neither self-respect, nor dignity, nor independence, nor the right to live as free people. But don't forget that even the most benevolent patience and the greatest obedience have a limit, that a single, last drop can bring the barrel to overflowing."*[203]

The British envoy wanted to defuse the situation and asked Skouloudis if the Greek government would agree to a sea transport. After consulting the

Above: "Gypsies in Bulgaria" in the background Lt. Voss and Lt. Kempf. (Slg Jörg Kempf, KG1)

cabinet, Skouloudis agreed to the new proposal. But in the meantime the Serbian envoy had raised the old demand again and pulled the Allied envoy back to his side. Again Skouloudis refused. Skouloudis instructed the Greek ambassadors in London, Rome and Petersburg to present the Greek position, but not in Paris because he knew that there would be no point in protesting there. Skouloudis 'arguments were accepted in London and an attempt was made to change Paris' mind through diplomatic pressure, which succeeded. But Prime Minister Briand retaliated against Greece on April 10th by refusing to pay out the first tranche of the 150 million gold francs loan that the Finance Commission had approved.[204]

On April 11th, Skouloudis complained that Paris was linking the transport of the Serbs and the disbursement of the loan. The claim that Greece wanted to have the transport paid for by paying out the loans was completely unjustified and pulled out of thin air. Greece urgently needed this money because mobilizing its army cost huge sums. Demobilization was rejected by the French. Therefore, the Greek government turned to Germany and asked for financial help. On April 20th, the imperial government informed the Greeks that it was ready to advance 40 million Reichsmarks in order to keep Greece neutral. Berlin transferred this sum. The French press later claimed that this was the price which Greece had received for clearing the half-finished fort on the Roupel-Straits at the end of May, a good four weeks later.[205]

On the question of the transport of the Serbian troops, an agreement was finally reached that took the Serbian troops on a specially secured sea transport to the Chalkidiki peninsula on May 30th. The troops were divided into three armies. Crown Prince Alexander became the new commander in chief. Voivode Putnik had resigned from office due to exhaustion. The 130,000 men had 34,000 transport animals, 9 battalions of field artillery, 6 battalions of anti-tank artillery, 6 batteries with 120 mm cannons and 300 machine guns. The Serbian government in exile also moved its seat to Saloniki.[206]

At that time there were still consulates of Germany, Austria-Hungary, Bulgaria and Turkey in Saloniki. The consulates regularly reported on the Allied troops arriving in Saloniki. Each of these

Above: The port in Saloniki in aerial photographs, stitched together by KG 1. Judging by the large number if ships, it can be seen that this armed force could have shot up any resistance in this neutral port. If the Kaiser's reich had done this, it would have been "canon-boat politics". One can also recognize it through the numerous military barracks with which the allies supported Venizelos. Slg Jörg Kempf

consulates employed agents and spies to gather information about the Allies. General Mahon wanted to have these spies arrested in one fell swoop and informed Sarrail about it. He did not answer, but on December 30th, 1915, Sarrail informed his British colleague that he would have the four consuls arrested in a few minutes and asked him to take part in the operation. Mahon invited a small British group to take part. He feared violent protests from neutral Greece, which actually did become loud. The next morning it was decided to arrest the spies in a joint operation.[207]

A total of 20 Austrian, 17 Turkish, 12 Bulgarian and 5 German consulate members were arrested.[208] The consulates were occupied and the files confiscated. A little later the Norwegian consul was also arrested.[209] The diplomats of the Central Powers were interned on a French warship and taken to France, from where they were later repatriated. Greece protested violently, but Sarrail did not care.[210] He did not find it necessary to inform the Greek government of his actions.[211] The arrest of foreign consuls on the soil of a neutral country was an outright breach of international law. But Sarrail went even further: he interfered in Greek domestic politics. He treated Greek royalists rudely and humiliated them, according to Mahon and the British ambassador.[212] Two Greeks who had flyers in their possession that were dropped out of German aircraft were sentenced to death and shot.[213]

Constantine was said to have sympathy for Germany because he was a brother-in-law of Wilhelm II. But nobody thought of accusing him of being kind to England because he was a nephew of the British King. He accused the Entente of dishonesty: *"It is the purest hypocrisy when England and France speak of a violation of Belgium's neutrality, after what they did and still do to me. They occupied Lemnos, Imbros, Mytilene, Castellorizo, Corfu, Saloniki, the Cha[l]kidike peninsula, a large part of Macedonia. Germany invaded Belgium under the pressure of military*

Above: Gotha G.II 208/16 flipped over upon landing. (Slg Jörg Kempf, KG 1)

necessity. But what was the military necessity to destroy the Demirhissar Bridge and to occupy Corfu? The Serbs could have been brought to Italy more easily. Is it not so that the Italians don't want the Serbs because they fear cholera? It's just as uncomfortable for us. [...] At the beginning of the war, eighty percent of the Greeks were favorably disposed [to the Allies], today it is not quite twenty percent."[214]

Constantine tried on various occasions to defend himself against the one-sided, completely distorted reports in the Allied press by offering interviews to the journalists. He corrected the facts, but Allied censorship did not allow the corrections to be published.[215] So Constantine turned to the American press, i.e. the AP correspondent Hibben. The German text above also comes from this interview. Regarding the demolition of the bridge, he said: *"What military reason compelled me to destroy the Demir-Hissar Bridge, which cost one and a half million drachmas and which was the only way by which I could feed my troops in East Macedonia? The bridge had been prepared to be blown up. It would have been possible to destroy it in one minute, if the enemy really came. But it is absolutely certain that there was not a single man of the enemy anywhere near the bridge. There is also no sign of a planned enemy advance. The only reason that remains is the clear intention to surrender the Greek troops surrounding Drama and Serres to starvation."*[216]

Hibben sought Constantine's authorization of the interview text, which he approved. But what followed was an unbelievable story.

Before I sent my text to America, I informed Mr. Guillemin [...], with whom I was on friendly terms, the content of King Constantine's admission. He reprimanded me as a friend of France for wanting to send such a message, which he described as German propaganda. At the same time, he sent an encrypted message to his government to have the censors stop the interview. When I found out about this, I sent the same telex via England because all messages from Greece to the United States must go through either France or England and are subject to Allied censorship. I discussed the case with Sir Francis Elliot, the British Ambassador to Greece."

"Nothing is gained by suppressing one side of the matter," I told the English diplomat. "The Allies have dozens of correspondents in Greece who are flooding the Entente. The American press is also here with its view of what is happening in the Middle East. I believe that King Constantine is sincere and I am concerned that he will not be treated fairly. If this telex is stopped, it will arrive later as a letter. You cannot suppress the truth. It would be better to let the message go through."

"It seemed that Sir Francis agreed and telegraphed his government to let the interview pass. The British censors, however, had a different opinion. The report was withheld in London. Finally, I informed M. Guillemin that I would also send the report to Berlin or that I would carry it to New York personally to circumvent the censorship. After the report was held back for six days, Prime Minister Briand decided it made more political sense to pass it on, but he made it a condition that a semi-official opinion of the French government

Above: The port of Saloniki with numerous ships and military barracks on land. These allied troops came for the most part from the unsuccessful Dardanelles battles in 1916. (Slg Jörg Kempf, KG 1)

would be published at the same time."[217]

Hibben's interview, accompanied by an official statement, was published a week later. It contained the usual justifications. The Allies had landed in Saloniki at the invitation of Greece. No land was occupied, but rather it was simply temporarily subjugated, a process that at the time had already taken two years. Guillemin and Elliot wrote a text that attacked the German propaganda. They demanded Hibben publish this text as well. The text ended with a statement that can only be described as cynical: *"The whole matter between the Entente and Greece is a question of trust. If Greece is loyal to the deal, it will not suffer."*[218]

When, at the end of May 1916, Bulgarian and German troops occupied the unfinished *Fort Roupel* on the Struma strait, this provoked two movements: Venizelos returned to active politics and Sarrail was at the helm of an even tougher occupation policy.

Venizelos wanted to return to power, but he knew that he would not be able to achieve this in Athens, because loyalty there and in ancient Greece was to the king. Therefore, he took a detour via Saloniki. There, he wanted to launch a revolutionary movement under the protection of Sarrail, which would allegedly be directed against the Bulgarians and not the King. Supporters of his were active among the troops in Saloniki, and took advantage of the general hate of Bulgaria, in order to draw them to Venizelos' side. He hoped to be able to bring the King and the country behind him as well.

In truth, Venizelos had spoken of treason in connection with *Roupel*. To the British envoy Francis Elliot, he expressed his suspicions that the German and the Greek general staff were working together. On May 29th he suggested to Elliot and the French envoy Guillemin that he should go to Saloniki with General Panagiotis Danglis, who was highly regarded in the army. There, he wanted to set up a provisional government, summon members of the parliament elected in June 1915, which in his opinion had been wrongly dissolved, and mobilize an army that was to attack Bulgaria together with the Allies. This was not directed against the King

Above & Below: The "Second crash of Lt. Eggers." Lt. Hans Herbert Eggers, born on January 31st 1894, died on August 19th, 1962. The aircraft was Gotha G.II 202/16. (Slg Jörg Kempf, KG 1)

Above: Braila in Romania, on the Danube (Slg Jörg Kempf, KG 1)

or the dynasty; on the contrary, he hoped that this would draw the King to the Allied side. For a while there would be two Greek states: the official Germanophile Greece and his pro-Allied one. He was certain that five or six regiments of the army in Macedonia would join him. He wouldn't strike until he had these troops on his side.[219]

A liaison officer was sent to Salonika to make the necessary arrangements with the Defense Committee and Sarrail. Venizelos asked Sarrail to postpone the planned imposition of a state of emergency so as not to turn the entire 11th Division against him in Salonika. Parts of that division would be sure to support the movement. For the same reason, Elliot urged London to postpone the peaceful blockade of the Greek ports that Admiral John de Robeck had requested. Addressing London, he added: *"Personally, I firmly believe that nothing can save this country from ruin, threatened by the stubbornness of the king and the hostility of his government towards us, other than a revolution based on the ideas of Monsieur Venizelos. As it currently stands, the revolution will not take place until after the war, when it is too late to be of any use to us. To be successful now, the revolution must be supported by us and France."*

But London was against this. On June 1st, Elliot was instructed to refrain from making commitments and to not encourage Venizelos' project. Venizelos dropped his plans for the moment. But he agreed that preparations should continue among the troops in Saloniki with Sarrail's support.[220] On May 31st, the French envoy Guillemin informed Sarrail that Venizelos was planning to establish a provisional government in Saloniki, but Britain *"had vetoed the undertaking"* so Venizelos and his French co-conspirators would have to postpone their plans.[221] Sarrail did not give up, but supported the Greek officers who were preparing the military coup.

Like Sarrail, the envoy Guillemin was also fixated

Cutout from the photo album of Johannes Keller. His military resume will be described in a later chapter. He flew in Macedonia in 1917 together with Lt. Rudolf von Eschwege.

Above: An emergency landing of a Gotha G.II which ended well. The local population is immediately there to marvel at the flying "monstrosity". (Slg Jörg Kempf, KG 1)

Above: This is Nieuport 1864 of Escadrille N.77, which was shot down near the 5th army, both crewmembers dead, on November 8th, 1916.

Above: AEG G.III (219/16) training aircraft. Eight mechanics stand around the aircraft. Judging by the clothing, it's November/December 1916. Second from the right is Lt. Fritz Kempf, who was training on this type in Bucharest-Pipera. (Slg Jörg Kempf, KG 1)

Above: "Kempf's aircraft" – registration Go G.II 20x/16; right is Go G.II 207 (?)/16, probably during the training in Bucharest-Pipera. (Slg Jörg Kempf, KG 1)

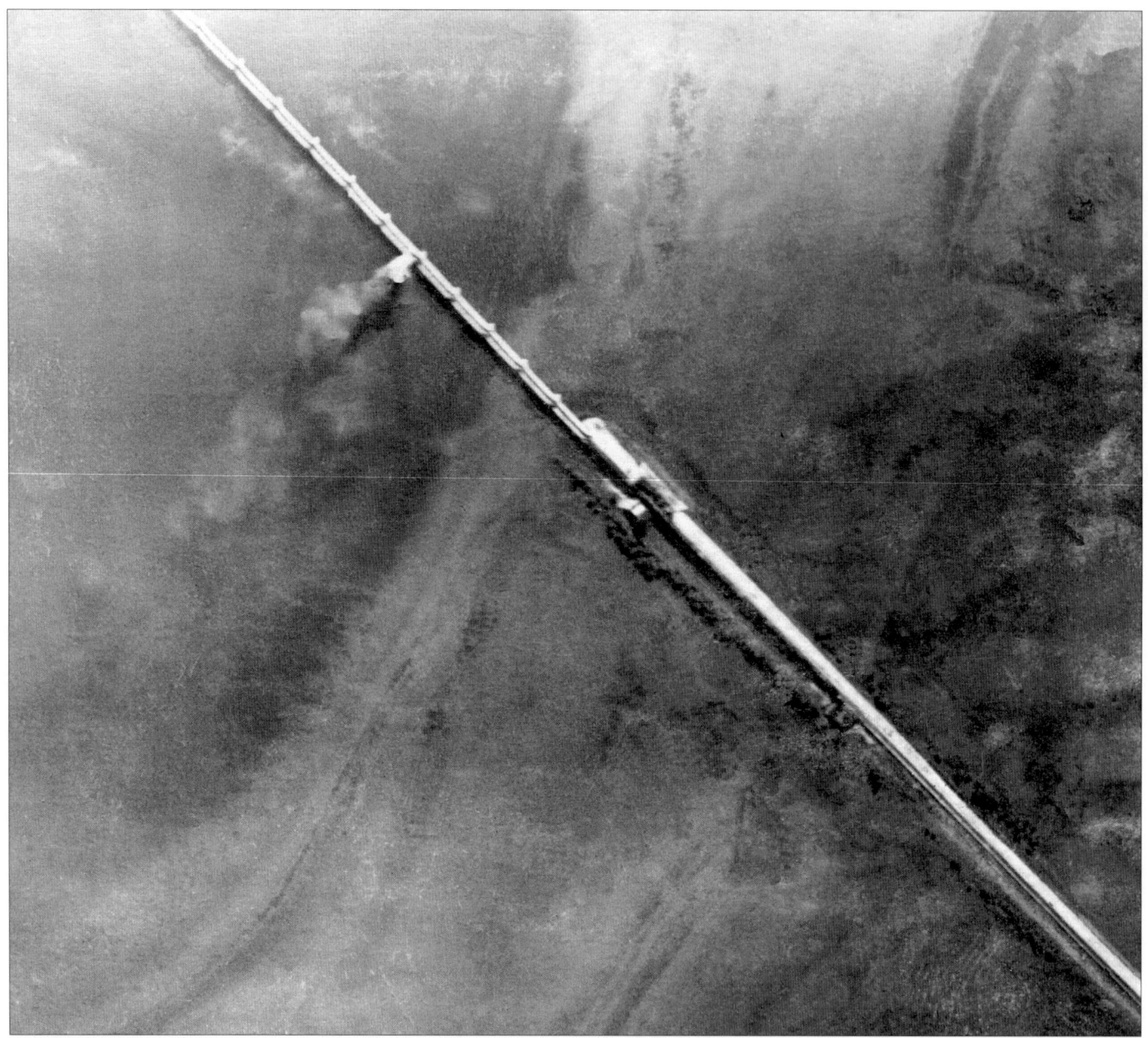

Above: Bombs hitting their target on the Cernovada bridge. (Slg. Jörg Kempf)

on Venizelos. The American correspondent Hibben characterized his attitude as follows: *"He did not once take into account the Greeks' view of their own situation. He was barely in contact with the Greeks, let alone the followers of Venizelos. His policy was based solely on the assumption that Venizelos represented the true opinion of the Greek people."*

In addition, he was of the opinion that because of the French help in the Battle of Navarino (1827) Greece was obliged to obey France. In Guillemin's eyes, Venizelos was the man for the Entente. Why should he keep in touch with anyone else? In contrast, the ambassadors of Italy and Russia, Count Alessandro de Bosdari and Prince Demidoff, distrusted Venizelos. They felt that *"he was playing his own political game in Greece, with the support of the Entente and at their expense."*[222]

In his apologetic account of Venizelos' politics, Georgios Ventiris denies that he (Venizelos) was involved in the conspiracy.[223] Another apologist, Neokosmos Grigoriadis, claimed that the Saloniki revolt was a spontaneous action by the Macedonian people. The organizers of the revolt did not have time to ask Venizelos. [224] The latter is probably true; this is also supported by Venizelos' reaction when the mutiny actually broke out. It is astonishing, however, that more recent Greek historiography

Above: "Birthday celebration of Lt. Peres, taken at 3:30 in the morning. Lt. Peres, Camin, Dörstling, Kempf. Bottom: Hauff, Jiergovitz, Kaestner." Dates of pilot Jiergorovitz: born on Novober 15th, 1887, died on March 20th, 1917 in Prilep. (Slg. Jörg Kempf, KG 1)

takes over this view uncritically, perhaps because one is now not able to speak French and cannot take note of the clear statements made by Sarrail.[225]

The surrender of *Fort Roupel* to the German-Bulgarian troops without a fight was of course grist on Sarrail's mill, who had wanted to strengthen his policy of *"oppression and coercion"*[226] for a long time. Joffre had held him back so far, but when he heard of the Roupel straits occupation, he gave Sarrail the go-ahead for a tougher course.[227] In fact, the only difference between the occupation of Roupel and the Karabournous was that the former was carried out by Bulgarians and the latter by French. Both occupied neutral territory in disregard of international law.[228] But Sarrail didn't care, he declared a state of siege in the Saloniki area on June 3rd, although Milne had wanted to wait at least a day,[229] because June 3rd was the name day of King Constantine, and the residents of Saloniki had planned to conduct a torchlight procession and to hold a ceremonial *Te Deum* in the cathedral in his honor. After all, Constantine, as Crown Prince, had led the army that had liberated Salonika from the Turks. The *"Republican"* Sarrail forbade this *"royalist"* demonstration. On his orders, the French gendarmerie took control of the railway lines, all postal and telegraph services, and set up a censorship center for all newspapers that appeared in Saloniki. The British commander disapproved of these measures, but in order not to disavow Sarrail, he placed the British military police under him. *"This made Saloniki an occupied city under foreign military administration, like Brussels, Warsaw or Belgrade. Unusually, there were demonstrations in Athens against the powers of the Entente."*[230]

Sarrail imposed a blockade on all Greek ports and decommissioned all Greek merchant ships. A squadron of the French Navy with an infantry brigade on board went to the Cyclades Islands to put pressure on Athens. On June 21st, the diplomatic representatives of France and England demanded the demobilization of the Greek army, which they viewed as a *"political weapon against Venizelos"*, the replacement of the allegedly pro-

This Page: In Sofia (Bulgaria) a "German inspection body" was established, which listens to the enemy radio transmissions in the Aegae and the Balkans. The man on the right is the later Lieutenant Protze.

Right: Lt. Pfeffer, unfortunately no further life dates are known. Slg. Jörg Kempf

Below: Lt. Adolf Gutknecht in the late fall of 1916 in a thick fur coat in the southeastern war theater. Born on September 12th, 1891 in Badingen, Kreis Stendal, died on June 15th, 1979 in Johannesburg, South Africa. Slg. Jörg Kempf

Above: Friedrichshafen G.II 109/16.

German Skouloudis government with a neutral one, new elections and the dismissal of certain police officers which had not sufficiently protected the diplomats of the Entente against the popular anger of the Athenians. Since the French fleet was anchored in the bay of Milos and could therefore appear in Athens in eight hours, the Greeks gave in. Skouloudis resigned and his predecessor Zaïmis formed a new government. The blockade was lifted, except in the case of the port of Kavalla, which was too close to the Bulgarian border. The fleet withdrew again. German propaganda picked up on the incident and announced that the Entente's promises to protect the freedom and rights of small states

Above: The airfield in Hudovar in the winter of 1916-1917, from KG1. In this winter landscape with much snow, cold and lodging was uncomfortable for the personnel and the aircraft. The photo was taken with a cold-resistant aerial camera.

Above: The Bucharest-Pipera airfield in a nice aerial photograph. Above the horizontal street a fighter site can be recognized. There are five aircraft tents and seven fighter aircraft. Left of the landing "T" an aircraft is coming in to land. In the lower part of the airport are 13 aircraft tents, as well as 11 large aircraft. The personnel of the former Kasta 20 retrained here on twin-engined bombers.

sounded rather hollow.[231]

Sarrail's actions in the wake of the Roupel occupation cut Saloniki off from the rest of the world. In order to communicate undisturbed with Prime Minister Skouloudis, General Moschopoulos had to go to Katerini on May 23rd. He demanded that Skouloudis give him command of the gendarmerie and that he should be in charge of Saloniki. Skouloudis agreed and urged him to pursue a policy of neutrality. When the Greek government protested against Sarrail's actions to the large neutral states like the United States, the Allies viewed it as a hostile act.[232]

The Allied intervention of June 21st turned the Greek public against them. It was viewed as a humiliation, even in some Venizelist circles. The anti-Venizelists capitalized on it when a telegram from Venizelos and Briand was also made public on June 24th. Venizelos had written in it: *"The seriousness of his address, the sincerity of his arguments and the complete separation between the Greek people and the previous government give him, more than any other, the character of his fatherly concern for the Greek people. The protecting powers act as parents with the entirety of their rights."* In other words, Venizelos justified the allied intervention.[233]

The British Chief of Staff Robertson rejected any offensive from Saloniki. On May 28th, he even ordered two British divisions to withdraw. He instructed Milne to assist the French only in the defense of Saloniki. When the British cabinet heard of the French advance, it was angry. On June 7th, an inquiry was made in Paris as to whether the French government had really approved the advance. Gray had adopted the view of the British military that nothing could be achieved in Macedonia for the moment. On the other hand, he also knew that a British refusal to cooperate could cause political trouble with the French.[234]

As the financial burden on the Greek budget grew, the King ordered the mobilization status to be

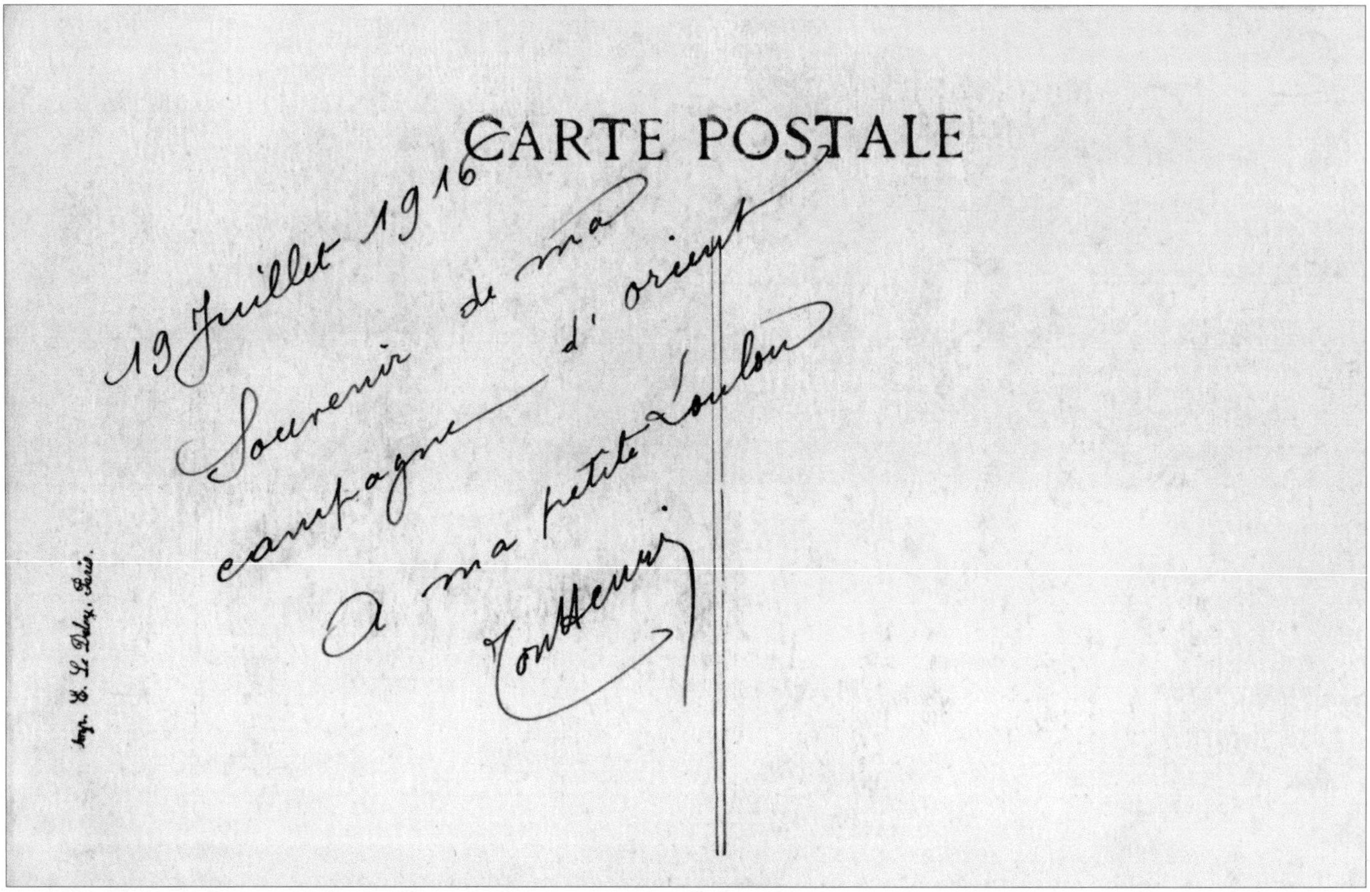
CARTE POSTALE
19 Juillet 1916
Souvenir de ma
campagne d'orient
A ma petite Loulou
Ton Henri
Imp. E. Le Deley, Paris.

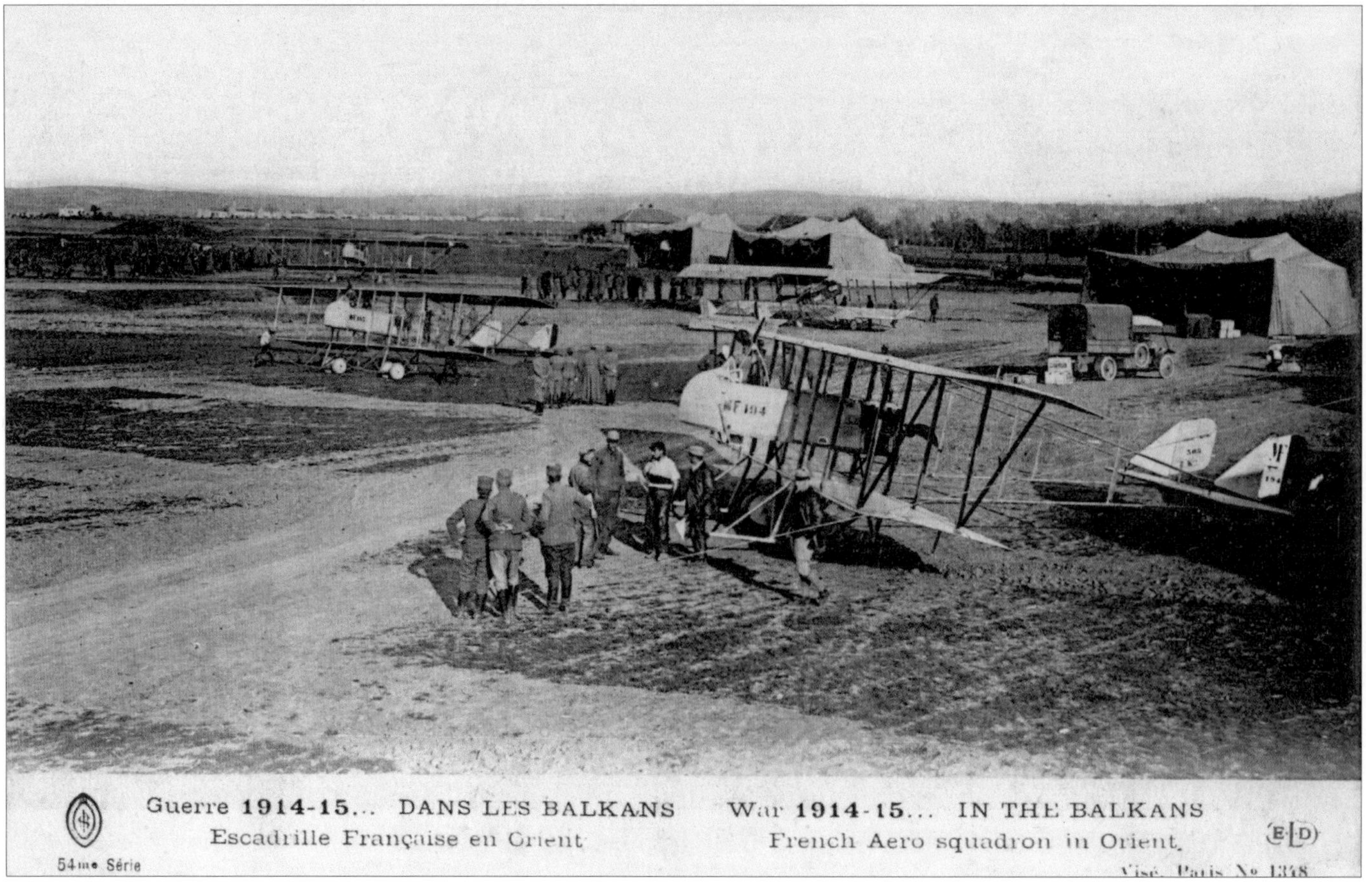
Guerre 1914-15... DANS LES BALKANS
Escadrille Française en Orient
War 1914-15... IN THE BALKANS
French Aero squadron in Orient.
54me Série
E.L.D.
Visé. Paris No 1348

Above & Right: Captured Farman F.40 F.3131.

Facing Page: An Allied post card of the era.

108

Left: "A car is stuck in the snow on the way to the Bucharest-Pipera airfield in January 1917." (Slg. Jörg Kempf)

Below: "The train of the KG 1 is totally covered in ice." With these massive amounts of snow, no air operations could take place. Photo: Slg. Jörg Kempf

reduced to *"partial mobilization"* on June 8th. This would significantly reduce the number of 300,000 conscripts. On June 11th, draftees from ancient Greece and Crete received a two-month vacation. With these measures, Constantine also wanted to show the seriousness of his policy of neutrality. How could such a scaled-down army be dangerous to anyone?[235] The Allies did not like it all. They wanted Skouloudis to resign and Greece to give certain guarantees about its position. Sarrail and the French envoy Jean Guillemin demanded the king's abdication. Only in this way would Greece be deemed favorable classification to the Allies. The British were against it at first, but then they joined the French. As mentioned, on June 21st, the envoys of the Entente presented the Greek Foreign Ministry a note calling for the full demobilization of the army, the resignation of the Skouloudis government and the establishment of an interim government, the dissolution of parliament, the holding of elections and the removal of some police officers. Skouloudis had found out about it and on the same day submitted his resignation, and the king had reappointed Zaïmis as prime minister.[236] Stavrianos rightly notes that *"seldom has an independent country been subjected to such humiliating treatment. And that was just the beginning. Greece [...] has become a "public place" where all sides did not hesitate to act according to their own ideas."*[237]

The new government was sworn in on June 22nd and new elections were scheduled for October 8th. The police officers, who the Entente did not like, were replaced and demobilization began. Despite all of the Allied demands being met, Sarrail remained suspicious. At the end of June, he asked the chief of the 3rd Army Corps that he be given command of the gendarmerie for security reasons. When Moschopoulos refused, Sarrail told him to leave Saloniki within 24 hours. There was a violent exchange of words. Finally, Sarrail apologized and dropped the demand.[238]

But Sarrail's attacks continued. The Allied-sponsored newspaper *Rizospastis*, a propaganda paper, published a series of inflammatory articles in July against the King, the government and some high-ranking officers. The officers felt provoked and 14 of them angrily stormed the newspaper's office and beat up the publisher. Moschopoulos had the 14 arrested and locked up. Sarrail demanded their extradition to the French military judiciary, although the crime against Greeks had taken place in neutral Greece. When Moschopoulos refused, Sarrail sent an armed patrol that broke into the military prison and kidnapped the 14 officers. They were to be tried in a French court martial for allegedly insulting the French flag. Moschopoulos turned to the government and achieved the release of the officers. They were then sentenced by a Greek court to two months' imprisonment, which they served in Akronafplia prison.[239]

Sarrail's advance northward on August 10th brought new movement into this issue. This and the demand for demobilization led to the Bulgarians attacking East Macedonia. That the weakening of the army had been carried out on the orders of the Allies did not interest Venizelos; for him demobilization was an act of the king to support the Germans. On August 16th, a rally was held in Athens, where Venizelos appeared again in public for the first time. He announced that he did not agree with the King's policy. The King was Germanophile, he would do something about it. The neutralists then claimed in speeches on August 19th that Venizelos wanted to overthrow the King from the throne on behalf of foreign interests. Most of the officer corps agreed.[240]

On August 27th Venizelos held a speech from the balcony of his house, in which he attacked the King directly: *"But, oh King, you have been the victim of unscrupulous councils trying to destroy the work of the 1909 revolution in order to restore the previous mismanagement and to satisfy the passionate hatred of the popularly elected leader. Hence, you have become the victim of military advisors with limited intelligence and oligarchic principles. Therefore, you became the victim of your admiration for Germany, in whose victory you believed. In the hope of pushing aside our free constitution through this victory and uniting the entire state power on yourself. [...] Instead of expanding into Asia Minor, Thrace and Cyprus, there was a Bulgarian invasion of Macedonia. Furthermore, valuable war material was lost. If no one listens to us, then we should advise what to do to save what can still be saved from the disaster that has befallen us."*[241]

Shortly afterwards, mutinies began in Crete and Samos. The rebels demanded Greece enter the war against Bulgaria, on the side of the Entente.[242]

Already at the end of 1915, a large group of Venizelist-oriented officers had formed the 3rd Army Corps, which wanted to overthrow the government and replace it with a revolutionary council. This council was supposed to bring the troops in Macedonia under its control, side with the Allies and take part in the war against Bulgaria. The group was led by the commander of the 10th Division, Major General Leonidas Paraskevopoulos, the commander of the 11th Division, Colonel Immanouil Zymprakakis, and Artillery Lieutenant

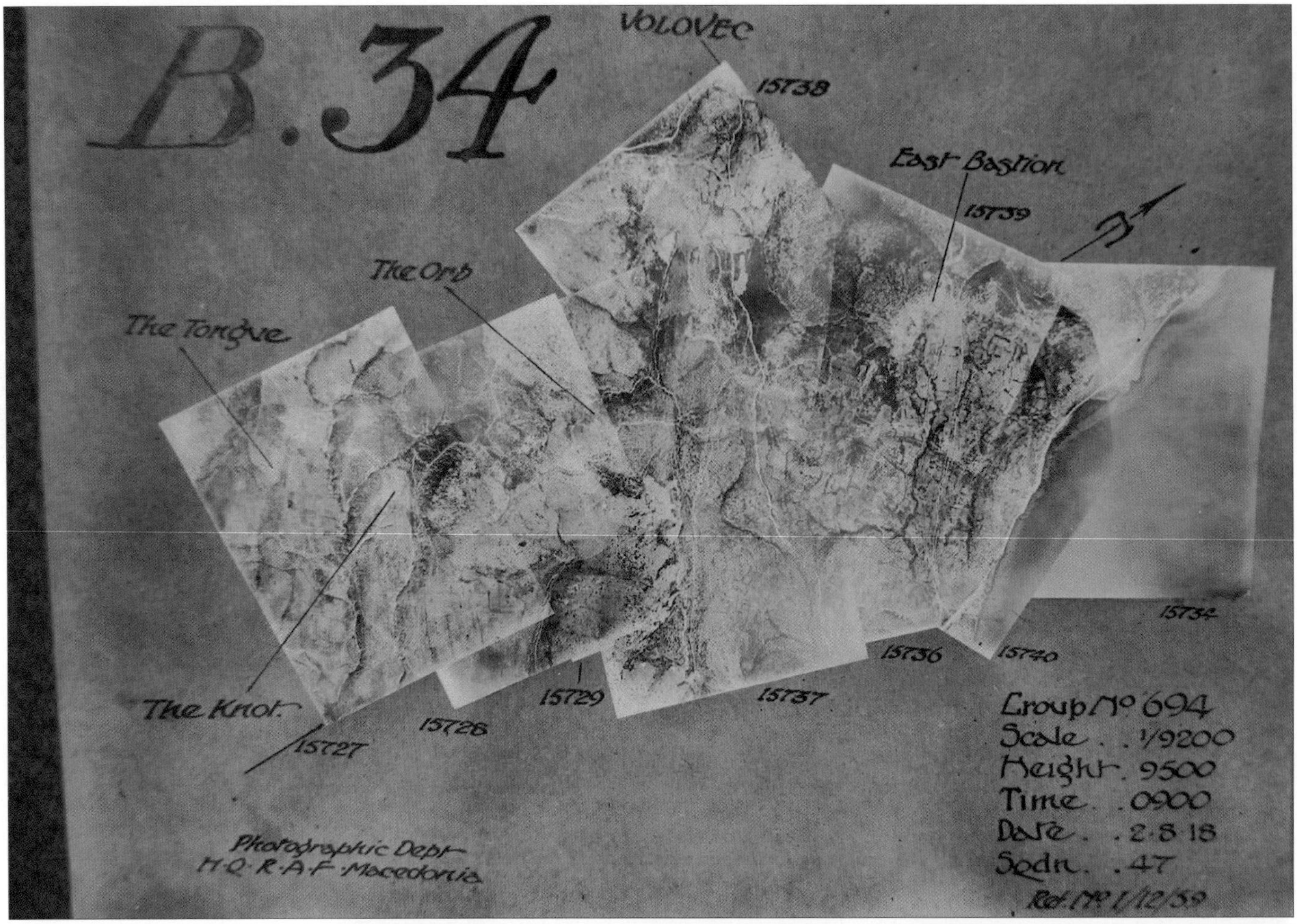

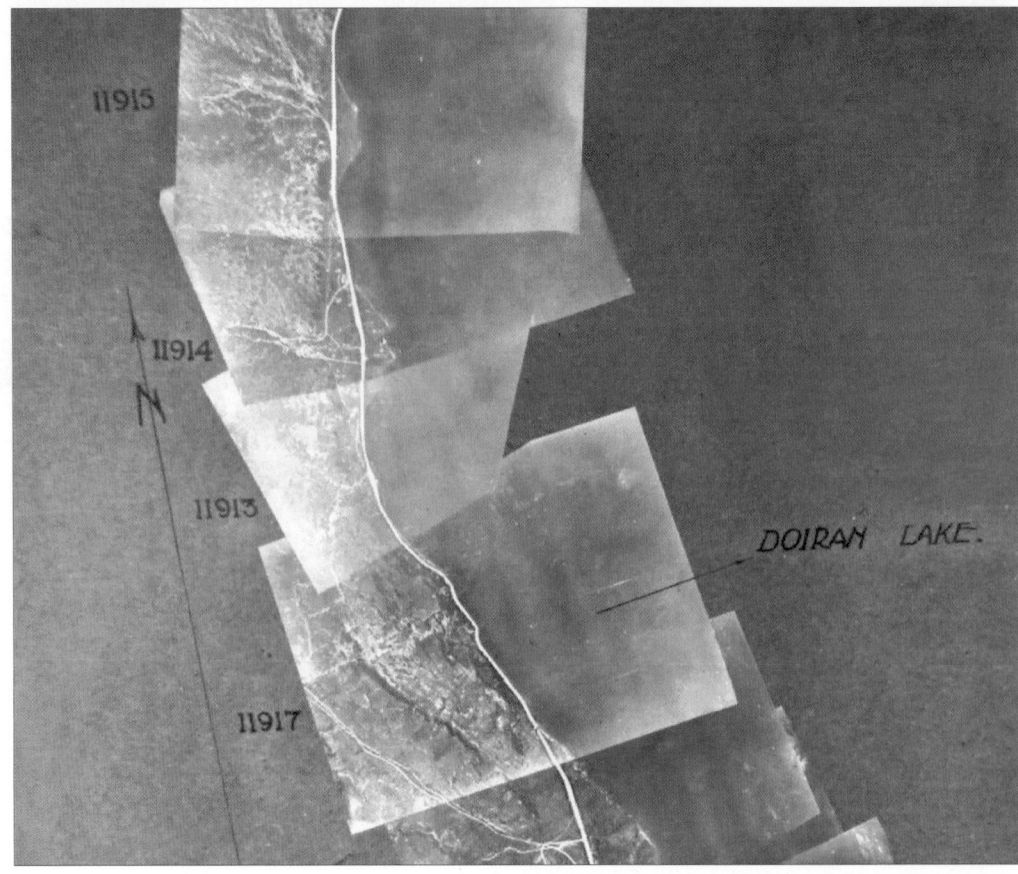

Above & Right: The RAF also wanted to be informed about the German activities in Macedonia and sent reconnaissance craft to find out what changes were taking place and to fight them directly.

Colonel Konstantinos Mazarakis. Sarrail supported this revolutionary movement, because he hoped to force the entry of Greece on the Allied side and the deposition of Constantine.[243]

On August 10th, 1916, Zymprakakis wrote to Venizelos that France was supporting their plans through Sarrail. Sarrail hated the monarchy and would be happy to participate in Venizelos' plan. In another letter to Venizelos, Periklis Argyropoulos made a similar statement. Further letters from followers of Venizelos confirm that Sarrail actively supported the coup in Saloniki. These letters were found in his house after Venizelos' departure from Athens and published in the newspapers.[244]

In addition to this military component, there was also a civil one. Saloniki had been simmering since the Allied invasion and the occupation of Eastern Macedonia by the Bulgarians. The Bulgarian riots caused many Greeks to flee to Saloniki. They were furious that Athens could not protect the new areas better. Palmer aptly described the further development: *"Angry patriots, sincere liberals and a handful of coffeehouse demagogues formed an "Alliance for National Defense" to give the will of the Macedonian Greeks a voice. With unconscious irony, the alliance established its headquarters in the same building from which the "Young Turks" wanted to revive the Ottoman Empire nine years earlier. A "Public Safety Committee" was formed, a gesture that Sarrail, friend of Danton's most enthusiastic biographer, undoubtedly welcomed."*[245]

When the government learned of the conspiracy, they ordered Moschopoulos to conduct an investigation, which, however, yielded no results. Only Zymprakakis was replaced by Major General Andreas Momferatos. At the end of July 1916, Moschopoulos was called to Athens for advice, precisely at the time when the German-Bulgarian invasion of the *Roupel-Strait* took place. On August 26th, the government learned that the Allies would soon demand the replacement of the Chief of Staff, Major General Viktor Dousmanis, and his deputy, Colonel Ioannis Metaxas. The latter was considered a friend of the Germans. To avoid an Allied demarche, Moschopoulos was put in place of Dousmanis and Dousmanis was given five years' compulsory leave. Colonel Xenofon Stratigos took the place of Metaxas, who was deactivated. The post

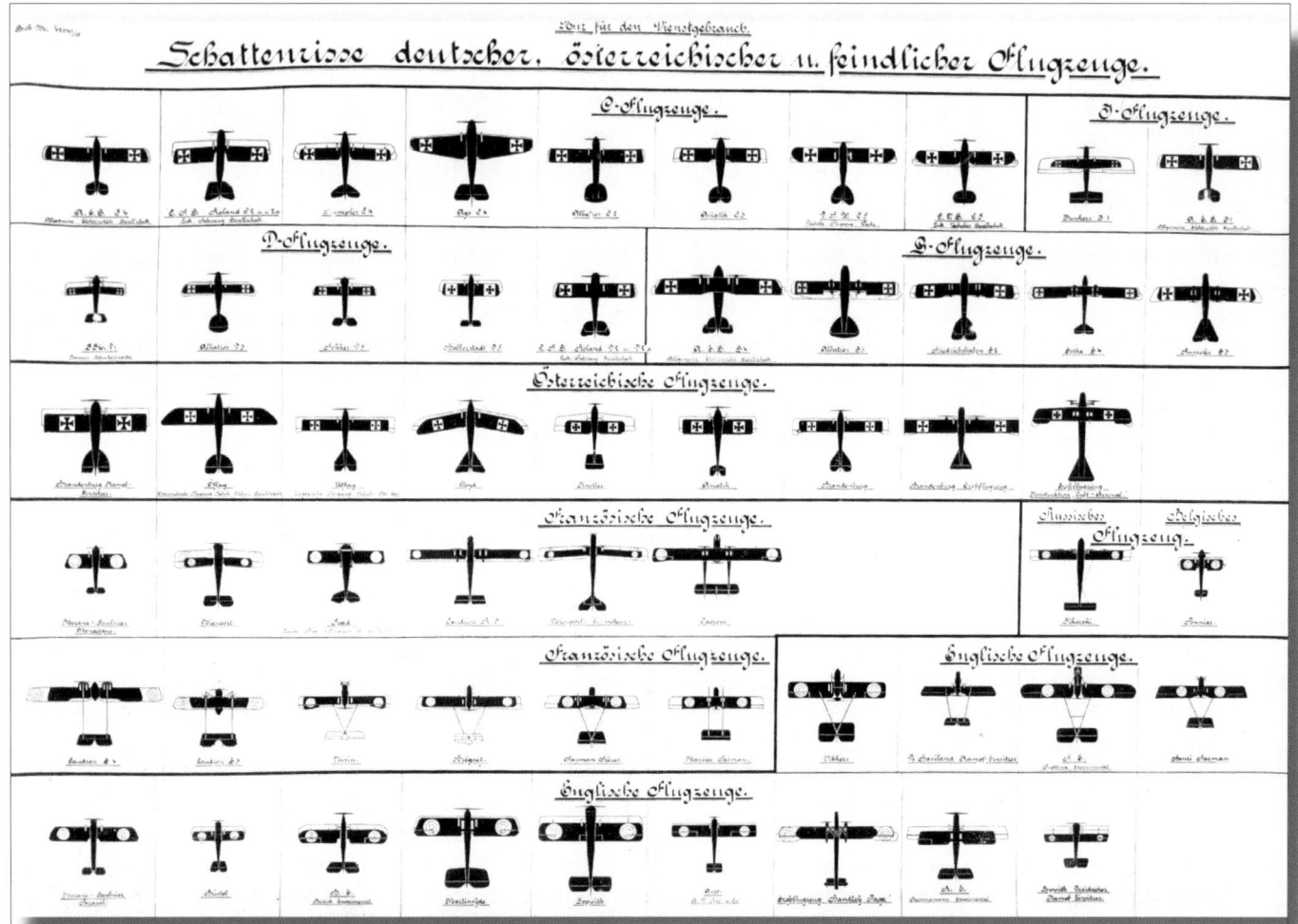

Above: "Parade of aircraft on the occasion of the visit of Excellency Lieutenant General Adolf von der Esch, Commander of the 101st Infantry Division in February 1917" in Hudovar. Three mechanics stand behind each of the crews, in front of the aircraft. The first aircraft on the right is the AEG GIII 225/16. Mr. Mückler identifies the aircraft beginning from right to left as: eight AEG G III, three Friedrichshafener G II, two AEG G III and one Gotha G II. [Mückler, "Deutsche Bomber im Ersten Weltkrieg:, Motorbuch-Verlag 2017]. Photo: Slg. Jörg Kempf, KG 1.

of commander of the 3rd Corps was taken over by the commander of the demobilized 11th Division, Colonel Nikolaos Trikoupis.[246]

This was the trigger for the following events. On the morning of August 30th, Trikoupis learned that Mazarakis had appeared at the Karabournou barracks and announced that he would set up a revolutionary government to organize the defense of Macedonia. He had asked the officers to join him, but only two had followed his call. When Trikoupis tried to attack Mazarakis and wanted to depose him, he discovered that he was under the protection of the French. Mazarakis published a leaflet allegedly from a National Defense Committee (Ethniki Amyna) calling on the army and civilians to join him. The gendarmerie defected to the mutineers. Those higher officials and clergy who were loyal to the King were arrested. Trikoupis' job was to put down the mutiny. But since he did not want to get into a conflict with the French, he sent the nomarch and the chief prosecutor to them. They learned from Sarrail that he would send an armed intervention if there were any disturbances. Nevertheless, in the course of the evening there were small clashes between the mutineers and troops loyal to the government.

The next morning, Sarrail informed Trikoupis that he would not tolerate armed clashes in the city and would use armed force against them. Trikoupis replied that he had enough troops to quell the mutiny unless Sarrail supported the mutineers. At about 4 p.m. French troops occupied the seaside promenade and Sarrail demanded that the troops of the 3rd Corps return to their barracks, otherwise he would attack them with artillery. Trikoupis fearlessly replied that Sarrail should shoot the mutineers and demanded that he repeat his requests in writing. After some back and forth, he received it. Since their rejection would have led to an armed conflict in the city, Trikoupis gave in, especially since he knew that Athens did not want an armed clash with the Allies. On the evening of August 31st, the loyal Greek officers were put on a French passenger steamer and sent to Piraeus. The teams were placed under French command. Sarrail proudly reported to Paris: *"Colonel Trikoupis [...] tried last night to prevent the coup that was reported yesterday. Three dead and five injured. Surrounded by the rebels and announced by me that order must be maintained in Thessaloniki, he surrendered to me and not to the Revolutionary Committee. All officers with their weapons and all disarmed soldiers are in our hands. Each of the officers and soldiers who want to fight with the Entente reinforces our Greek units. The others are already there, the soldiers are in the evacuation camp, the officers are on a steamer in the port with which they will be brought to Piraeus on Saturday, unless you give contrary orders. The Greek division in Thessaloniki no longer exists."*[247]

A short time later, Sarrail received a note stating

Above: Enemy bombing attack on the airfield in Hudovar. Bombs exploding! Photo: Slg. Jörg Kempf, KG 1.

Above: "Excellency Lieutenant General von der Esch visits the KG 1 in Hudovar. Hptm. von Römer, Hptm. Kastner, February 1917." The 101st Infantry division was on the Greek border from March 3rd, 1916 until October 27th 1917 and constantly had contact to the pilots. Von der Esch was their commander from November 19th, 1915 until May 23rd, 1917.

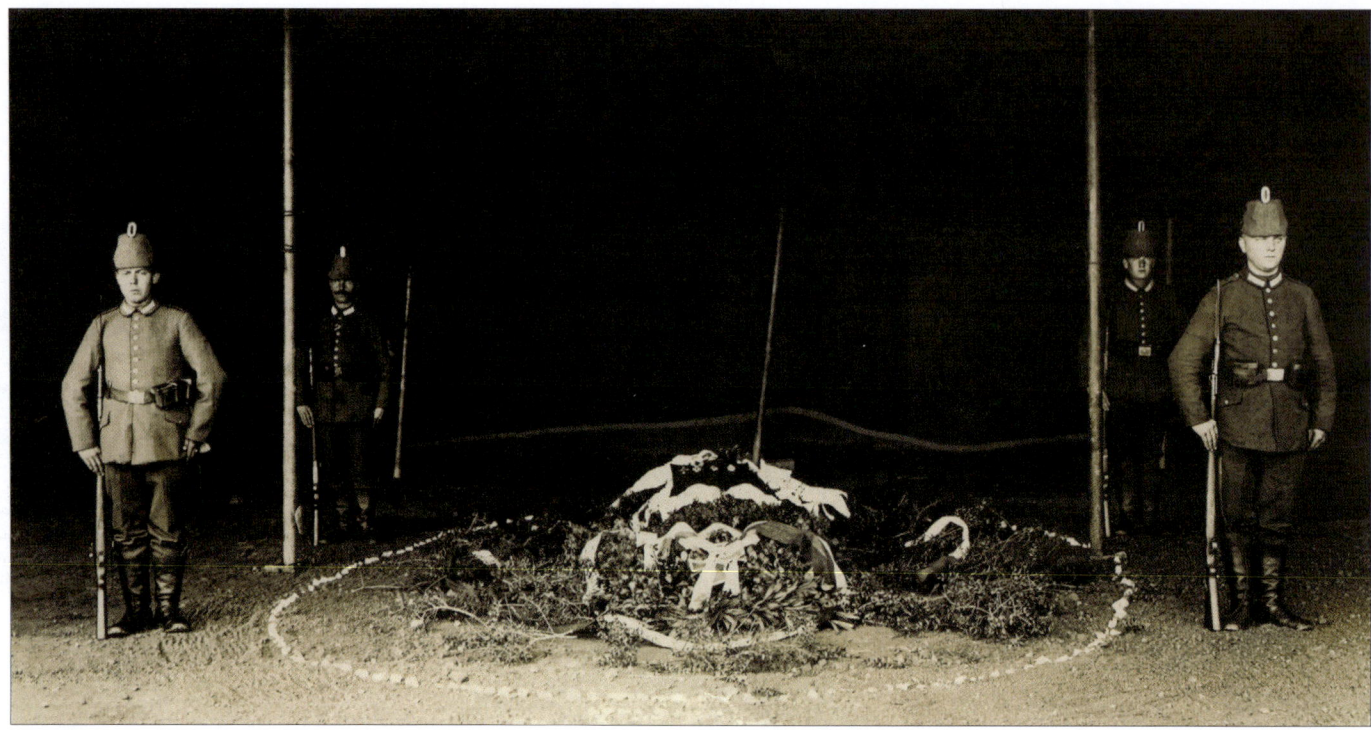

Above: Oblt. Freiherr Bodo von Lyncker, death honors. "An English aircraft shot down, a second was rammed and fatally crashed. Fl.-Abt. 30. A good comrade. Honor his memory." Freiherr Bodo von Lyncker, born on October 22nd, 1898 in Berlin, died on February 18th 1917 in Piravo (Macedonia). Photo Johann Keller.

Above: An Allied postcard of the era.

Right: Fort Broscarei (Fortul XIV Broscarei) about 5 kilometers south of Bucharest.

Below: 50kg and 12 ½kg bombs in front of a Gotha G.II. Slg. Jörg Kempf

Below Right: Detail of the marking on a 50kg bomb. Slg. Jörg Kempf

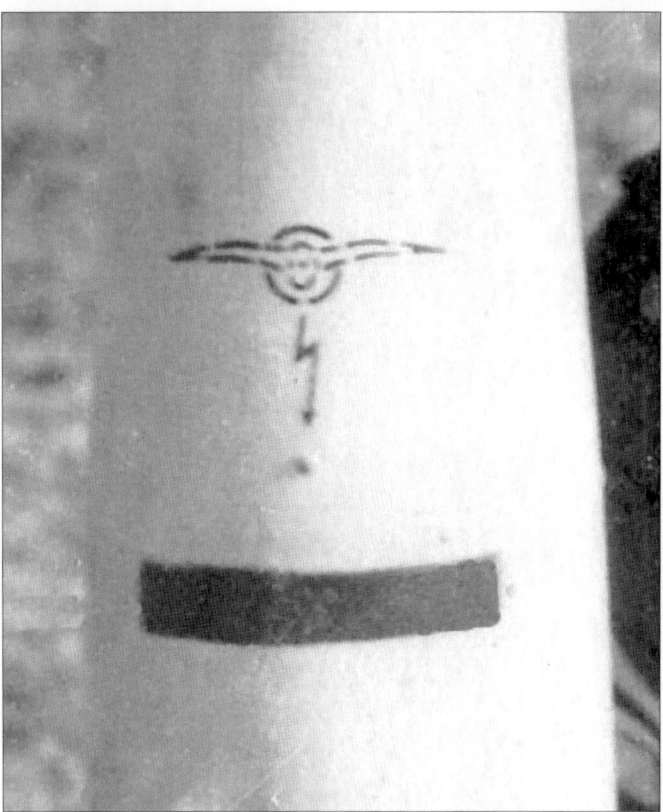

Left: "Oblt. Kaestner and his Bursche (orderly or batman)." Oblt. Kaestner prepares for an enemy flight and he needs help with his leather clothing. Slg. Jörg Kempf

Facing Page, Above: Two AEG G.III bombers of KG 1 on a bombing raid to Saloniki.

Facing Page, Below: "Second English aircraft shot down." Slg. Jörg Kempf

Below: "Before take-off: MG fighter Fw Lehmann, Lt. Dörstling, Oblt. Kaestner." The crew stand prepared in front of their Gotha G.II. Slg. Jörg Kempf

Above: Emergency landing in the Vardar. The AEG G.III is slowly sinking. Unfortunately, there is no way to find out what date this is. In any case, one can consider it lucky that the emergency landing went so smoothly and a soldier with a camera was nearby. "Landing in the Vardar by Lt. Rentsch and Lt. Pfeffer." The two crewmembers were saved and walked over the pier, soaking wet, to dry land. Lt.d.R. Werner Rentsch, born on August 30th, 1896 in Herford (Westphalia), died October 3rd, 1963 Düsseldorf. Slg. Jörg Kempf.

Above: Gotha G.II 207/16: "Large fighter aircraft of Hptm. von Römer and Lt. Kempf." In the background one can see the living and sleeping train, but since it is 3rd class, it must be the team's carriage.

Right: "At the Dojran Lake: Lt. Kempf, Lt. Rentsch, Lt. Löhr, Lt.d.R. Fritz Lankisch." (born on July 9th, 1894 Woizichau, Kreis Jarotschin, died June 10th, 1917 Ingelmünster, fallen near KG 1/1), Lt. Schröder in March 1917. Lt. Lankisch had captured two land turtles, that were being admired. Slg. Jörg Kempf

Below: "At the Dorian Lake: Lt. Lankisch, Lt. Rentsch, Lt. Kempf, front Fw. Müller, Lt. Löhr." (Lt.d.R. Friedrich Löhr, born on January 26th, 1890 in Halle a.d.S, died on April 8th, 1917 in a fighting squadron.) Slg. Jörg Kempf

Above: Crew and flying personnel in front of their aircraft, Gotha G.II 207/16. "From left to right: mechanics, MG-fighter Kluge, Hptm. von Römer, Lt. Kempf, Uffz. Mock, Mechanic Blattman in Hudova, April 1917." Slg. Jörg Kempf

Above, Below, & Facing Page, Below: Prize photos from a Russian crash (estate of Wehmeyer). These photos were taken from an air officer who was captured in the Romanian campaign. These photos show a flight school of the czar in Russia. The aircraft types are French Farman aircraft that were built under license.

Above: Prize photo from a Russian crash (estate of Wehmeyer). This photo shows a flight school of the czar in Russia. The aircraft types are French Farman aircraft that were built under license.

Above & Facing Page, Below: Russian prize photos (loot) from a crash (estate of Wehmeyer): Russian captive balloon.

Above: Air maneuvers at a Russian flight school and training (prize photos) These rare images gave clues as to the status of the czar's aviation capabilities from 1914.

Above & Below: Air maneuvers at a Russian flight school and training (prize photos) These rare images gave clues as to the status of the czar's aviation capabilities from 1914.

Right: Here are the graves of BO Lt. d.R. Fritz Herrlich, born on April 16th, 1889 in Berlin, died on January 23rd, 1916 in Demirkapu (Macedonia), FFA 1, and FF Offiz. Stellvertreter Otto Witt, born on June 9th, 1891 in Grimmau, died on January 23rd, 1916 Demirkapu (Macedonia), FA (A) 246.

Below Right: Hptm. Erwin von Römer, born on February 8th, 1885 Bad Elster (Kreis Oelsnitz), died February 25th, 1948 as a Soviet prisoner of war. The half-squadron KG1 was transferred back to the western front from Hudova in May 1917. (Slg. Jörg Kempf)

Above & Below: Deadly crash of observer Lt. Walter Crome (born on August 13th, 1893 in Goslar), on May 5th, 1917 with Alb C.III from Flugpark 11, with pilot Flg. Karl Mayer (born on July 6th, 1895 in Hofen) in Üsküb.

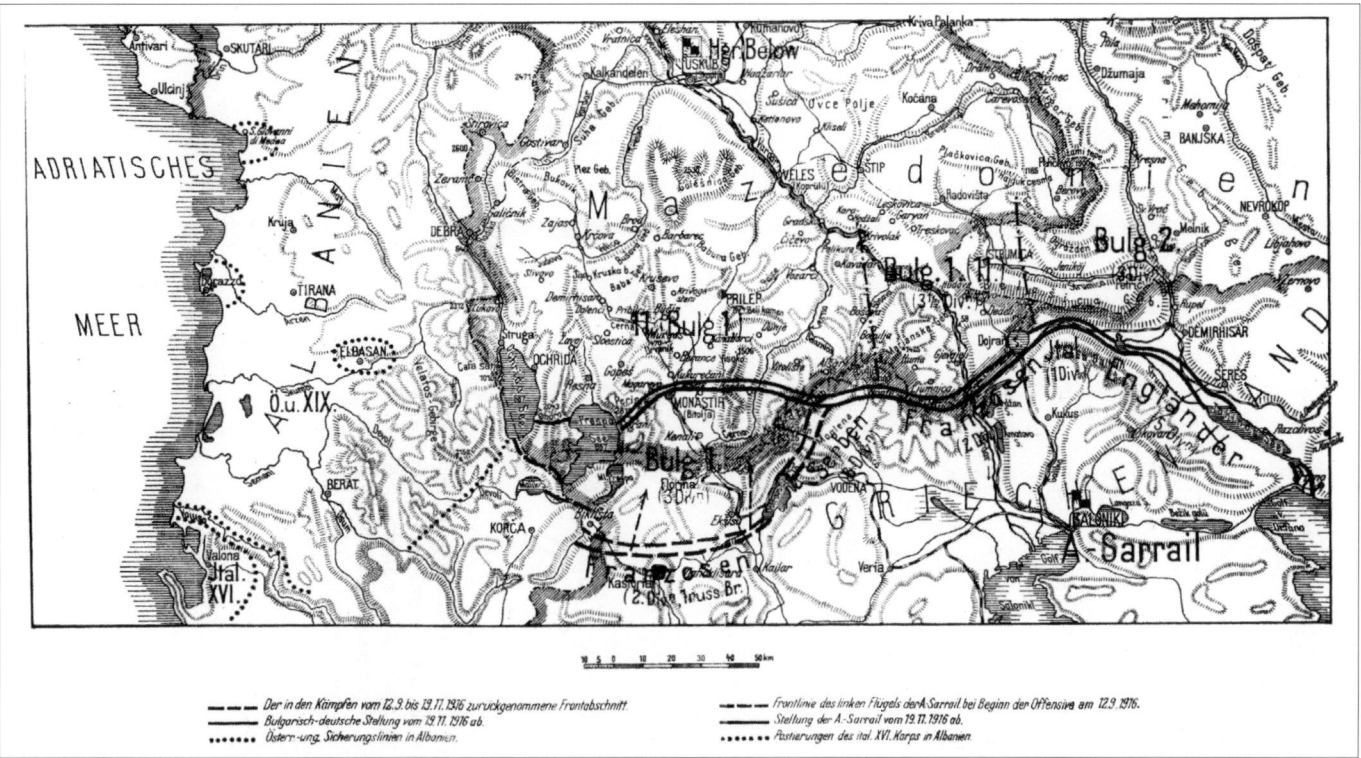

Above & Following Page, Above: Deadly crash of Vzfw. Walter Gröppel (born on September 21st, 1894 in Leipzig) as pilot near the AFP 11 in Üsküb on May 16th 1917.

Left: From top to bottom, in Hudovar: Lt. Löhr. Lt. Hunzinger, Lt. Schröder, Lt. Rentsch, Lt. Lankirch, Lt. Kempf, Oblt. Gutknecht, Hptm. von Römer. The pictured wagon is a luxury carriage of the French "Compagnie Internationale des Wagon-Lits", a sleeper car that was built in 1908 by MAN (Series No. 1772–1783), Slg. Jörg Kempf, KG1.

Facing Page, Below: Before the air unit 20 (Flieger-Abteilung 20) with Lt. Wehmeyer is sent to the Romanian war theater, it flew reconnaissance over French airfields. Here is the French airfield Corcieux, 36 kilometers west of Colmar, on June 15th, 1917, from an altitude of 3,800 meters, photographed by pilot Vfw. Fruhner and the observer Lt.d.R Weihs. The deployment of the FA 20 to Macedonia occurred only in December 1917, via Focsani to Macedonia.

VK Fl. 30 · 1419 · 31.5.17 · 10–12 v. Lager Strasse Orljak –

F. 20 – 5350 – 15.6.17. H. 3800 – B. 50 Flugpl. Corcieux – F. Vicef. Fruhner – B. Lt. d. R. Weihs

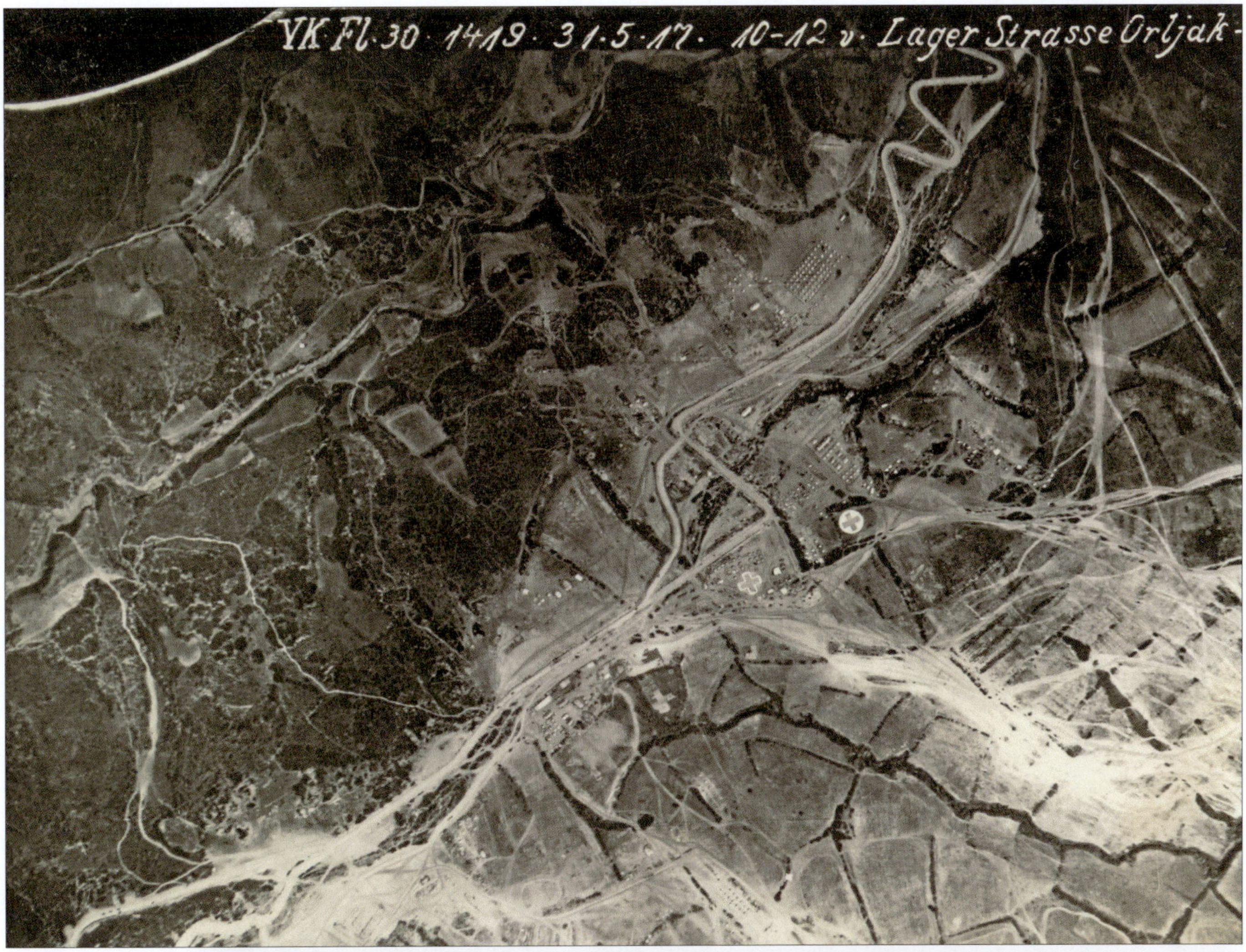

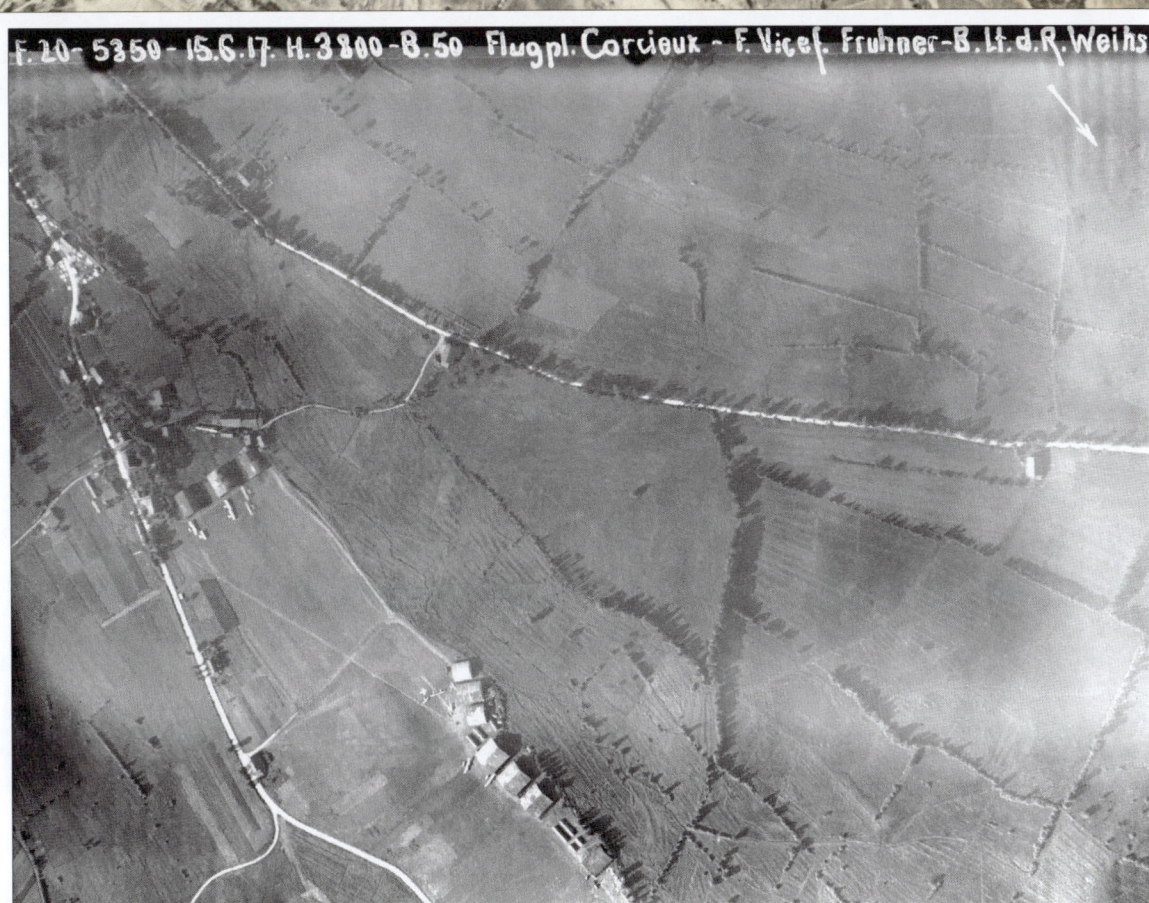

Left: "The inhabitants of Serbia" Keller writes on the back side of this photograph.

Facing Page, Above: "Serbia 1917" in front of an Albatros C.VIIs (1916). In the Serbia campaign the German and Austrian air forces lost about 104 aircraft, but only two due to engagement with the enemy. The rest crashed on take-off or landing, or were lost due to weather.

Facing Page, Below: The "iron gate" near Üsküb. Keller writes: "The 'iron gate' in Serbia, which was very difficult to fly over." This narrows in the Danube between Serbia and Romania near the Mraconia monastery is no longer visible today, due to a dam.

Below: Aerial photograph near Salonika (Thessaly) with allied ships.

Das Eiserne Tor in Serbien

that Venizelos would come to Saloniki on September 9th to take command of the movement. This was the beginning of the division of Greece, the beginning of the so-called *"Ethnikos Dichasmos"*.

The mutiny was sharply criticized in Athens and ancient Greece, and even by many Venizelists. Venizelos himself criticized it as too early and irresponsible, but he didn't condemn it, after all, he himself was behind it. The cooperation with the French also worked extremely well in Athens. On August 26th, the French admiral Louis Dartig du Fournet, who in 1915 had rescued 4,000 Armenians on Musa Dag on his own initiative, received the order to set up a special squadron and go to Greece, according to the plans drawn up in June. He left Malta with four ships and reached the island of Milos on August 28th, where 39 British ships were waiting for him. With this armada they could prevent a Bulgarian-German advance to the south, but also the royalist resistance to the Venizelian putsch. The departure of the fleet was delayed again by four days because the British had scruples, but on September 1st the fleet appeared in the Saronic Gulf.[248] This was the beginning of the Allied intervention in favor of Venizelos' action.

But the Venizelists were not the only ones who were organizing, their opponents did the same. Demobilization in the face of the occupation of the Roupel-Strait and eastern Macedonia had violated the patriotism of the common soldiers. They felt that they were being sent home and that Macedonia was being left to the enemy. In their eyes the Allies and Venizelos were to blame. When the first railway trains with demobilized soldiers arrived in Athens, there were demonstrations against the Allies and Venizelos. Tensions were high as the anti-Venizelists in the army organized themselves. A league of reservists had already been set up in April, which quickly developed, under the leadership of royalist officers, into a paramilitary organization that controlled the low country. It formed the counterbalance to the Venizelist groups of the National Reservist League, which had been founded by Danglis in mid-July. The general staff was behind the league. It was also the existence of the league that led to the postponement of the elections, which had originally been scheduled for August, to October.[249]

London was appalled by the French action in Greece. Attempts were made in Paris to convince Briand to exercise more moderation, but the latter was urged to take the tough course by his naval minister, who came from a ship-owner's family. But the British did not want a major argument with the French government as the Battle of the Somme reached its climax. So they gave in again and grudgingly accepted the French course in Greece. In contrast to the British politicians, the English King George V, who otherwise did not interfere in foreign policy, was no longer willing to accept the French behavior. On September 4th, in a letter to Asquith, he complained about the behavior of the Allies towards a *"neutral and friendly country"*.

"I cannot shake the feeling that we have let France rule too much on the Greek issue. As a republic it is probably a little intolerant, if not even trying, to abolish the monarchy in Greece. But that's not my government's policy. Nor is it that of the Russian ruler who wrote to me a few days ago: I am concerned about the internal affairs of Greece. It seems to me that the protecting powers are interfering too much in internal affairs as they try to protect their own interests with regard to Greece's neutrality. This to the detriment of the King. I cannot hold back from expressing my astonishment

and regret at General Serrail's unauthorized behavior towards the troops. They showed themselves to be loyal to the King and the government and refused to join the revolutionary movement in Saloniki. Couldn't it be possible to send a protest note to the French government taking a stand against Sarrail's actions? Greek public opinion, as well as the attitude of the King, is obviously changing. If the Allies treated the Greeks kindly and not, if I may put it so, in an oppressive spirit, then in all likelihood they would join."[250]

For all of the weeks that followed, George V and the Tsar were suspicious of French policy, particularly that of Sarrail. Constantine, for his part, tried to appeal to monarchical solidarity by sending his brothers Andreas and Nikolaos to London and Petersburg, respectively. In Paris, Marie Bonaparte, the wife of Constantine's son Georg, used her charm in a similar sense towards Briand, who was not immune to it.[251]

Right: Lt. Rudolf von Eschwege (born February 27th, 1885 in Homburg v.d.H, died November 21st, 1917 in Orljak). (Photo: Slg. Lance Bronnenkant)

Macedonia under French Occupation

Normally, the following period in the English and Greek historiography would be described as *The Entrenched Camp of Salonika*.[252] In reality, this and other parts of Greece experienced a kind of occupation.

Sarrail enjoyed inserting himself into politics. "During his entire life, Maurice Sarrail had an insatiable hunger for political intrigues."[253] His future place of activity, Greece, or rather Saloniki, offered the right breeding ground for this obsession, as we will see later. In his political views and judgments, Sarrail did not differentiate. His maxim was: whoever is not for me is against me; Or to put it another way: whoever is not for France is an enemy, and this also applied to neutralists. In his eyes, the neutral King Constantine was a friend of Germany who had to be fought. Furthermore, Sarrail was a fanatical Republican who had wanted the King's removal from the moment he arrived in Greece. *"Acting as proconsul on the part of Greek territory that he had brought under his control a few months ago, he is neglecting ancient Greece. He went to Athens in April 1916 to make a courtesy visit to King Constantine. It made him feel "employed" in the Salonika region"*[254]

The occupation began in Corfu. This island was given to Greece by Great Britain as part of an international contract in 1864, and was declared forever neutral. France was also a signatory. Corfu therefore had an internationally-accepted neutral status, just like Belgium. But now some French military ships appeared off the island, and an infantry division occupied it on January 1st, 1916. But the French government explained that it was not an occupation. The country justified this with the claim that Corfu was a German submarine base, probably because Wilhelm II had a summer residence (Achilleion) there.

Admiral du Fournet[255] wrote: *"The intelligence service had portrayed Achilleion as a perfectly organized submarine base, with a mooring that channeled the fuel to the sea. The occupation of Corfu allowed all of these legends to disappear. The quay was made for the palace's boats; The oil supplies were simply a little gasoline for automobiles. The pipeline was just the main sewer."*[256]

The whole thing was an untenable claim. In fact, the occupation was against international law. It was justified to the Greek government as having humanitarian motives. Then, beginning on January 14th, the last Serbian troops were brought to the island, which also became the seat of the Serbian government in exile.[257]

An Entente declaration spoke of a "pure duty of humanity". The "conquest" of the Achilleion as a submarine base was portrayed in the French press as a great feat, although the only people there were the castle administrator and a gardener.[258] A total of 133,000 Serbs were brought to Corfu, of which 3,681 died of exhaustion, typhoid and dysentery. When the survivors recovered, six divisions were formed. France supplied the weapons, the artillery and the animals to be transported. In mid-March the Serbian units were ready for transport.[259]

Another occupation of Greek territory had preceded this. On December 28th, 1915, French sailors from the cruiser *Jeanne d'Arc* occupied the island of Kastellorizo. The French consul had incited a group of French-friendly residents of the island into a revolt. The official justification for the occupation, however, was to prevent possible Turkish reprisals. When the Greek Evzones wanted to land, the French blocked them. The Turkish artillery then promptly shelled the island because of the French occupation.[260] The real reason for the occupation of the island was that the French wanted to establish a naval base on Kastellorizo to fight German U-boats.[261]

The fear of German submarines in Greek waters reached proportions that bordered on hysteria. Mainly in the French press, Greece was accused of *"treason"*: Greece was allegedly a hornet's nest of German submarines. The Greek coast was full of hiding places for German submarines. The submarines were said to have used the Greek ports without the slightest hindrance. In December 1915 it was reported that – in Saloniki of all places – a supply depot for the submarines had been discovered. Allegedly, the Allies found countless oil depots on the coast, and these would be tracked down again and again by the vigilance of the Allies. *"Some newspapers published Jules Verne-style inventions: one swore that there was a workshop on Athens' popular Phalerum beach. This diabolical, underground and underwater facility allowed submarines to approach submerged in full daylight.*

They could connect themselves to a pipeline 500 meters offshore and collect large quantities of oil. Queen Sophie (emphasis added) often came to Phalerum for "teatime" to watch this."[262]

Every day the press brought new "revelations" of this kind. Gullible readers were appalled and angry at the Greeks and their King who committed such crimes. In the eyes of readers in Paris, the Greeks were worse than the (German) "Huns". There was not a shred of truth in all of the reports: they came from the poisonous kitchen of the French secret service under the leadership of the Naval Attaché de Roquefeuil[263] and the Venizelist press, according to a reply by Admiral Dartige to an article in the *Revue des deux Mondes* of March 1917, which commented negatively on Greece: *"The revelations of the Venizelist press about the supply of German submarines in Greece are a tangle of absurd legends. Despite the most repeated checks, not a single one of these "facts" was true. A number of mischievous villains, however, made a living from trading in fake*

Below: Reconnaissance photo of Skala-Stavros taken 7-6-1917.

Above: From the pay book of Johannes Keller. Stamp from Army Flight Park 11.

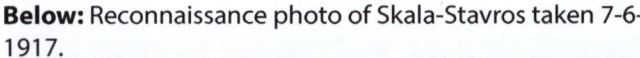

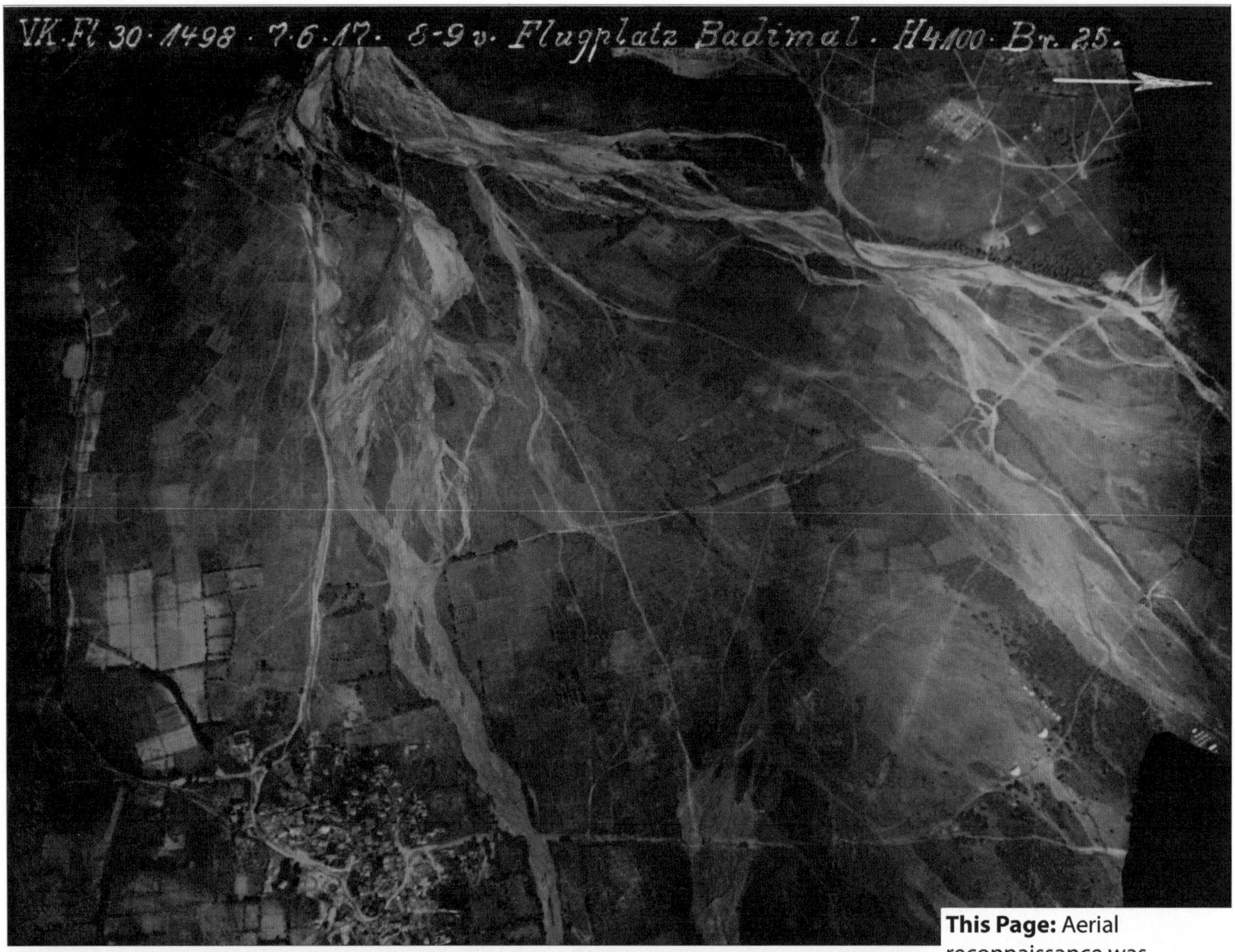

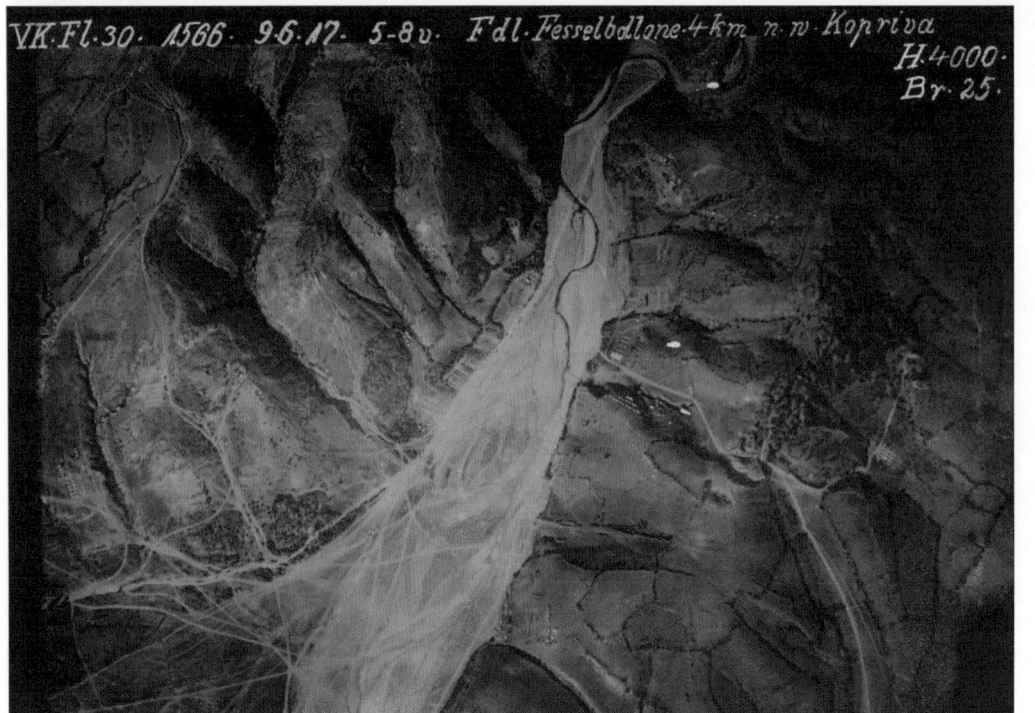

This Page: Aerial reconnaissance was the pilots' daily bread. The additional war with Romania called on all operating resources, which were already stretched to the limit.

Facings Page: Johannes Keller: of the advance party of Flieger-Abteilung 30, probably August 1917: 9.
1. Oberlt. Haupt
2. Lt. von Eschwege, who shot down his 11th enemy
3. Lt. König
4. Lt. Siebold
5. Lt. Rottka [with bandage] suffered a broken nose due to a hard landing [August 1st, 1917]
6. Lt. Lenz
7. Assistant Doctor
8. Lt. Becker

Above: August 1st, 1917, crash of Rumpler C.Ia (Hann) C.4713/16 with 180 hp Argus As.III. Pilot Gefr. Matull, bruised ribs and flesh wound, Observer Lt. Rottka, broken nose." Johann Keller.

news and abused the naivety or negligence of our secret service. A large number of people who wanted to be informed, in their ignorance, repeated this nonsense. We have cited a few examples; they were numbered in the hundreds. Tanks, towers that were mistaken for submarines and places for refueling that were absolutely inaccessible are just a few examples, etc. etc."[264]

It can be assumed that the fear of the U-boats also influenced the planning of French policy in Greece, since at the beginning of January 1916 the Quai d'Orsay was rife with talk about occupying Piraeus, in order to bring Athens under its own control. The neutral King should be chased out and Venizelos brought to power by force, if needed.[265] But so far, these were just considerations.

In order to secure his position in Macedonia against possible attacks from the east, be it from the Greeks or the Bulgarians, Sarrail had the railway bridge of Demir Hissar (Sidirokastro) over the Struma (Strymonas) as well as the bridges of the same railway line south of Lake Dojran on a branch line, blown up.[266] In doing so, he ultimately signaled to the enemy that he was not planning an attack to the east. The Greek *Fort Roupel* under construction on Struma-strait was about five kilometers north of the bridge, and through the demolition, it was cut off from supplies from the west and east of Macedonia. So it was at the mercy of Bulgaria. Sarrail paid no attention to proposals from either the Greek or French officers to take possession of the fort.[267]

On January 22nd, 1916, the British troop transport *Norseman* was torpedoed by a German submarine at Cape Karabournou. The captain succeeded in running the ship aground so that neither the people nor the 500 mules were harmed. But the attack convinced the French that some of the Greek officers in the Karabournou fortress had had contact with the Germans, and he decided to put an end to it.[268]

At the end of January, Sarrail had the *Karabournou Fort* at the entrance to the port of Saloniki and the battery at the mouth of the Vardar, occupied by a French battalion. French warships took over artillery security. Sarrail asked General Mahon, who was still not under his command, to occupy the eastern outskirts of Saloniki with two battalions in order to stop any Greek protest marches to the fort. Fortunately for Sarrail, the fort's crew did not fight back, and there was no unrest in Salonika. Mahon criticized Sarrail's behavior towards London: the bridges being blown up and the occupation of the fort would drive even the Entente-friendly Greek officers into the arms of the friends of the Central Powers. Something should be done to appease the Greeks.[269]

Briand and Joffre warned Sarrail to treat the Greeks a bit more kindly but that did not help. In January and February, more and more French naval units landed on certain islands and arrested foreigners and Greeks who had been blacklisted by the French secret service as friends or agents of the Central Powers. Petrol tanks belonging to private merchants were blown up in the ports because their owners were suspected of selling the fuel to German submarines.[270] Obviously, the Sarrail officers had no idea that the German submarines were powered by

Above: Reconnaissance photograph taken 20-6-1917. Note the crack in the glass plate negative. Allied camp near Kartasliderbend.

diesel engines. Such measures naturally turned the population against the Entente.

But it got even worse. At the end of December, at French instigation, imports of grain, flour, minerals, coal and the like were severely restricted. The reason given was that the stockpiling of supplies in neutral states bordering on those of the enemy should be prevented. However, it would be ensured that necessary imports could pass unhindered. But exactly this did not happen, and so *"normal"* goods came into the country in greatly reduced quantities. The result was that they became scarce and the corresponding stores raised their prices. This of course primarily affected the little people, who were outraged. The Allies turned the Greeks against them more than the Central Powers' propaganda could ever have done. When the Prime Minister and King protested violently, this turned the Allies against them and they reacted harshly. This, in turn, was grist on the mill of the Venizelists, who began to stir up against the King and the royalists. A national split loomed on the horizon.[271]

The British and Russians watched developments with concern and asked Paris to hold Sarrail back. Briand therefore admonished Sarrail and the French ambassadors on various occasions, to hold themselves back more, and not take any measures which would anger the Greeks. In this context, Sarrail and Constantine met on February 22nd. Briand had given Sarrail clear instructions. He should tell the king that there was no intention of humiliating him or his country. Certain measures

Above & Below: Two reconnaissance photos from the area of fighting on the Struma, on June 28th, 1917. One can see the excellently-built English positions near the mouth of the Struma. Fighting rarely occurred, but most losses – on both sides – were due to illnesses such as fever and malaria.

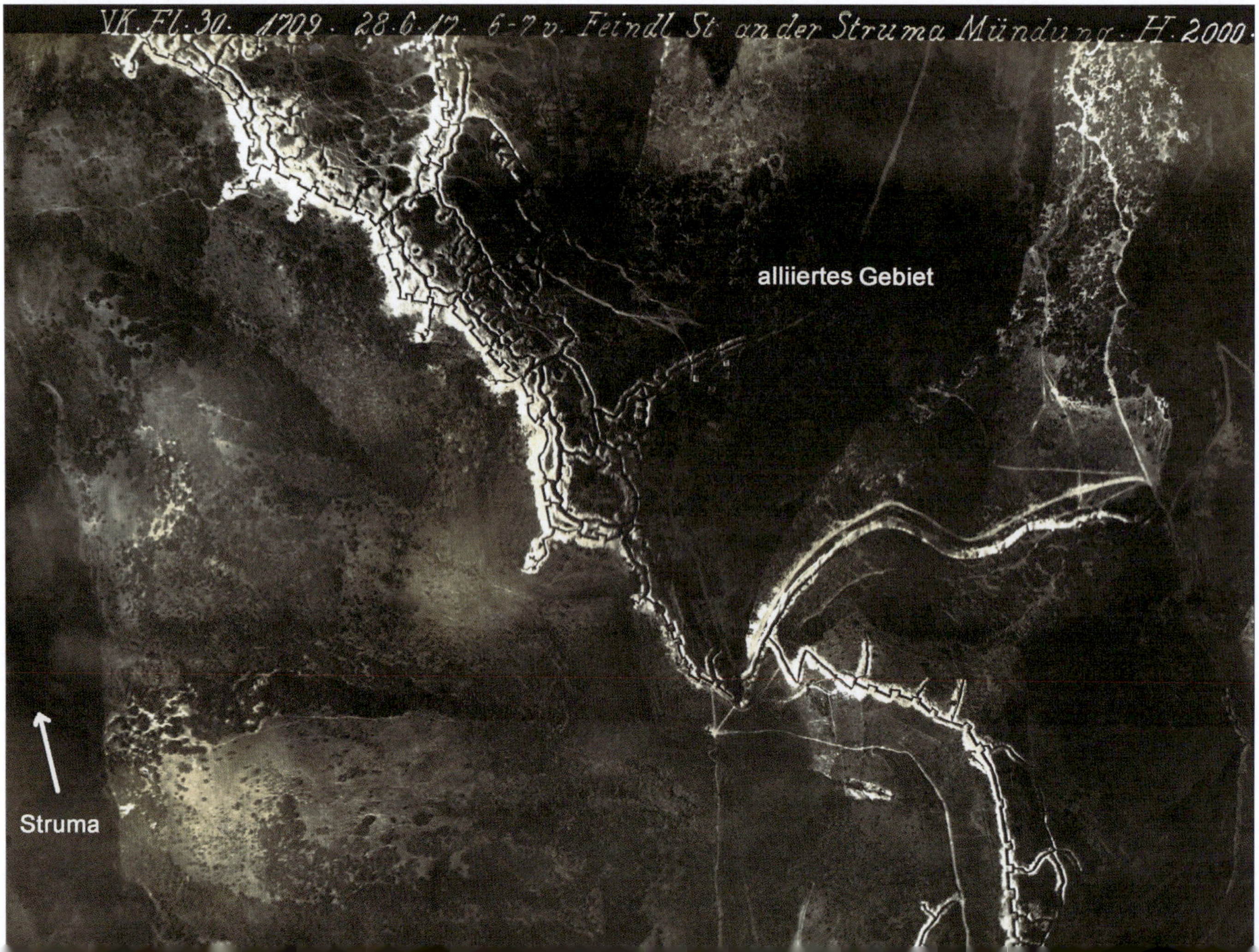

Above: Reconnaissance photograph taken 29-6-1917 of the bridge near Komarjan.

Below: Lt. König sits in his working space. In the background a map of the reconnaissance sectors.

Above: Enlargement of the reconnaissance sectors.

Left: Lt. König.

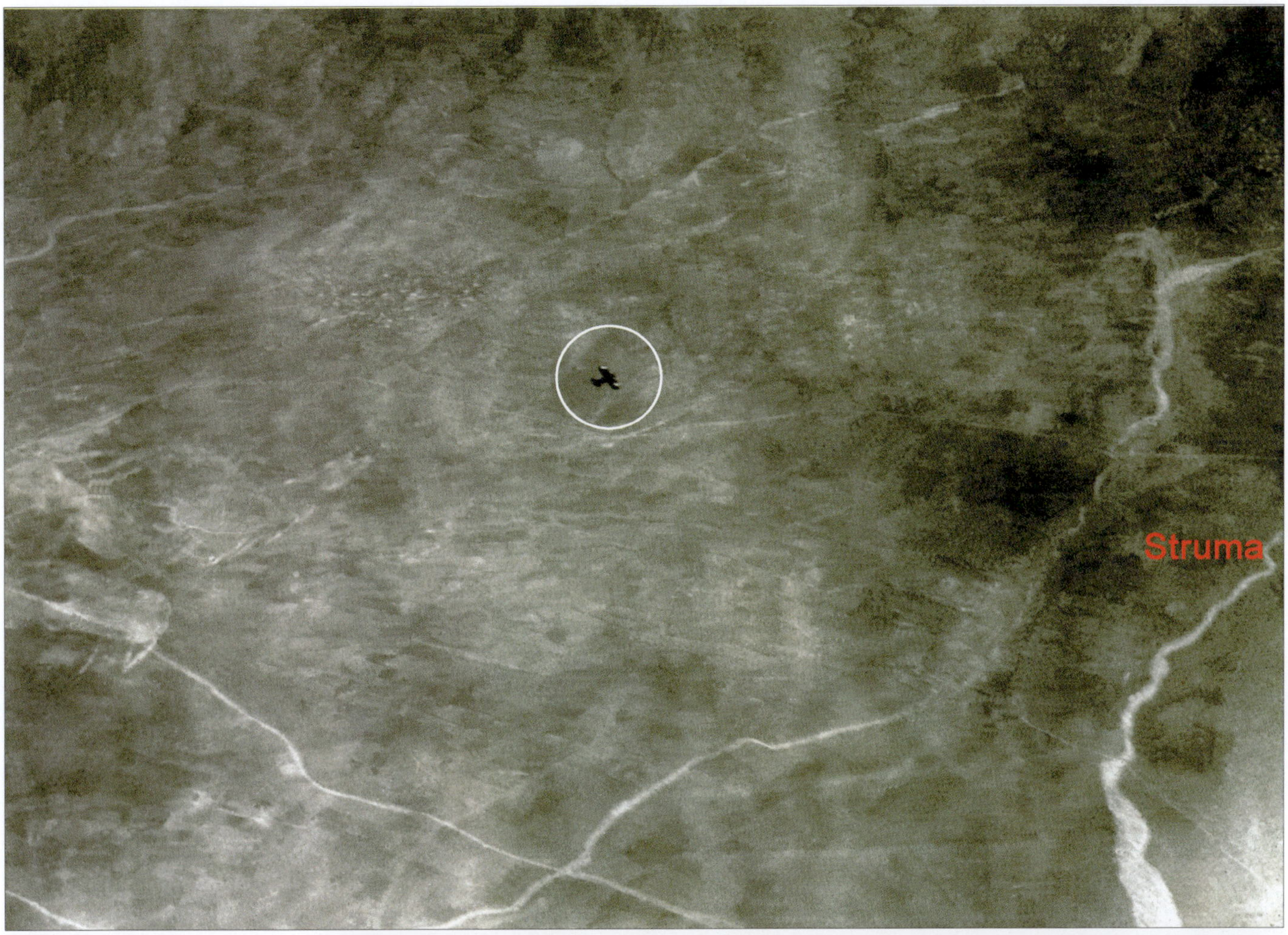

Above: Johann Keller: With an English aircraft B.E. dogfight on July 17th, 1917 near Lakana. Altitude 4,500m."

Right: Enlargement of the Ailled aircraft in the photo above.

144

Facing Page Aboe: Reconnaissance photograph of the English positions.

Facing Page, Below: The swamps near the Struma mouth, into the Golf of Orfano.

Right: Telegram-Thank you note from the Bulgarian High Command 2 for excellent reconnaissance results: "To the advance command of Fl.-Abt. 30: I would not like to fail to express my best thanks and very special appreciation to you for the extremely successful and valuable image of the enemy positions at Neohori. Rittmeister von Bardeleben, German liaison officer at the Bulgarian High Command 2."

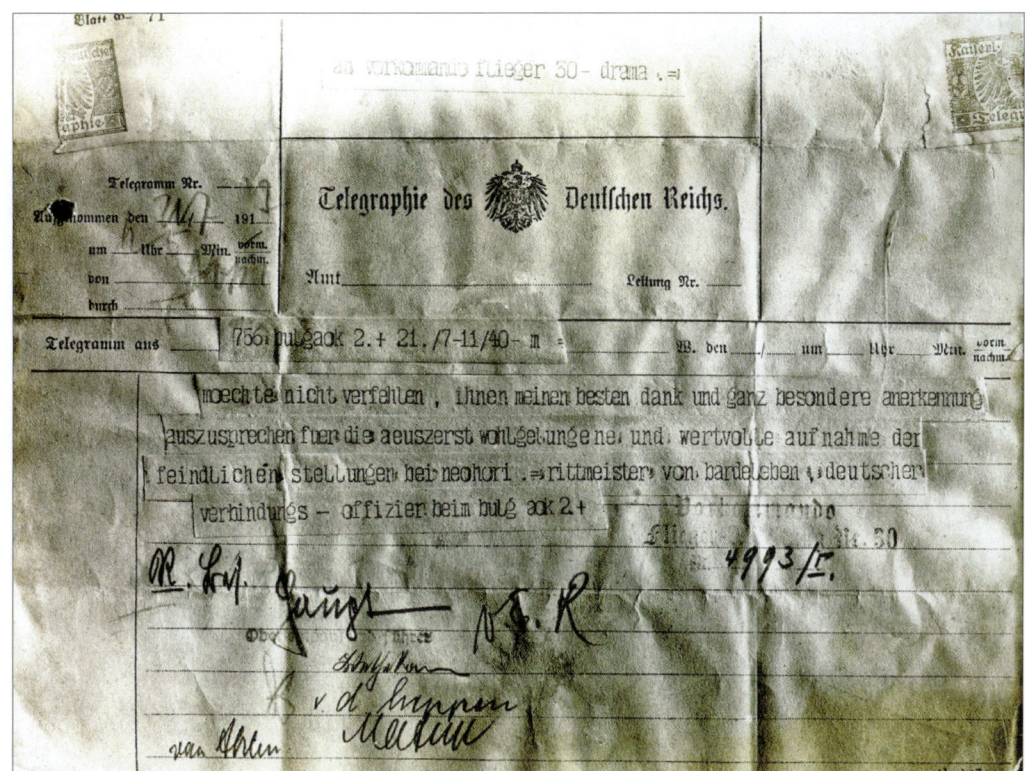

Above: Group photo of the advance command 30: No. 1 Oblt. Haupt, No. 2 Fw. Johann Keller, No. 3 Oblt. von Eschwege.

were militarily necessary. They respected Greece's neutrality, but welcomed the cooperation. The conversation itself was harmonious and Sarrail relented on a few small points.[272]

Skouloudis used this temporary relief to address the bond issue in February. The International Finance Commission had approved a loan of 150 million gold francs for Greece, which had not yet been disbursed. Greece urgently needed this money, because the government was on the verge of running out of money due to the mobilization. London and Paris were ready to pay it out if Greece made concessions. The British envoy wanted the Greek police to report to the Allies and his French colleague supported the demobilization of the Greek army. When Venizelos found out about this, he suggested that Greece enter the war. Demobilization was not an option for him. The Allies postponed the problem for some time.[273]

In the meantime, opposition to the Saloniki operation in the British General Staff increased massively. Chief of Staff William Robertson was preparing for the Battle of the Somme and wanted to withdraw all British troops from Saloniki and as many as possible from Egypt. On February 12th, he submitted a memorandum to the *War Committee*

with his ideas: Britain must gain the diplomatic upper hand in the Entente. Peace should be made with Turkey, with Russia forgoing Constantinople. If the latter was not possible, Bulgaria should be broken out of the Central Powers' front by offering it all of Macedonia from the Turkish border to Saloniki in the west and Veles in the north if Bulgaria becomes neutral again. The area could immediately pass into Bulgaria's possession. If Bulgaria accepted this offer, the supply of Turkey from Germany would be blocked. The 200,000 men could be withdrawn from Saloniki. The only state that would not benefit from this is Greece, but it did not support Serbia either. Although the *War Committee* did not agree with the ideas of a separate peace, it fully agreed with his ideas regarding Saloniki. There should be no offensive in the Balkans.[274]

France then tried in vain to get Italy to send troops to Saloniki. Afterwards, the Saloniki question was more or less forgotten until June because of the

Left: Oberleutnant Walter Wehmeyer born July 7th, 1887 in Spandau, died March 31st, 1938 in Berlin. Came from the Garde-Fuss-Artillerie Regiment and during the war was an observer, including at the FAA 283, FA 20 and FA 38. Many of the documents in this book are from him. Here Lt. Wehmeyer sits in his room in Romania.

Below & Facing Page, Above: My tent and machine, destroyed by the hot sand storm. Johann Keller

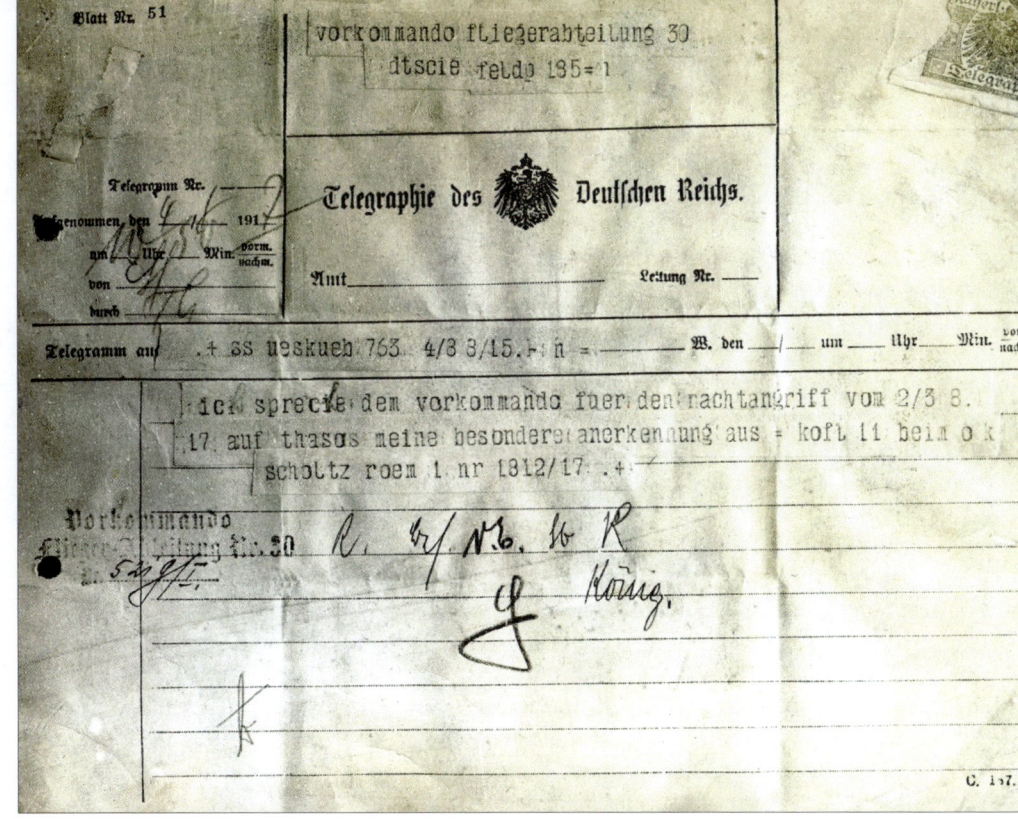

Right: Telegram: August 4th, 1917, Üsküb: "To the advance command of Flieger-Abteilung 30: I express my special appreciation to the advance detachment for the night attack on August 2nd / 3rd, 1917 on Thasos. Kofl 11 of OK Scholtz." [Commanding officer of aviation at the chief command 11 of the Army Group Scholtz].

German offensive against Verdun. At a meeting on March 14th, it was decided that the Serbian troops should be transferred from Corfu to Saloniki.[275]

On April 5th, 1916, the French and British ambassadors visited Prime Minister Skouloudis and informed him that the Serbian troops were operational. They were to be brought to Patras by ship and transported to Thessaloniki by train via Athens and Larissa. The diplomats hoped that the Greek government would agree. Skouloudis protested violently. The transport of 100,000 men across Greece meant that civil rail traffic and thus the country's economic life would practically be shut down for two months. In addition, such a transport violated the neutrality of Greece. He, Skouloudis, must emphatically reject the

Left: Telegraph stamp: "Kaiserliche Deutsche Telegraphie." ("Imperial German Telegraphy.")

project: *"No, gentlemen, we will not allow such a thing. I will officially declare this to you." "Our government,"* replied the French minister, *"has not instructed us to ask your permission, but to inform you of the decision."* […] Skouloudis said:*" I declare that it is the decision of my government to not to allow the overland passage and further inform you that I see myself compelled to blow up the railroad "*. With increasing anger, he continued: *"You have left nothing for our country - neither self-respect, nor dignity, nor independence, nor the right to live as free people. But don't forget that even the most benevolent patience and the greatest obedience have a limit, that a single, last drop can bring the barrel to overflowing."*[276]

The British envoy wanted to defuse the situation and asked Skouloudis if the Greek government would agree to a sea transport. After consulting the cabinet, Skouloudis agreed to the new proposal. But in the meantime the Serbian envoy had raised the old demand again and pulled the Allied envoy back to his side. Again Skouloudis refused. Skouloudis instructed the Greek ambassadors in London, Rome and Petersburg to present the Greek position, but not in Paris because he knew that there would be no point in protesting there. Skouloudis 'arguments were accepted in London and an attempt was made to change Paris' mind through diplomatic pressure, which succeeded. But Prime Minister Briand retaliated against Greece on April 10th by refusing to pay out the first tranche of the 150 million gold francs loan that the Finance Commission had approved.[277]

On April 11th, Skouloudis complained that Paris was linking the transport of the Serbs and the disbursement of the loan. The claim that Greece wanted to have the transport paid for by paying out the loans was completely unjustified and pulled out of thin air. Greece urgently needed this money because mobilizing its army cost huge sums. Demobilization was rejected by the French. Therefore, the Greek government turned to Germany and asked for financial help. On April 20th, the imperial government informed the Greeks that it was ready to advance 40 million Reichsmarks in order to keep Greece neutral. Berlin transferred this sum. The French press later claimed that this was the price which Greece had received for clearing the half-finished fort on the Roupel-Straits at the end of May, a good four weeks later.[278]

On the question of the transport of the Serbian troops, an agreement was finally reached that took the Serbian troops on a specially secured sea transport to the Chalkidiki peninsula on May 30th. The troops were divided into three armies. Crown Prince Alexander became the new commander in chief. Voivode Putnik had resigned from office due to exhaustion. The 130,000 men had 34,000 transport animals, 9 battalions of field artillery, 6 battalions of anti-tank artillery, 6 batteries with 120 mm cannons and 300 machine guns. The Serbian government in exile also moved its seat to Saloniki.[279]

At that time there were still consulates of Germany, Austria-Hungary, Bulgaria and Turkey

Above: Lt. d.R. Walter Voigt burned in his aircraft. The crash location is Braila in Rumania (Voigt, born on June 18th, 1893, died on August 14th, 1917).

in Saloniki. The consulates regularly reported on the Allied troops arriving in Saloniki. Each of these consulates employed agents and spies to gather information about the Allies. General Mahon wanted to have these spies arrested in one fell swoop and informed Sarrail about it. He did not answer, but on December 30th, 1915, Sarrail informed his British colleague that he would have the four consuls

Above: Romanian prisoners are led away

Above: Lt. Steinbrück before take-off in a DFW C.V.
Below: Lt. Protzek and Lt. Matte in Drama.

arrested in a few minutes and asked him to take part in the operation. Mahon invited a small British group to take part. He feared violent protests from neutral Greece, which actually did become loud. The next morning it was decided to arrest the spies in a joint operation.[280]

A total of 20 Austrian, 17 Turkish, 12 Bulgarian and 5 German consulate members were arrested.[281] The consulates were occupied and the files confiscated. A little later the Norwegian consul was also arrested.[282] The diplomats of the Central Powers were interned on a French warship and taken to France, from where they were later repatriated. Greece protested violently, but Sarrail did not care.[283] He did not find it necessary to inform the Greek government of his actions.[284] The arrest of foreign consuls on the soil of a neutral country was an outright breach of international law. But Sarrail went even further: he interfered in Greek domestic politics. He treated Greek royalists rudely and humiliated them, according to Mahon and the British ambassador.[285] Two Greeks who had flyers in their possession that were dropped out of German aircraft were sentenced to death and shot.[286]

Right: Aerial reconnaissance of the FA 20 on August 19th, 1917 from 3,700 meters' altitude. Photographed at 10:30am, a Romanian military stockpile 2½ kilometers east of Anghelesti (west of Bucharest), by observer Oblt Wehmeyer.

Below: Aerial reconnaissance of the FA 20 on August 29th, 1917 over the train station of Badiana (?) from an altitude of 800 meters. One can clearly recognize the wagons on the turnout, as well as the people, casting long shadows.

Above: Aerial reconnaissance of FA 20 on August 29th 1917 from an altitude of 800 meters over the train station in Balca; many details can be recognized.

Constantine was said to have sympathy for Germany because he was a brother-in-law of Wilhelm II. But nobody thought of accusing him of being kind to England because he was a nephew of the British King. He accused the Entente of dishonesty: *"It is the purest hypocrisy when England and France speak of a violation of Belgium's neutrality, after what they did and still do to me. They occupied Lemnos, Imbros, Mytilene, Castellorizo, Corfu, Saloniki, the Cha[l]kidike peninsula, a large part of Macedonia. Germany invaded Belgium under the pressure of military necessity. But what was the military necessity to destroy the Demirhissar Bridge and to occupy Corfu? The Serbs could have been brought to Italy more easily. Is it not so that the Italians don't want the Serbs because they fear cholera? It's just as uncomfortable for us. [...] At the beginning of the war, eighty percent of the Greeks were favorably disposed [to the Allies], today it is not quite twenty percent."*[287]

Constantine tried on various occasions to defend himself against the one-sided, completely distorted reports in the Allied press by offering interviews to the journalists. He corrected the facts, but Allied censorship did not allow the corrections to be published.[288] So Constantine turned to the American press, i.e. the AP correspondent Hibben. The German text above also comes from this interview. Regarding the demolition of the bridge, he said: *"What military reason compelled me to destroy the Demir-Hissar Bridge, which cost one and a half million drachmas and which was the only way by which I could feed my troops in East Macedonia? The bridge had been prepared to be blown up. It would have been possible to destroy it in one minute, if the enemy really came. But it is absolutely certain that there was not a single man of the enemy anywhere near the bridge. There is also no sign of a planned enemy advance. The*

Above: A further photograph of Balca train station (?). Extensive piles of goods can be recognized. A series of photos was likely taken.

only reason that remains is the clear intention to surrender the Greek troops surrounding Drama and Serres to starvation."[289]

Hibben sought Constantine's authorization of the interview text, which he approved. But what followed was an unbelievable story.

Before I sent my text to America, I informed Mr. Guillemin [...], with whom I was on friendly terms, the content of King Constantine's admission. He reprimanded me as a friend of France for wanting to send such a message, which he described as German propaganda. At the same time, he sent an encrypted message to his government to have the censors stop the interview. When I found out about this, I sent the same telex via England because all messages from Greece to the United States must go through either France or England and are subject to Allied censorship. I discussed the case with Sir Francis Elliot, the British Ambassador to Greece."

"Nothing is gained by suppressing one side of the matter," I told the English diplomat. "The Allies have dozens of correspondents in Greece who are flooding the Entente. The American press is also here with its view of what is happening in the Middle East. I believe that King Constantine is sincere and I am concerned that he will not be treated fairly. If this telex is stopped, it will arrive later as a letter. You cannot suppress the truth. It would be better to let the message go through."

"It seemed that Sir Francis agreed and telegraphed his government to let the interview pass. The British censors, however, had a different opinion. The report was withheld in London. Finally, I informed M. Guillemin that I would also send the report to Berlin or that I would carry it to New York personally to circumvent the censorship. After the report was held back for six days, Prime Minister Briand decided it made more political sense to pass it on, but he made it a condition that a semi-official opinion of the French government

154

Fl.20.183.29.8.17. H.800. B.50. 8V. Brücke bei Balta

Fl.20.181.29.8.17. H.600. B.50. 8V. Adjudul-Nou - Sägewerk

Above: Enemy airfield is photographed by the FA 20 on August 29th, 1917 from an altitude of 800 meters. This is a Romanian airfield without fighter or flak defenses.

Facing Page, Above: Aerial reconnaissance of the FA 20 on August 29th, 1917, from 800 meters' altitude over the train bridge over the Sohodol bridge near Balta (Romania). Since the river is flooded only in the springtime with run-off from the Carpathians, tracks of other river crossings can be seen in the gravel.

Facing Page, Below: Aerial reconnaissance of the FA 20 on August 29th, 1917 over a sawmill near Adjudul-Nou from an altitude of 600 meters.

would be published at the same time."[290]

Hibben's interview, accompanied by an official statement, was published a week later. It contained the usual justifications. The Allies had landed in Saloniki at the invitation of Greece. No land was occupied, but rather it was simply temporarily subjugated, a process that at the time had already taken two years. Guillemin and Elliot wrote a text that attacked the German propaganda. They demanded Hibben publish this text as well. The text ended with a statement that can only be described as cynical: *"The whole matter between the Entente and Greece is a question of trust. If Greece is loyal to the deal, it will not suffer."*[291]

When, at the end of May 1916, Bulgarian and German troops occupied the unfinished *Fort Roupel* on the Struma strait, this provoked two movements: Venizelos returned to active politics and Sarrail was at the helm of an even tougher occupation policy.

Venizelos wanted to return to power, but he knew that he would not be able to achieve this in Athens, because loyalty there and in ancient Greece was to the king. Therefore, he took a detour via Saloniki. There, he wanted to launch a revolutionary movement under the protection of Sarrail, which would allegedly be directed against the Bulgarians and not the King. Supporters of his were active among the troops in Saloniki, and took advantage of

Above: Kaiser Wilhelm II as he visits the southeastern war theater. Behind the driver are two guns, for possible defense. After the end of the war in 1918 the commander in chief fled to exile in Holland. He did not want the war, but also did not stop it.

the general hate of Bulgaria, in order to draw them to Venizelos' side. He hoped to be able to bring the King and the country behind him as well.

In truth, Venizelos had spoken of treason in connection with *Roupel*. To the British envoy Francis Elliot, he expressed his suspicions that the German and the Greek general staff were working together. On May 29th he suggested to Elliot and the French envoy Guillemin that he should go to Saloniki with General Panagiotis Danglis, who was highly regarded in the army. There, he wanted to set up a provisional government, summon members of the parliament elected in June 1915, which in his opinion had been wrongly dissolved, and mobilize an army that was to attack Bulgaria together with the Allies. This was not directed against the King or the dynasty; on the contrary, he hoped that this would draw the King to the Allied side. For a while there would be two Greek states: the official Germanophile Greece and his pro-Allied one. He was certain that five or six regiments of the army in Macedonia would join him. He wouldn't strike until he had these troops on his side.[292]

A liaison officer was sent to Salonika to make the necessary arrangements with the Defense Committee and Sarrail. Venizelos asked Sarrail to postpone the planned imposition of a state of emergency so as not to turn the entire 11th Division against him in Salonika. Parts of that division would be sure to support the movement. For the same reason, Elliot urged London to postpone the peaceful blockade of the Greek ports that Admiral John de Robeck had requested. Addressing London, he added: *"Personally, I firmly believe that nothing can save this country from ruin, threatened by the stubbornness of the king and the hostility of his government towards us, other than a revolution based on the ideas of Monsieur Venizelos. As it currently stands, the revolution will not take place until after the war, when it is too late to be of any use to us. To be successful now, the revolution must be supported by us and France."*

But London was against this. On June 1st, Elliot was instructed to refrain from making commitments

Above: "Excellency Generalfeldmarschall August von Mackensen with his staff on a visiting ride." The German army leaders already have a premonition of the coming defeat when the Americans entered the war in 1917. The economic potential of the allies crushed the German air forces, despite all their courage. The determined fighting will have no influence on the outcome of the war.

and to not encourage Venizelos' project. Venizelos dropped his plans for the moment. But he agreed that preparations should continue among the troops in Saloniki with Sarrail's support.[293] On May 31st, the French envoy Guillemin informed Sarrail that Venizelos was planning to establish a provisional government in Saloniki, but Britain "*had vetoed the undertaking*" so Venizelos and his French co-conspirators would have to postpone their plans.[294] Sarrail did not give up, but supported the Greek officers who were preparing the military coup.

Like Sarrail, the envoy Guillemin was also fixated on Venizelos. The American correspondent Hibben characterized his attitude as follows: "*He did not once take into account the Greeks' view of their own situation. He was barely in contact with the Greeks, let alone the followers of Venizelos. His policy was based solely on the assumption that Venizelos represented the true opinion of the Greek people.*"

In addition, he was of the opinion that because of the French help in the Battle of Navarino (1827) Greece was obliged to obey France. In Guillemin's eyes, Venizelos was the man for the Entente. Why should he keep in touch with anyone else? In contrast, the ambassadors of Italy and Russia, Count Alessandro de Bosdari and Prince Demidoff, distrusted Venizelos. They felt that "*he was playing his own political game in Greece, with the support of the Entente and at their expense.*"[295]

In his apologetic account of Venizelos' politics, Georgios Ventiris denies that he (Venizelos) was involved in the conspiracy.[296] Another apologist, Neokosmos Grigoriadis, claimed that the Saloniki revolt was a spontaneous action by the Macedonian people. The organizers of the revolt did not have time to ask Venizelos.[297] The latter is probably true; this is also supported by Venizelos' reaction when the mutiny actually broke out. It is astonishing, however, that more recent Greek historiography takes over this view uncritically, perhaps because one is now not able to speak French and cannot take note of the clear statements made by Sarrail.[298]

The surrender of *Fort Roupel* to the German-Bulgarian troops without a fight was of course grist on Sarrail's mill, who had wanted to strengthen his policy of "*oppression and coercion*"[299] for a long time. Joffre had held him back so far, but when he heard of the Roupel straits occupation, he gave Sarrail the go-ahead for a tougher course.[300] In fact, the only difference between the occupation of Roupel and the Karabournous was that the former was carried out by Bulgarians and the latter by French. Both occupied neutral territory in disregard

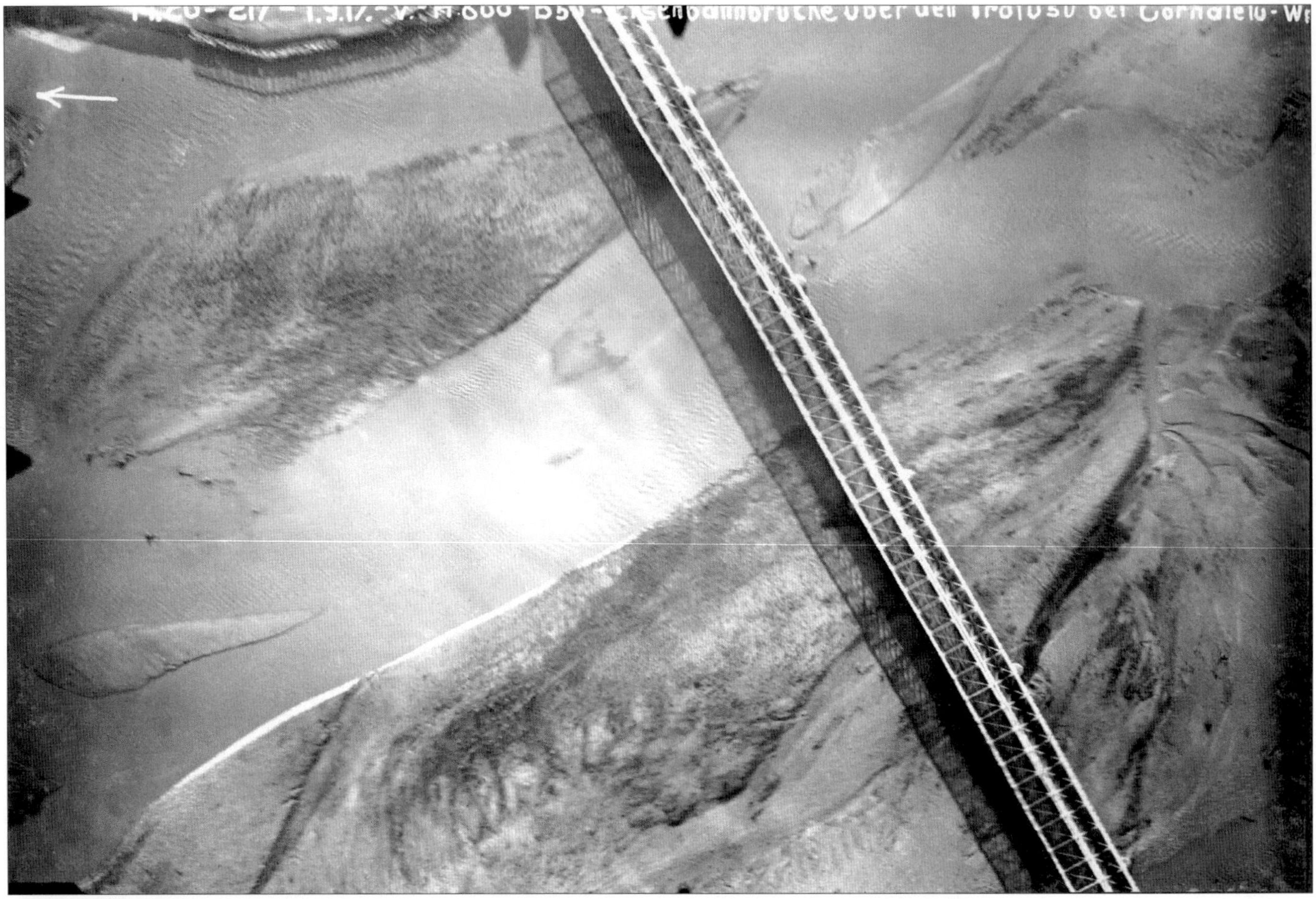

Above: FA 20 – Photo 217 – September 1st, 1917 – Altitude 800, train bridge over the Trotusu near Cornalelu.

Left: FA 20, Altitude 50 meters, shot down enemy aircraft west of Ciorani. Text Wehmeyer: "Air victory over a Farman from Vzfw. Ehlers and Oblt Wehmeyer on the morning of September 1st, 1917." In the afternoon, Vzfw. Ehlers is already dead. Celebration about an air victory and one's own death are sometimes very close to one another.

of international law.[301] But Sarrail didn't care, he declared a state of siege in the Saloniki area on June 3rd, although Milne had wanted to wait at least a day,[302] because June 3rd was the name day of King Constantine, and the residents of Saloniki had planned to conduct a torchlight procession and to hold a ceremonial *Te Deum* in the cathedral in his honor. After all, Constantine, as Crown

Above & Below: "DFW crash from an altitude of 30 meters, stalled at take-off, pilot Vzfw. Karl Ehlers (born on August 29th, 1892 in Celle – died September 1st, 1917 in Focsani) dead, observer Oblt Wehmeyer wounded, in the afternoon of September 1st, 1917 in Focsani (Romania).

Prince, had led the army that had liberated Salonika from the Turks. The *"Republican"* Sarrail forbade this *"royalist"* demonstration. On his orders, the French gendarmerie took control of the railway lines, all postal and telegraph services, and set up a censorship center for all newspapers that appeared in Saloniki. The British commander disapproved of these measures, but in order not to disavow Sarrail, he placed the British military police under him.

"This made Saloniki an occupied city under foreign military administration, like Brussels, Warsaw or Belgrade. Unusually, there were demonstrations in Athens against the powers of the Entente."[303]

Sarrail imposed a blockade on all Greek ports and decommissioned all Greek merchant ships. A squadron of the French Navy with an infantry brigade on board went to the Cyclades Islands to put pressure on Athens. On June 21st, the

Above: Grave of Vzfw. Karl Ehlers in Focsani. The text on the wreath ribbons reads: "The Officers of Fliegerabteilung 20, to their comrade Ehlers" and "To the brave comrade [Ehlers], the officer corps of the Flieg.-Abt. 42".

diplomatic representatives of France and England demanded the demobilization of the Greek army, which they viewed as a *"political weapon against Venizelos"*, the replacement of the allegedly pro-German Skouloudis government with a neutral one, new elections and the dismissal of certain police officers which had not sufficiently protected the diplomats of the Entente against the popular anger of the Athenians. Since the French fleet was anchored in the bay of Milos and could therefore appear in Athens in eight hours, the Greeks gave in. Skouloudis resigned and his predecessor Zaïmis formed a new government. The blockade was lifted, except in the case of the port of Kavalla, which was too close to the Bulgarian border. The fleet withdrew again. German propaganda picked up on the incident and announced that the Entente's promises to protect the freedom and rights of small states sounded rather hollow.[304]

Sarrail's actions in the wake of the Roupel occupation cut Saloniki off from the rest of the world. In order to communicate undisturbed with Prime Minister Skouloudis, General Moschopoulos had to go to Katerini on May 23rd. He demanded that Skouloudis give him command of the gendarmerie and that he should be in charge of Saloniki. Skouloudis agreed and urged him to pursue a policy of neutrality. When the Greek government protested against Sarrail's actions to the large neutral states like the United States, the Allies viewed it as a hostile act.[305]

The Allied intervention of June 21st turned the Greek public against them. It was viewed as a humiliation, even in some Venizelist circles. The anti-Venizelists capitalized on it when a telegram from Venizelos and Briand was also made public on June 24th. Venizelos had written in it: *"The seriousness of his address, the sincerity of his arguments and the complete separation between the Greek people and the previous government give him, more than any other, the character of his fatherly concern for the Greek people. The protecting powers act as parents with the entirety of their rights."* In other words, Venizelos justified the allied intervention.[306]

The British Chief of Staff Robertson rejected any offensive from Saloniki. On May 28th, he even ordered two British divisions to withdraw. He instructed Milne to assist the French only in the defense of Saloniki. When the British cabinet heard of the French advance, it was angry. On June 7th, an inquiry was made in Paris as to whether the French government had really approved the advance. Gray had adopted the view of the British military that nothing could be achieved in Macedonia for the moment. On the other hand, he also knew that a British refusal to cooperate could cause political trouble with the French.[307]

As the financial burden on the Greek budget grew, the King ordered the mobilization status to be reduced to *"partial mobilization"* on June 8th. This would significantly reduce the number of 300,000 conscripts. On June 11th, draftees from ancient Greece and Crete received a two-month vacation. With these measures, Constantine also wanted to show the seriousness of his policy of neutrality. How could such a scaled-down army be dangerous to anyone?[308] The Allies did not like it all. They

Above: The Gulf of Orfano with a British warship.

wanted Skouloudis to resign and Greece to give certain guarantees about its position. Sarrail and the French envoy Jean Guillemin demanded the king's abdication. Only in this way would Greece be deemed favorable classification to the Allies. The British were against it at first, but then they joined the French. As mentioned, on June 21st, the envoys of the Entente presented the Greek Foreign Ministry a note calling for the full demobilization of the army, the resignation of the Skouloudis government and the establishment of an interim government, the dissolution of parliament, the holding of elections and the removal of some police officers. Skouloudis had found out about it and on the same day submitted his resignation, and the king had reappointed Zaïmis as prime minister.[309] Stavrianos rightly notes that *"seldom has an independent country been subjected to such humiliating treatment. And that was just the beginning. Greece [...] has become a "public place" where all sides did not hesitate to act according to their own ideas."*[310]

The new government was sworn in on June 22nd and new elections were scheduled for October 8th. The police officers, who the Entente did not like, were replaced and demobilization began. Despite all of the Allied demands being met, Sarrail remained suspicious. At the end of June, he asked the chief of the 3rd Army Corps that he be given command of the gendarmerie for security reasons. When Moschopoulos refused, Sarrail told him to leave Saloniki within 24 hours. There was a violent exchange of words. Finally, Sarrail apologized and dropped the demand.[311]

But Sarrail's attacks continued. The Allied-sponsored newspaper *Rizospastis*, a propaganda paper, published a series of inflammatory articles in July against the King, the government and some high-ranking officers. The officers felt provoked and 14 of them angrily stormed the newspaper's office and beat up the publisher. Moschopoulos had the

Above: Bulgarian headquarters of the south front in Küstendil.

14 arrested and locked up. Sarrail demanded their extradition to the French military judiciary, although the crime against Greeks had taken place in neutral Greece. When Moschopoulos refused, Sarrail sent an armed patrol that broke into the military prison and kidnapped the 14 officers. They were to be tried in a French court martial for allegedly insulting the French flag. Moschopoulos turned to the government and achieved the release of the officers. They were then sentenced by a Greek court to two months' imprisonment, which they served in Akronafplia prison.[312]

Sarrail's advance northward on August 10th brought new movement into this issue. This and the demand for demobilization led to the Bulgarians attacking East Macedonia. That the weakening of the army had been carried out on the orders of the Allies did not interest Venizelos; for him demobilization was an act of the king to support the Germans. On August 16th, a rally was held in Athens, where Venizelos appeared again in public for the first time. He announced that he did not agree with the King's policy. The King was Germanophile, he would do something about it. The neutralists then claimed in speeches on August 19th that Venizelos wanted to overthrow the King from the throne on behalf of foreign interests. Most of the officer corps agreed.[313]

On August 27th Venizelos held a speech from the balcony of his house, in which he attacked the King directly: *"But, oh King, you have been the victim of unscrupulous councils trying to destroy the work of the 1909 revolution in order to restore the previous mismanagement and to satisfy the passionate hatred of the popularly elected leader. Hence, you have become the victim of military advisors with limited intelligence and oligarchic principles. Therefore, you became the victim of your admiration for Germany, in whose victory you believed. In the hope of pushing aside our free constitution through this victory and uniting the entire state power on yourself. [...] Instead of*

Above: Wilderness and high mountains: A horror for every potential emergency landing of an aircraft!

Right: A fountain in Xanthi.

expanding into Asia Minor, Thrace and Cyprus, there was a Bulgarian invasion of Macedonia. Furthermore, valuable war material was lost. If no one listens to us, then we should advise what to do to save what can still be saved from the disaster that has befallen us."[314]

Shortly afterwards, mutinies began in Crete and Samos. The rebels demanded Greece enter the war against Bulgaria, on the side of the Entente.[315]

Already at the end of 1915, a large group of Venizelist-oriented officers had formed the 3rd Army Corps, which wanted to overthrow the government and replace it with a revolutionary council. This council was supposed to bring the troops in Macedonia under its control, side with the Allies and take part in the war against Bulgaria. The group was led by the commander of the 10th Division, Major General Leonidas Paraskevopoulos, the commander of the 11th Division, Colonel Immanouil Zymprakakis, and Artillery Lieutenant Colonel Konstantinos Mazarakis. Sarrail supported this revolutionary movement, because he hoped to force the entry of Greece on the Allied side and the deposition of Constantine.[316]

On August 10th, 1916, Zymprakakis wrote to Venizelos that France was supporting their plans through Sarrail. Sarrail hated the monarchy and would be happy to participate in Venizelos' plan. In another letter to Venizelos, Periklis Argyropoulos made a similar statement. Further letters from

Above: A spy is hanged by the Bulgarian military in Drama.

followers of Venizelos confirm that Sarrail actively supported the coup in Saloniki. These letters were found in his house after Venizelos' departure from Athens and published in the newspapers.³¹⁷

In addition to this military component, there was also a civil one. Saloniki had been simmering since the Allied invasion and the occupation of Eastern Macedonia by the Bulgarians. The Bulgarian riots caused many Greeks to flee to Saloniki. They were furious that Athens could not protect the new areas better. Palmer aptly described the further development: *"Angry patriots, sincere liberals and a handful of coffeehouse demagogues formed an "Alliance for National Defense" to give the will of the Macedonian Greeks a voice. With unconscious irony, the alliance established its headquarters in the same building from which the "Young Turks" wanted to revive the Ottoman Empire nine years earlier. A "Public Safety Committee" was formed, a gesture that Sarrail, friend of Danton's most enthusiastic biographer, undoubtedly welcomed."*³¹⁸

When the government learned of the conspiracy, they ordered Moschopoulos to conduct an investigation, which, however, yielded no results. Only Zymprakakis was replaced by Major General Andreas Momferatos. At the end of July 1916, Moschopoulos was called to Athens for advice, precisely at the time when the German-Bulgarian invasion of the *Roupel-Strait* took place. On August 26th, the government learned that the Allies would soon demand the replacement of the Chief of Staff, Major General Viktor Dousmanis, and his deputy, Colonel Ioannis Metaxas. The latter was considered a friend of the Germans. To avoid an Allied demarche, Moschopoulos was put in place of Dousmanis and Dousmanis was given five years' compulsory leave. Colonel Xenofon Stratigos took the place of Metaxas, who was deactivated. The post of commander of the 3rd Corps was taken over by the commander of the demobilized 11th Division, Colonel Nikolaos Trikoupis.[319]

This was the trigger for the following events. On the morning of August 30th, Trikoupis learned that Mazarakis had appeared at the Karabournou barracks and announced that he would set up a revolutionary government to organize the defense of Macedonia. He had asked the officers to join him, but only two had followed his call. When Trikoupis tried to attack Mazarakis and wanted to depose him, he discovered that he was under the protection of the French. Mazarakis published a leaflet allegedly from

Right: A Turkish soldier.

Below: The German airfield Focsani in Romania.

Above: A group of soldiers pose in front of the Vardar river.

Left: German aircraft from the Vorkommando Flieg.-Abt. 30 that had crashed and burned.

a National Defense Committee (Ethniki Amyna) calling on the army and civilians to join him. The gendarmerie defected to the mutineers. Those higher officials and clergy who were loyal to the King were arrested. Trikoupis' job was to put down the mutiny. But since he did not want to get into a conflict with the French, he sent the nomarch and the chief prosecutor to them. They learned from Sarrail that he would send an armed intervention if there were any disturbances. Nevertheless, in the course of the evening there were small clashes between the mutineers and troops loyal to the government.

The next morning, Sarrail informed Trikoupis that he would not tolerate armed clashes in the city and would use armed force against them. Trikoupis replied that he had enough troops to quell the mutiny unless Sarrail supported the mutineers. At about 4 p.m. French troops occupied the seaside promenade and Sarrail demanded that the troops of the 3rd Corps return to their barracks, otherwise he would attack them with artillery. Trikoupis fearlessly replied that Sarrail should shoot the mutineers and demanded that he repeat his requests in writing. After some back and forth, he received it. Since their rejection would have led to an armed conflict in the city, Trikoupis gave in, especially since he knew that Athens did not want an armed clash with the Allies. On the evening of August 31st, the loyal Greek officers were put on a French passenger steamer and sent to Piraeus.

Above: English dud bomb. Hptm. Haupt-Heydemarck describes the explosion in his book. He had it detonated and placed an aircraft tent over the crater so that the effect of the bomb is not recognizable on English aerial photographs.

The teams were placed under French command. Sarrail proudly reported to Paris: *"Colonel Trikoupis [...] tried last night to prevent the coup that was reported yesterday. Three dead and five injured. Surrounded by the rebels and announced by me that order must be maintained in Thessaloniki, he surrendered to me and not to the Revolutionary*

Above: The detonation of the English bomb in Hudovar.

On the transport from Schlettstadt to Macedonia, the train makes a stop for a meal near the "KUK Eisenbahnverköstigungsstation" (train meal station), followed by this in the Hungarian language. The name of the train station is "ARUKEZELESI IRODA"

Piatra train station in Romania.

The photographers of the Abteilung 20 on their transport wagon from the west.

Lt. Wehmeyer stands in the middle of this group photograph. To his left is also an observer, to his right a pilot.

Friends.

Above: Lt. Danneberg and Uffz. Maret in the Albatros. The aircraft is an Albatros C.VII with 200-HP Benz D.IV engine. Lt.d.R. Artur Danneberg [born on October 19th, 1885 in Halle, died on May 4th, 1918 in Orljak, South Macedonia]. Lt. Danneberg was the observer and Uffz. Maret was his pilot. Here there is the old problem again: Die correct timing/indexing of these photos. Uffz. Emil Maret, born on November 27th, 1895 in Kuhmen, was taken prisoner on July 13th, 1917 vetweeb La Hollande and Nompatelize.

Left: The mess of the Flieger-Abt. 20 in Braila, Romania.

Above: On the Danube near Kilia, barges are being prepared for a river crossing. These floating bridge pieces are connected in threes, and were implemented near Ismail.

Above: A single track train bridge in Romania was bombed. The iron piece has been cleared and the new construction of the missing bridge is in full gear.

Above: The Army baggage train.

Below: Participants of the preliminary peace negotiations in Focsani led by Excellency von Morgen.

Above: Körner, Schwabedissen, Danneberg, Wehmeyer, von Morgen, Grube, Pohle, Kaskeline, Schulz, Felmy, Stern, Walter, Bechta.

Below: Excellency von Kosch and Excellency Hilmi Pasha with their officers together with Fl.-Abt. 28 as guests at Fl.-Abt. 20 in Braila.

Above: Lt.d.R. Arthur Danneberg, (*19.10.1885 in Halle - † 04.05.1918 near Orljak [South Macedonia]) and Vzfw. Maret in front of their Albatros C.VII.s

Committee. All officers with their weapons and all disarmed soldiers are in our hands. Each of the officers and soldiers who want to fight with the Entente reinforces our Greek units. The others are already there, the soldiers are in the evacuation camp, the officers are on a steamer in the port with which they will be brought to Piraeus on Saturday, unless you give contrary orders. The Greek division in Thessaloniki no longer exists."[320]

A short time later, Sarrail received a note stating that Venizelos would come to Saloniki on September 9th to take command of the movement. This

was the beginning of the division of Greece, the beginning of the so-called "*Ethnikos Dichasmos*".

The mutiny was sharply criticized in Athens and ancient Greece, and even by many Venizelists. Venizelos himself criticized it as too early and irresponsible, but he didn't condemn it, after all, he himself was behind it. The cooperation with the French also worked extremely well in Athens. On August 26th, the French admiral Louis Dartig du Fournet, who in 1915 had rescued 4,000 Armenians on Musa Dag on his own initiative, received the order to set up a special squadron and go to Greece, according to the plans drawn up in June. He left Malta with four ships and reached the island of Milos on August 28th, where 39 British ships were waiting for him. With this armada they could prevent a Bulgarian-German advance to the south, but also the royalist resistance to the Venizelian putsch. The departure of the fleet was delayed again by four days because the British had scruples, but on September 1st the fleet appeared in the Saronic Gulf.[321] This was the beginning of the Allied intervention in favor of Venizelos' action.

Left: Vzfw. Otto Matte (*1895 in Leißling) fell on 01.12.1917 in Galatz.

Below: Before the offensive at Focsani.

But the Venizelists were not the only ones who were organizing, their opponents did the same. Demobilization in the face of the occupation of the Roupel-Strait and eastern Macedonia had violated the patriotism of the common soldiers. They felt that they were being sent home and that Macedonia was being left to the enemy. In their eyes the Allies and Venizelos were to blame. When the first railway trains with demobilized soldiers arrived in Athens, there were demonstrations against the Allies and Venizelos. Tensions were high as the anti-Venizelists in the army organized themselves. A league of reservists had already been set up in April, which quickly developed, under the leadership of royalist officers, into a paramilitary organization that controlled the low country. It formed the counterbalance to the Venizelist groups of the National Reservist League, which had been founded by Danglis in mid-July. The general staff was behind the league. It was also the existence of the league that led to the postponement of the elections,

Facing Page, Above: In Romania, a "Kühlstein Torpedo Eindecker" was found and repainted in German markings.

Facing Page, Below: Aerial view of the city of Braila in Romania on the Danube River.

Left: Lt. Graul was an observer.

Below: Aircraft hangars at Sierenz airfield.

Above: Once again, a general view of the casino of Flieger-Abt. 20 in Braila, Romania.

which had originally been scheduled for August, to October.³²²

London was appalled by the French action in Greece. Attempts were made in Paris to convince Briand to exercise more moderation, but the latter was urged to take the tough course by his naval minister, who came from a ship-owner's family. But the British did not want a major argument with the French government as the Battle of the Somme reached its climax. So they gave in again and grudgingly accepted the French course in Greece. In contrast to the British politicians, the English King George V, who otherwise did not interfere in foreign policy, was no longer willing to accept the French behavior. On September 4th, in a letter to Asquith, he complained about the behavior of the Allies towards a *"neutral and friendly country"*.

*"I cannot shake the feeling that we have let France rule too much on the Greek issue. As a republic it is probably a little intolerant, if not even trying, to abolish the monarchy in Greece. But that's not my government's policy. Nor is it that of the Russian ruler who wrote to me a few days ago: I am concerned about the internal affairs of Greece. It seems to me that the protecting powers are interfering too much in internal affairs as they try to protect their own interests with regard to Greece's neutrality. This to the detriment of the King. I cannot hold back from expressing my astonishment and regret at General Serrail's unauthorized behavior towards the troops. They showed themselves to be loyal to the King and the government and refused to join the revolutionary movement in Saloniki. Couldn't it be possible to send a protest note to the French government taking a stand against Sarrail's actions? Greek public opinion, as well as the attitude of the King, is obviously changing. If the Allies treated the Greeks kindly and not, if I may put it so, in an oppressive spirit, then in all likelihood they would join."*³²³

For all of the weeks that followed, George V and the Tsar were suspicious of French policy, particularly that of Sarrail. Constantine, for his part, tried to appeal to monarchical solidarity by sending his brothers Andreas and Nikolaos to London and Petersburg, respectively. In Paris, Marie Bonaparte,

Above: Austro-Hungarian monitors on the Danube.

Below: In the officers' mess there was a lot of excitement. Oblt. Wehmeyer wrote: "Flieger 28 und Ordonanzoffiziere des Generalkommandos bei der Flg.-Abt. 20 in Braila."

the wife of Constantine's son Georg, used her charm in a similar sense towards Briand, who was not immune to it. [324]

Left: Macedonian Mother's Happiness.

Below: Camel caravan at Dojran Lake. However, the scene, which seems archaic for a modern war, reflected the reality of 1917: this means of transport was faster than trucks.

Above: Observer Oblt. Fahrig of FA 244" (Oblt. Hermann Fahrig, *11.07.1890 in Bad Liebenstein - † 13.02.1947), with his unit on the way to the Eastern Front.

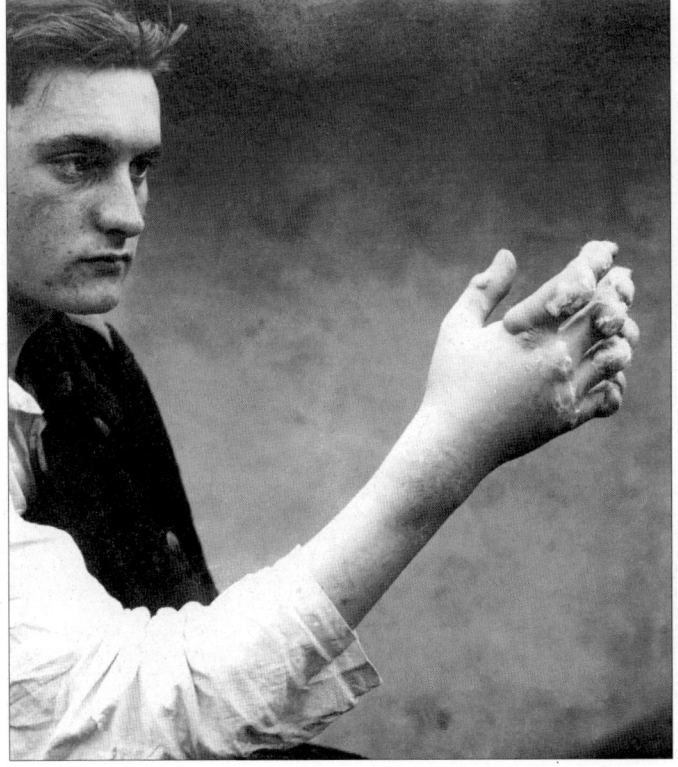

Above: DFW C.V destroyed by aerial bombing. Unfortunately, no identification marks are visible.

Left: Lt. Duderstadt with his hand frozen in aerial combat.

Above: Cozy little corner in the Greek cemetery in Drama.

Below: Macedonian ways. On the hood there is still the following inscription: "F.Fl.A. 30". The license plate has the inscription: "XI 5148", which indicates the army air park 11. The officer sitting in the back apparently expects the soldiers to get him out of this awkward situation with dry feet.

Above: Casino Flg.-Abt. 20 in Braila (Romania). The officers are having a good time.

Right: Lt. Hellmuth Felmy, born on May 28th, 1885 in Berlin, died on December 14th, 1965 in Darmstadt. He was a brave officer and was abundantly decorated. On his left sleeve he wears the insignia of the FA 51.

Above: The swampy lowlands on the river Vardar.

Below: Kavalla, the Greek city on the Aegean Sea. The English pilots from the island of Thasos often encountered the German pilots here and engaged them in fierce air battles.

Above: The English buried Lt. Rudolf von Eschwege (*27.02.1885 in Homburg v.d.H. – † 21.11.1917 in Orljak) with military honors. Photographs were taken of this funeral, which were dropped by the English over the German airfield Hudova. Text by Johann Keller: "Englishmen bring my dear friend von Eschwege to grave. Own photograph after ceasing fire. Ich hatte einen Kamerad. (I had a comrade.) Johann Keller".

Below: Text by Johann Keller: "The grave in the English lines of my dear friend von Eschwege, who went with me to the air battle. Rest gently in foreign soil."

Above: Lt. von Eschwege.

Above: Text by Johann Keller: "The grave in the English lines of my dear friend von Eschwege, who went with me to the air battle. Rest gently in foreign soil."

Left: Text by by Johann Keller: "Bulgarian General at the dedication of the memorial stone for von Eschwege near Drama" The landscape around Drama belongs to the Greek territory today. In Bulgaria, however, he is still commemorated because he alone with his single-seater fighter kept about 70 km of the front free from British planes.

Right: Text by Johann Keller: "Bulgarian General at the dedication of the memorial stone for von Eschwege near Drama" The landscape around Drama belongs to the Greek territory today. In Bulgaria, however, he is still commemorated because he alone with his single-seater fighter kept about 70 km of the front free from British planes.

Above: Hudova airfield.

Above: Keller wrote: "My squadron". He himself stands on the far right and has drawn a cross of beams over many of his fallen comrades of Fl. Abt. 20.

Left & Facing Page: Funeral of airplane pilot Vzfw. Otto Matte, *23.03.1895 in Leißling, Weißenfels – † 01.12.1917 in Galatz, northeast of Braila in Romania. The field chaplain gives a speech. The wreath ribbon on the left is inscribed "Ruhe sanft – Unteroffizierskorps Fl. Abt. 20", the 2nd ribbon next to it is inscribed "Das Offizierskorps Fl. Abt. 20", next to it is the wreath ribbon from the "Offizierskorps des Generalkommandos 5".

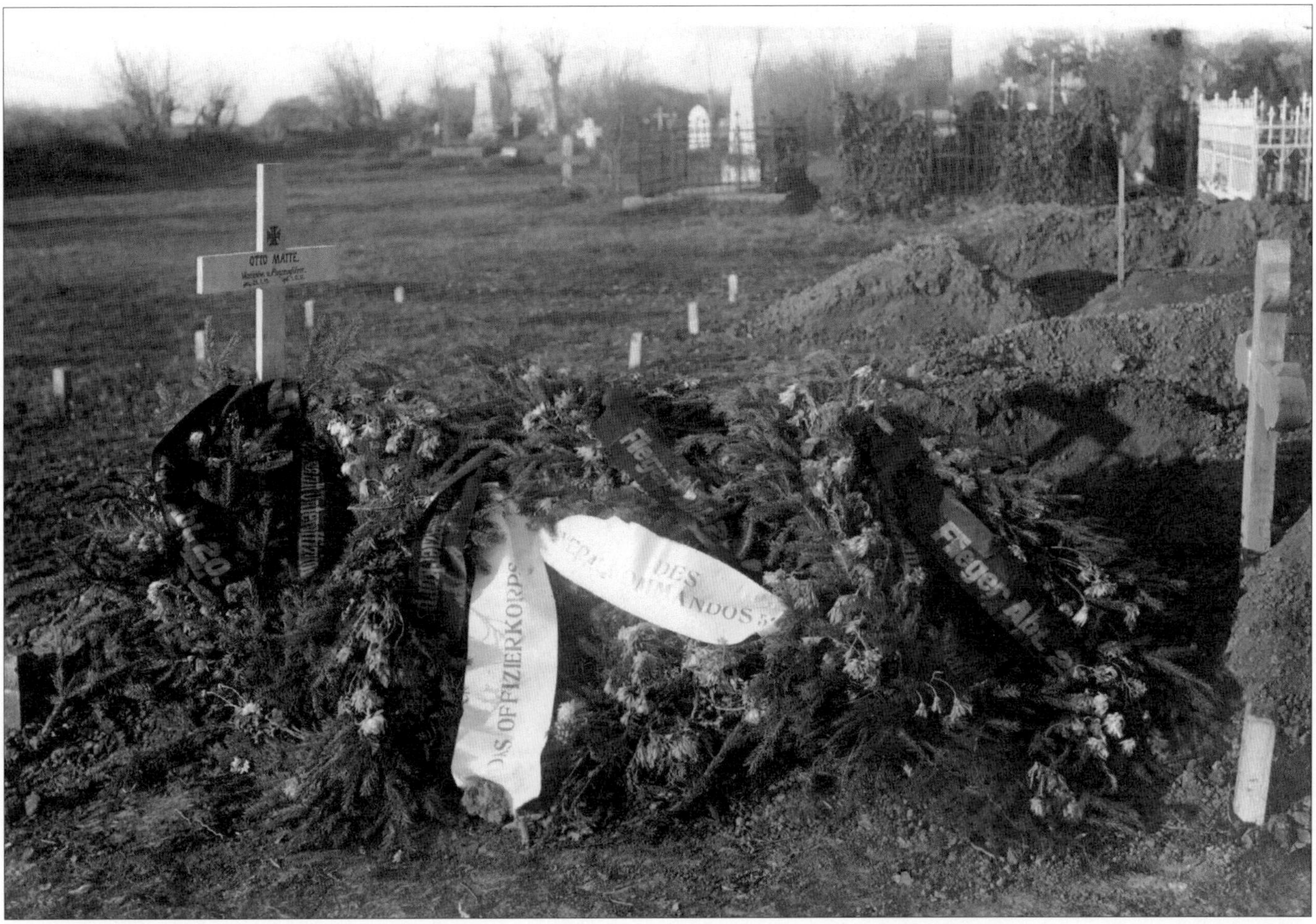
OTTO MATTE
DES
DES OFFIZIERKORPS

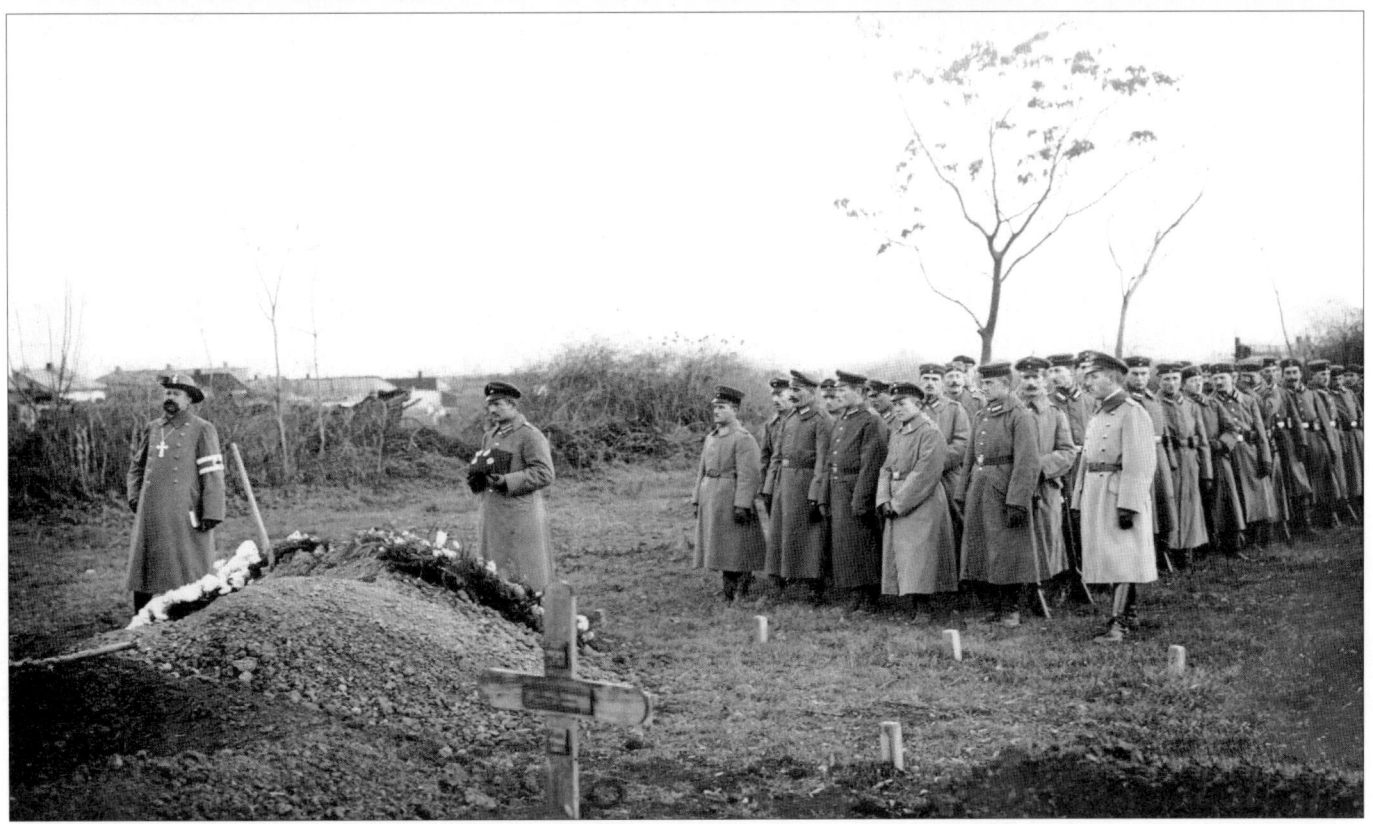

The Entrenched Camp of Saloniki

Above: Christmas 1917 at the Cavadia House, Braila."Lt. Wehmeyer lists the names in 5 rows: 1st row: Lt. Severdin Bey - Mr. Krysoveloni. 2nd row: Mr. Goldberg and father - Consul Suidas - Miss Galati - Lt. Stephan - Hptm. Reisch. 3rd row: Oblt. von Morgen - Lt. Metzger - Fr. Karentino - Mj. Nasim Bey - Fr. Drakulis - Fr. Pantusopulo - Fr. Vlasopulo - Fr. Kelaidites - Fr. Galati - Fr. Karentino - Fr. Skender - Fr. Krysoveloni - Fr. Vlasopulo - Fr. Dimopuli. 4th row: Mr. Mavromati - Mrs. Cavadia - Excellency Hilmi Pasha - Mrs. Sulioti - Mr. Cavadia. 5th row: Miss Skender - Miss Galati - Miss Cavadia - Miss Ooanga.
For the German pilots it was the last glorious Christmas for a very long time!

At the talks at the end of November 1915, the Greeks had agreed that the Allies could build defensive structures near Saloniki. Fearing attack by the Central Powers, Sarrail and Mahon decided to turn Saloniki and its hinterland into a defensible position. [325] Due to the small number of troops, a relatively small area was initially fortified. It began in the west on the lower reaches of the Axios River, moved north around the 561m-high Koryphi Heights to the Choriates mountains in the north of the Chalkidiki peninsula. But they soon discovered that the area was far too small to be used as a starting point for major offensive operations. Therefore, it was decided in December 1915 to push the border further. In the west, the new fortified line began again on the Axios south of Kastanas. But then it made a greater detour around the *561 Heights*, which included the villages of Monolofos, Drymos, Liti, the Langadas and Volvi lakes and reached the Gulf of Orfano at the gorge of Rintina. To the west of Monolofos was the French sector, to the east of it the British. By the end of the year, this line had been expanded into a system with three successive trenches with cross connections. Concrete machine gun positions that covered for each other secured the camp. There were barbed wire barriers in front of the trenches. At the end of January 1916 the camp was fully constructed.[326]

If reinforcements came in greater numbers, the line to be defended was to begin in the west in the Vermion Mountains, through the Paikon Mountains to Lake Dojran, and from there through the Dysoron Mountains to the southeast. In the Vertiskos Mountains, it should pivot east to Orfanos Bay.[327]

Three French and five British divisions were later located in the *"entrenched camp"*. Four

Lt. John McGilchrist (1893 - 1977) processed his war experiences in WWI as a balloon observer for the RAF and was shot down 4x in this capacity. Here on this colored lithograph, two Sopwith Camels are attacking a German balloon whose occupants are trying to save themselves by jumping out of the basket. The yellowish color of the balloon comes from the rubber impregnator, which was only available in this shade.

"The unexpected attack"
(Sopwith Camel & K.B.)

Above: A German observer of the Cologne "Stollwerkbrigade" landed happily and only slightly injured from his burning balloon with the singed parachute. As a souvenir, a photo is taken of the observer - with head bandage - and the charred parachute.

of the British divisions came from the Western Front. These were the 22nd, 26th, 27th and 28th Divisions. The 22nd and 26th Divisions were re-formations that had never been in combat. The 27th and 28th Divisions were regular units that had been in action at Ypres and Loos. The 10th Division came from Gallipoli and was still pretty battered.[328] The six Serbian Divisions were added to the eight Allied Divisions. Tremendous amounts of food and supplies were brought in. But neither the Germans nor the Bulgarians felt compelled to transfer large numbers of troops to the Greek border. German journalists spoke contemptuously of the largest internment camp in the world, to which the Allies had voluntarily gone. But the situation in the camp was much worse, because the whole area was infested with malaria. Closed units fell victim to the mosquitoes. The camp became a huge hospital. The troops at Saloniki did not draw any German forces from the western front, nor did they support the Russians or weaken the Turkish front. Their stay was completely pointless.[329]

Both general staffs opposed the launch of larger offensives from Saloniki. There were two alternatives. The troops could be kept strong enough so that they could repel a Bulgarian advance that would be supported by small forces of the Central Powers. For this, the number of troops could be reduced. The British General Staff tended towards this alternative. The second alternative would have been an attack on European Turkey or a breakthrough through the Balkan Mountains

> Guerre 1914-15-16... EN ORIENT
> Baie de SALONIQUE. Amiral Guéprate bras croisés. Général Baumann derrière
>
> War 1914-15-16... IN ORIENT
> SALONICA bay. Amiral Guéprate crossed arms. General Baumann is behind him

Above: A postcard showing the bay of Salonica.

Right: A sweet tooth pig. For the funny moments in this cruel time the photographer still had time.

Above: Kaiser's Birthday Celebration 1918 in Drama. Emperor Wilhelm II was born on 27.01.1859. Accordingly, this photograph is from 27.01.1918 in Drama. Lt. Wehmeyer lists 4 rows of names in the caption, all from left to right: "Top row: Vzfw. Käppeler, Dr. Fränze. Second row: Lt. von Morgen, Lt. Lenz, myself [Wehmeyer], Col. Seveljeff, Hptm. Pohle, Lt. Griebsch, Lt. Sonneff. Third row: Vzfw. Keller [with cigarette in mouth], Lt. Fröhner, Lt. Schwabedissen, Prefect Angelow, Uffz. Bechta. Bottom row: Lt. Westphal.

and the defeat of the Bulgarian army. This would have created a land bridge with Romania and Russia. The French general staff wavered between the alternatives. But in order to even be able to implement the second alternative, heavy artillery suitable for mountain use and the corresponding ammunition would have been necessary, and they were missing. French political leadership believed in the second alternative. Joffre, however, tended towards the first alternative. He wanted to win on the Western Front. But his position had been weakened since Verdun.[330]

At the end of February, the Allies learned that the Germans were moving troops north from the Balkans. Probably to prevent a further withdrawal and their relocation to Verdun, Sarrail decided to advance his troops to the Greek frontier in order to give the impression that an offensive was imminent. Therefore, he moved four French and two British divisions forward. The Greek units who were there stepped aside as ordered, and General Moschopoulos dutifully protested that Sarrail was violating the sovereignty of Greece. The Bulgarians responded and moved 11 divisions up to the border from Monastir (Bitola) in the west to Alexandroupolis in the east. But not only the Bulgarians.[331]

Because on March 14th, Chief of Staff Falkenhayn informed the Greek government that German and Bulgarian troops would occupy the Kerkini and Angistron mountain ranges west and east of the Struma Strait as far as the village of Neo Petritsi at the exit of the Strait.[332] The war ministry in Athens ordered the commanders of Army Groups C and D and the forts at Saloniki and Kavalla to withdraw their security troops as soon as the invasion began, because they feared an armed clash with the Bulgarian and German troops. The forts were also to be evacuated. It was assumed that the attacks would

Above: A DFW C.V.

Above: A Russian airplane with two crew members has been shot down. The two airmen lie dead in front of their plane.

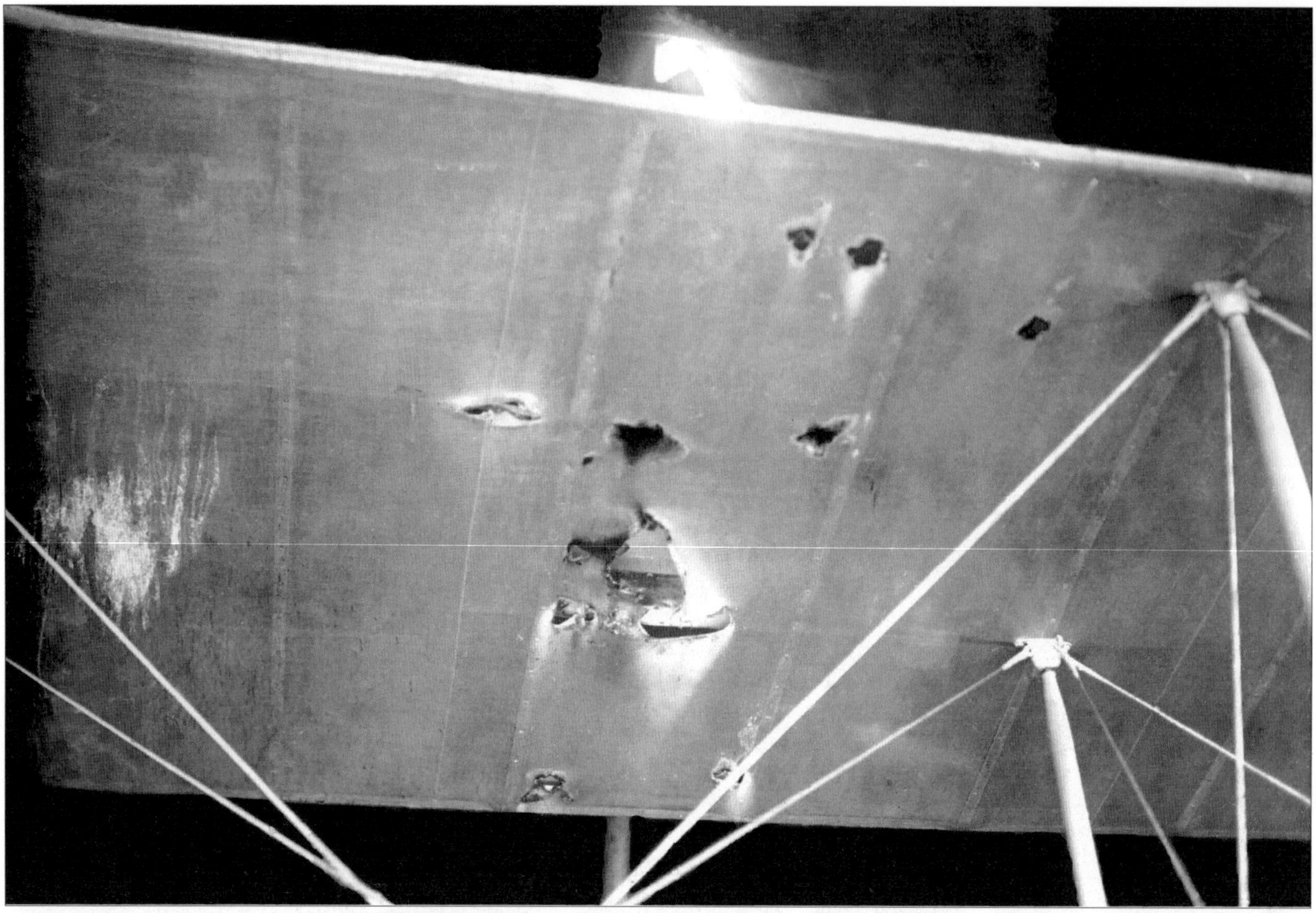

196

Above: Wreckage of Sergeant Wethekam with an Albatros C.III or C.VI. The landing gear has been completely pressed against the fuselage before the rollover.

Facing Page: The splintering effect of an English bomb through the roof of a wooden aircraft hangar damaged the aircraft, but primarily the fabric covering.

Below: Sergeant Fritz Stattaus shot down near Thasos. Rollover on landing at Kavalla." The aircraft is presumably an Albatros C.VII. Fw. Stattaus (Vfw. Fritz Stattaus, *02.08.1891 in Sauskeppen, Krs. Insterburg, date of death unknown), was shot down in France on 09.09.1916 and ended up in French captivity. He escaped across the Swiss border back to Germany and came to Drama to FA 30. During this flight his observer was Lt. Lenz, who shot down an attacking English aircraft.

Above: A very battered Halberstadt fighter at an airfield.

take place at Gevgeli, Lake Dojran, the Struma or in the Angistron Mountains. On May 7th, Falkenhayn informed the Greeks that, in view of the Allied advance towards the Greek-Bulgarian border, they considered it necessary to take over the *Roupel-Strait*. On May 10th, Athens' above-mentioned order was revoked and the troops ordered to defend themselves from a penetration-depth of 500 meters. Before that happened, local commanders should contact the enemy commanders, asking them to stop until a diplomatic solution was found. No fort was to be abandoned, but incidents were to be avoided if at all possible.[333]

On May 10th, the German military attaché in Athens informed the Greek government that the German General Staff considered the occupation of *Roupel-Strait* and the mountains to the left and right of it to be necessary because the Allies had crossed the Struma. Athens protested against this in Berlin, but on May 14th the German Foreign Ministry in Berlin informed the Greek ambassador that, in light of the Allies' advances, the occupation of the *Roupel-Strait* could no longer be postponed. Could they kindly instruct the Greek troops on site accordingly? On May 22nd, 1916, the ambassadors of Germany and Bulgaria informed the Greek government that their armed forces would occupy the *Roupel-Strait*. Should there be any destruction, the damage would be replaced and at the end of the war they would withdraw from the narrows. The Greek government saw no alternative and took note of these assurances.[334] When the occupation actually took place, Greece formally protested the violation of its neutrality, similar to what it had done when the Allies had occupied Saloniki.[335] According to the Bulgarian Prime Minister Radoslavov, there is even said to have been a written agreement between the Greek and Bulgarian governments.[336]

On May 24th, the 6th Greek Division was informed from its outpost in the village of Strymoniko that two large Allied columns were approaching. On May 26th the Greek outposts at Fort Roupel reported that strong Bulgarian-German troops were returning to the narrows. The commander of the fort, Major Ioannis Mavroudis, alerted his crew and sent his adjutant to the Germans as a parliamentarian: if they advanced more than 500 meters into Greek territory, he would have to stop them by force of arms. The German commander then ordered a cavalry detachment that had already advanced further to withdraw. At the same time, he informed the parliamentarian that he had orders to occupy the *Roupel-Strait*. He would do this soon too. Major Mavroudis reported this to his division commander. He alerted his troops and ordered everyone to resist. At the same time the German commander was informed that he had consulted Athens and was waiting for an answer. If the Germans chose to not wait for the answer as well, there would be fighting.[337]

The Germans nevertheless advanced on the two mountain ridges to the left and right of the narrows. In the strait itself, the German-Bulgarian units pushed forward to the left and right of the Struma. According to his orders, the commander of the fort had the tip of the German-Bulgarian column bombarded, which did not return fire but withdrew back to the border. Meanwhile, defense preparations

Right: "To his pilot, forever remembered: Otto Voigt." Keller writes: „My loyal mechanic Voigt."

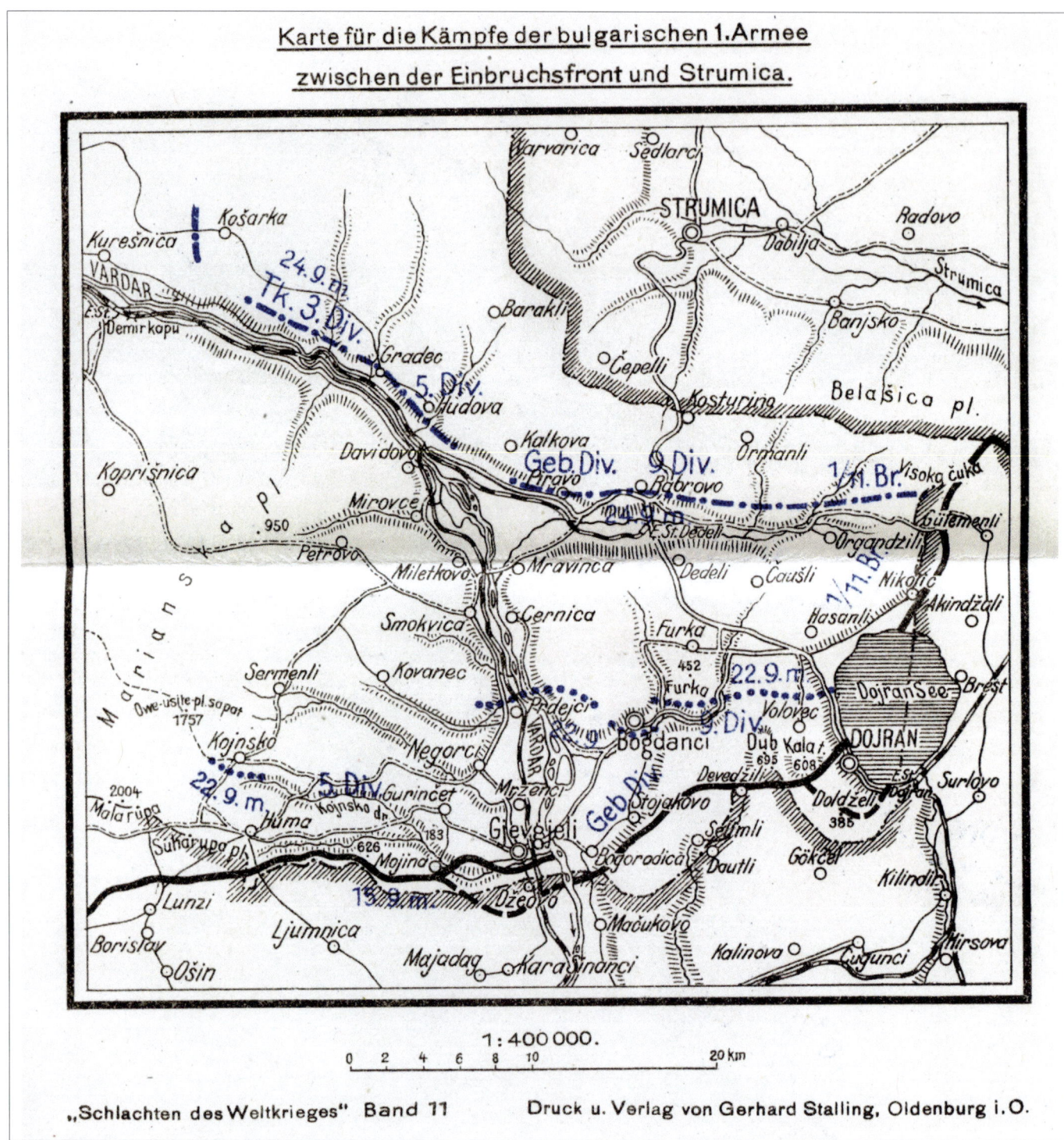

continued. But around 3 p.m. an order from the Ministry of the Army in Athens arrived, prohibiting resistance - the troops were to withdraw without a fight. However, the commander should protest against the German-Bulgarian invasion, which he did. But the fort's crew did not want to surrender and it took until the next day before Fort Roupel was surrendered without a fight upon the orders of Athens. Then the German-Bulgarian units crossed the border and advanced to Siderokastro at the exit of the *Roupel-Strait*. At the same time they occupied the ridges of the two mountain ranges. On May 27th, 1916, the representatives of the Greek government protested in Berlin, Vienna and Sofia against the violation of Greek neutrality and demanded the withdrawal of the German-Bulgarian troops.[338]

Facing Page & Right: Maps from the „Reichsarchiv, Schlachten des Weltkrieges", Volume 11, Verlag Stalling, 1923: Looking at the data of the retreat one can think it was more of a fleeing or escape.

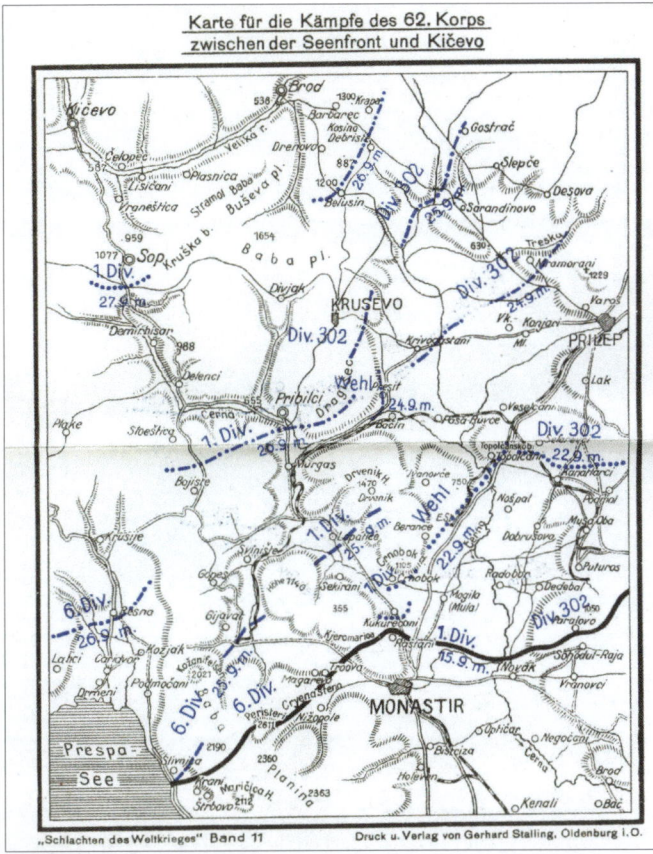

Below: Recognition chart showing silhouettes of German and Allied aircraft.

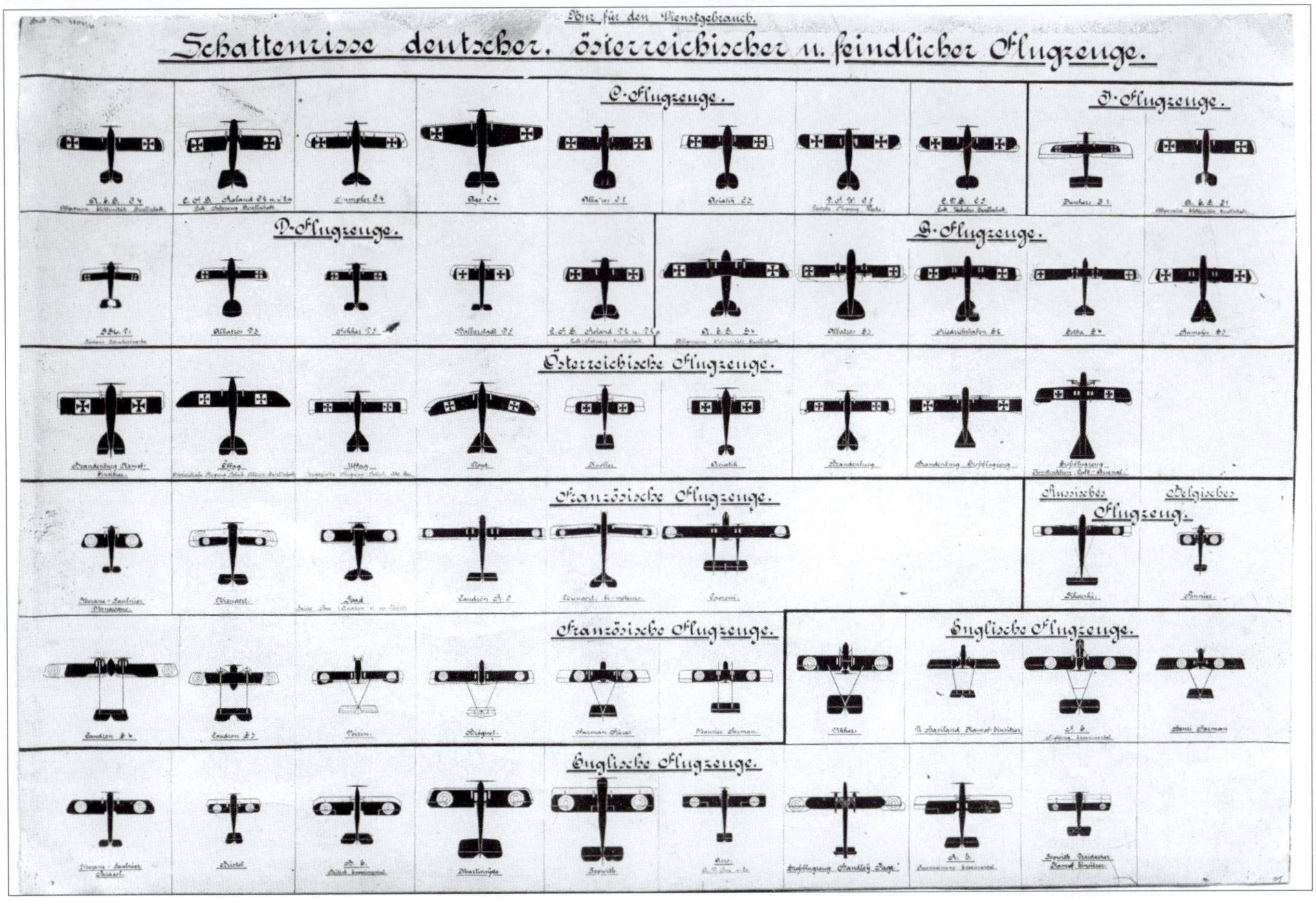

Above: The inscription on the back: "Two of my best friends, Knaurig and Kiselowski." Vzfw. Fritz Knaurig, born on July 31st, 1884 in Berlin, died on May 4th, 1918 in Orljak and Vzfw. Hans Kiselowski, born on January 17th, 1899, died on September 16th, 1918 in Macedosnia. Vzfw. Keller from FA 20 stands to the right.

Facing Page: Right, Vzfw. Johann Keller with an unknown colleague.

Above & Below: The aircraft of the crew of Lenz/Hermstedt, which fell during the air war, on February 5th, 1918 in Drama. This crew was on a ferry flight and was shot down by surprise by an Englishman.

Above: The funeral of the crew Lenz/Hermstedt which fell during the air war on February 5th, 1918 in Drama. Observer Lt. Wilhelm Lenz, born on December 5th, 1895 in Berlin, died on February 5th, 1918 in Topoljani near Üsküb and Lt. Friedrich Hermstedt, born on March 5th, 1892 in Erfurt, died on February 5th, 1918 in Topoljani near Üsküb.

Above: The On the way to the laying-out.

Below & Following Page, Top: "Rest in peace in foreign soil!" The funeral of the fallen crew Lenz/Hermstedt on February 5th, 1918 in Drama. Next to the coffins is the honor guard. The medal pillow of the observer Lt. Lenz with EK II and EK I as well as the observers' medal. Lenz and Hermstedt were to switch out an aircraft in the rear territory. On this flight, they were surprised by an enemy aircraft and shot down.

Wilhelm Lenz.

Friedrich Hermstedt.

Below: "English airport Orljak. Bomb attack by …..Keller." [Slg Keller, Album 2, photo 158]

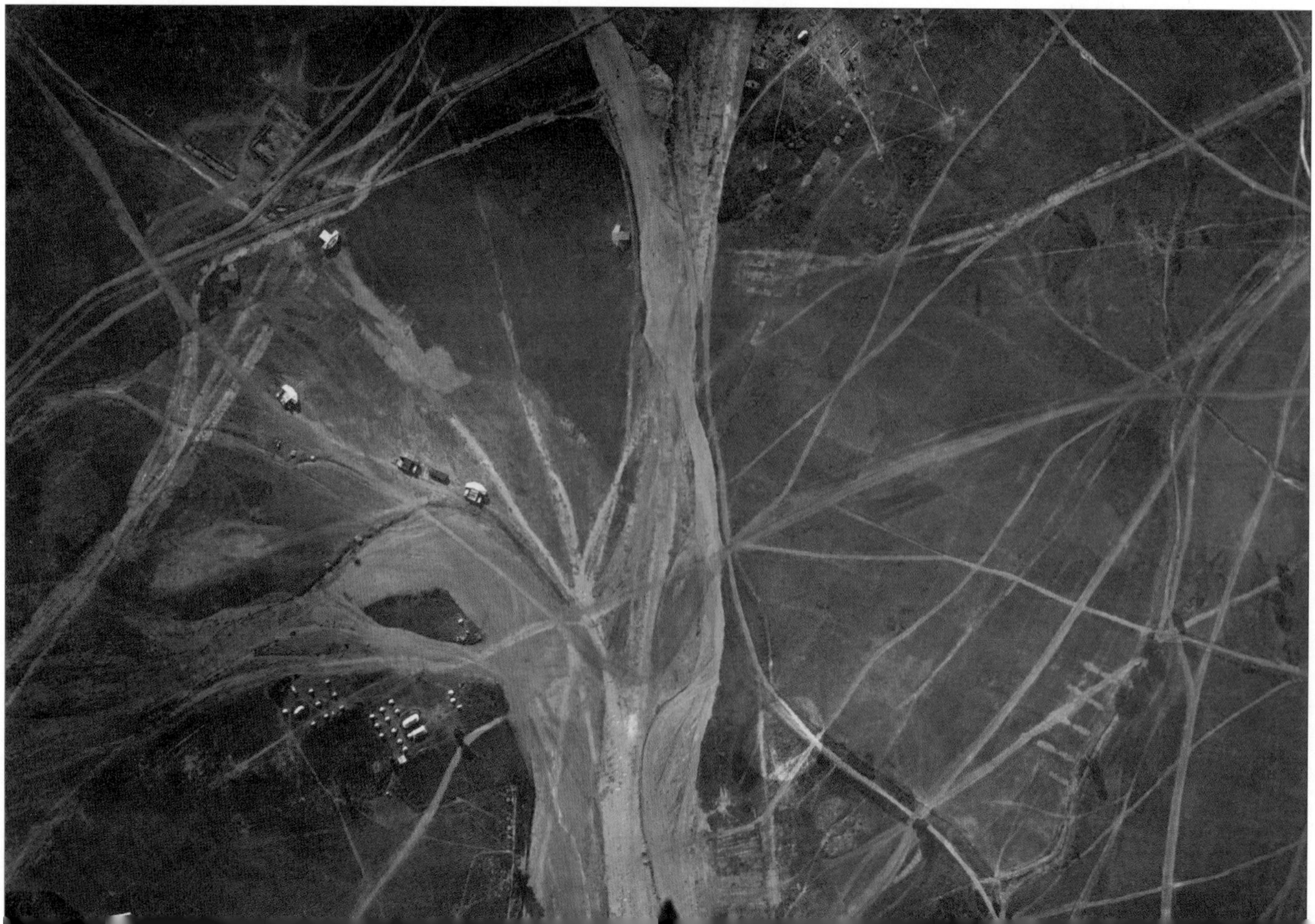

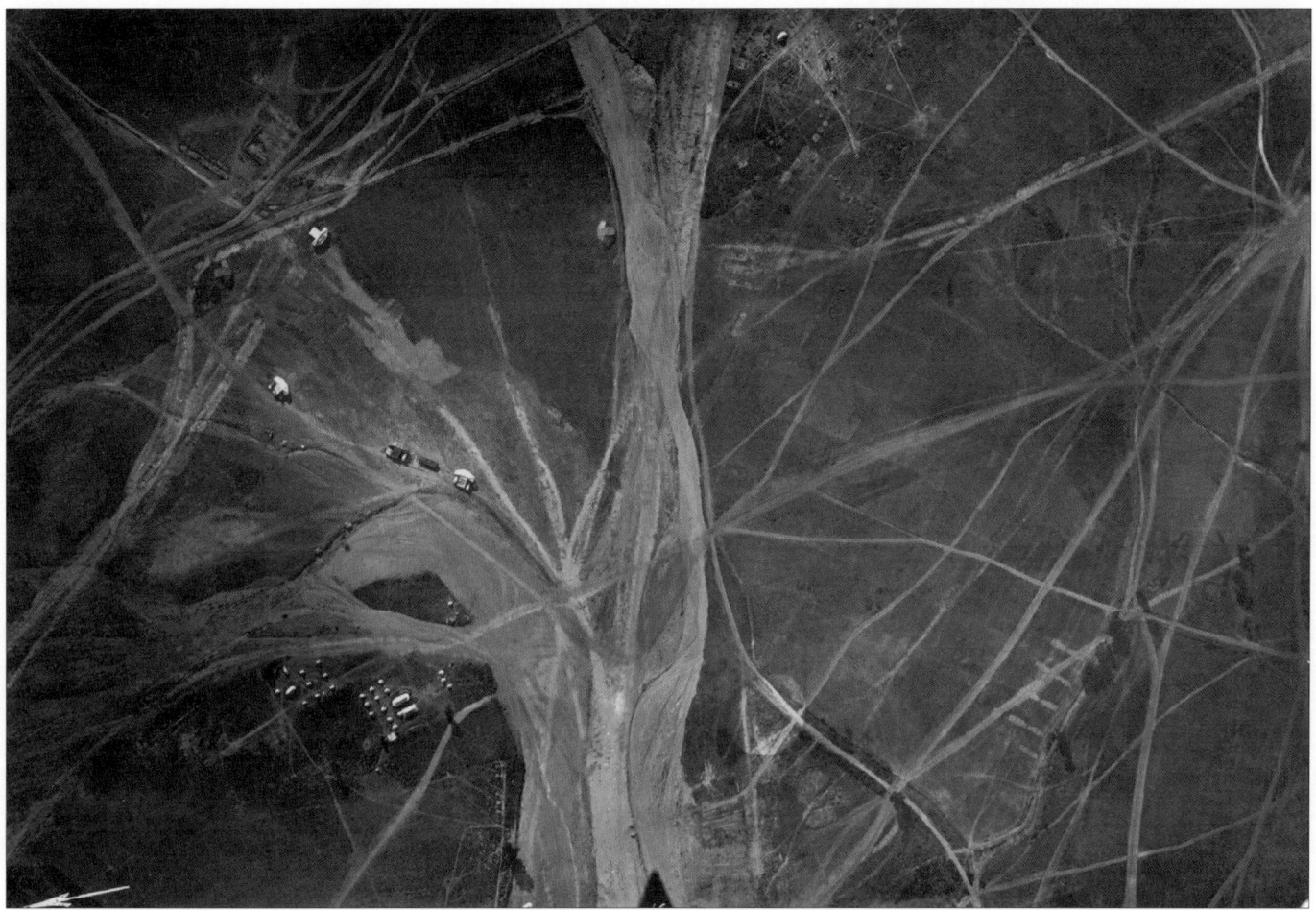

Above: "English airport Orljak. 3000m altitude. This image almost cost me my life. Keller." ." [Slg Keller, Album 2, Bild 160]

On the next day, the German-Bulgarian troops occupied some villages around Siderokastro. Their positions ran about 15 km south of the border. The main gateway to Bulgaria, the Struma Valley, was now in the hands of the Central Powers. No more was planned and their troops did not advance any further. [339] For Paris, the occupation of the *Roupel-Strait* meant a violation of the assurance of the benevolent neutrality of Greece towards the Allies.[340]

Strictly speaking, the occupation of the *Roupel-Strait* was the same under international law as the occupation of the Saloniki bridgehead by the Allies almost a year earlier. Both happened on the soil of a neutral state that did not resist. The difference was this: the Allies had created a starting point for offensives against Bulgaria, and the Central Powers had brought one of the few starting points for such offensives into their hands. The occupation of the *Struma-Roupel-Strait* was ultimately a defensive measure by the Central Powers. The Entente had believed that the Greek government would enter the war the moment the first Bulgarian soldier set foot across the frontier. The Greeks were outraged, but they understood that the occupation of the *Roupel-Strait* was an inevitable consequence of the Allied landing in Saloniki. But the occupation of the *Roupel-Strait* gave the Allies a powerful propaganda weapon. Venizelos did his best to reinforce the Allies' impression of a treacherous interplay between the Greek government and the Central Powers.

On June 5th, Skouloudis protested in parliament against the assertion made by the Allied propaganda that the *Roupel-Strait* had been surrendered to the German-Bulgarian troops on the basis of a secret agreement. On June 6th he addressed the French government through the Greek envoy and repeated this. The Allies should not be misled by false information. But French censorship banned the publication of the statement before parliament. Instead, an extremely violent press campaign against Greece began. At the same time, Sarrail banned Greek merchant ship traffic without justification, against which Skouloudis protested in vain in the capitals of the Entente.[341]

On April 25th Joffre had written to Robertson that the Bulgarians should be attacked as soon as the Serbian troops had arrived in Saloniki. The Bulgarians operated during the war only half-heartedly, and a victory over them would probably bring Romania to war on the side of the Entente. A victory in Macedonia would be more significant than a local one in the west. If nothing was done, the troops gathered in Saloniki would likely be reduced. In addition, attacks in Macedonia would force the Germans to send troops there. One might have expected the British to respond to these arguments, but Robertson was still in favor of withdrawing troops anyway. He replied evasively in early May. In mid-May, Joffre presented the British with an offensive plan and asked them to move two more divisions to Saloniki. The British rejected this with indignation. On May 17th, the War Committee discussed the French proposals and flatly rejected them. They informed the French. The arguments put forward had made it clear that the British government had no interest in operating in a region of Europe which was not in its traditional area of interest. At a May 26th meeting between Joffre and Robertson it became clear that a thorough discussion between the two governments was necessary.[342]

So Briand and Joffre came to London on June 9th, 1916. Joffre tried to persuade the British to send more troops to Saloniki, knowing that the British were planning the Somme Offensive. Because of the ongoing battle for Verdun, he was unable to send French troops. But he wanted a diversion attack on the Saloniki front to force the Germans to transfer forces from the Western Front to the east. He knew that his units in Saloniki were not fully equipped. Lloyd George wrote: *"It was one of the most cynical performances I have ever heard. Given the inevitable loss of brave human life that such a futile undertaking would entail, it would be dire if he had not counted on us to reject his proposal. I understood that he didn't mean it and that an offensive near Saloniki would fail without the necessary cannons and ammunition."*

Lloyd George replied that, like Briand, he believed in an offensive from Saloniki, but only when the necessary equipment was available. *"A failed offensive would affect all further attacks on this flank [front]. No government would try a second time after a first failure"* So it remained at the Battle of the Somme.[343] Grey's determination that the French were prepared to risk a disaster for internal political reasons is completely correct.[344]

When it became clear in April that the French absolutely wanted to attack in Saloniki, the British switched out their commander in Saloniki. In their eyes, Mahon was not steadfast enough towards Sarrail. Therefore, Mahon was transferred to Egypt. He left Greece on May 9th and when he arrived in Egypt suffered heat stroke. He was succeeded by Lieutenant General George Francis Milne. Milne had been in Saloniki since January as chief of the 27th Division and commander of the 16th Corps. He knew Sarrail's methods, and London hoped he could stop him.[345]

Robertson had informed Milne that the British troops were subordinate to Sarrail only in the defense of Saloniki and that the *Imperial General Staff* opposed any offensive against the Bulgarians. On June 3rd, the General Staff explicitly ordered him not to take part in offensives under any circumstances. Sarrail knew that Joffre was as angry as he was at the surrender of *Fort Roupel* to the Bulgarians. So he ignored Milne's admonitions to treat the Greeks moderately and announced that he would attack the Bulgarians with or without British support. But Milne did not allow himself to be impressed. He followed his orders and withdrew the British troops to the south. Sarrail continued his preparations for an offensive in August.[346]

The British and French soldiers had been in Saloniki for more than eight months and, strictly speaking, had not contributed anything to the course of the war. Their presence formed a kind of economic development program for the city of Saloniki. The region had new streets and quays. There was even an entertainment industry. The Saloniki plans called on withdrawing men and supplies from the other theaters of war. In Paris, in view of the losses in Verdun, voices rose to recall the former victor of Verdun. On June 20th, Briand spoke to the Chamber of Deputies about his Balkan policy and ended by stating that the prospects for an offensive in Macedonia were getting better and better.[347]

Former Foreign Minister Delcassé countered, revealing that he had resigned over the Saloniki business. Then he said: *"I have heard that if we weren't in Saloniki, the Germans would be here immediately. Everyone should voice their thoughts, I'm here to share mine. And I say I wish to heaven they were here. And not just 200,000 or 250,000 of them, like our army, but 400,000 or 500,000. Then, gentlemen, at least they would not be at the front in France right now."*[348]

But there was also biting criticism, especially from Clemenceau, who exposed the aim of his criticism: *"What are Sarrail's men doing there? Digging! Then they should be known in France and Europe as "The Gardeners of Saloniki".*[349] In Germany, people also made fun of them: *"German *humorists laughed themselves to death. Prisoners of war, they said, are locked in barbed wire cages when they are first captured. But there is no known example in history of an army going into the self-made cage and locking the door."*[350]

The advance of the Allied troops from the *Entrenched Camp* towards the border on August 20th was also due to the fact that Romania was about to enter the war on the side of the Allies. The attack on that day, by Serbian and French troops, in western Macedonia against the Kajmakçalan Mountains, was successful. The first foothills could be occupied, from which the offensive against Monastiri (Bitaloa) began later. But the advance also resulted in the occupation of the *Roupel-Strait*. The consequences for East Macedonia were far more serious. The 4th Army Corps was in Kavalla. Due to the demobilization, the corps only numbered 600 officers and 8,500 men. Many of the officers, including the corps commander, were on compulsory leave. The commander of the 7th Division, Colonel Ioannis Chatzopoulos, led the corps. The telephone and telegraph connections with the capital had been cut off by the French. One could only communicate with Athens via Radio Kavalla. The fortifications around Kavalla were not finished and even the few that were finished were inadequately armed. The forts were manned by just 35 officers and 250 soldiers.[351]

At the end of July, Chatzopoulos learned that the Bulgarians were advancing to the border. He asked Athens for instructions. The army minister replied that under no circumstances should he surrender or vacate the positions at the border without his order to do so, which he could request if the worst came to pass. In the meantime, the Greek government was negotiating with the Allies about possible support and guarantees. Since the Allies did not respond to the Greek wishes, the minister on August 15th ordered the evacuation of the garrison of the Kavalla fortress with all of their weapons. This happened in the following days. On August 18th, the security companies were withdrawn from the border. In the evening, the following order came from Athens: If the Bulgarian troops advanced, the units of the 4th Corps should withdraw to their home bases. Since armed rioters who terrorized the Greek population were advancing at the same time as the Bulgarian troops, the inhabitants fled the country to the cities of Serres, Drama and Kavalla. Prime Minister Zaïmis was informed and asked the governments of Germany and Bulgaria to take the necessary measures to protect the civilian population in the areas occupied by their troops. Germany replied on August 2nd, assuring him that the German-Bulgarian troops would not enter the cities in question and that the captured Greek soldiers would be released.[352]

But some Bulgarian units, especially Bulgarian *Komitatschis* (irregulars) did not keep these promises and attacked Drama, Serres and Chrysoupolis, so that the inhabitants fled to Kavalla in a panic. There were murders, rape and looting.[353] The commander of the 4th Corps asked Athens for instructions; he would mobilize the reservists and defend Kavalla. On August 21st, Athens stopped this. Any use of force should be avoided. Between August 19th and 22nd, the Bulgarians occupied all of Eastern Macedonia. The Greek units withdrew to Kavalla. Since the Allies, like the Bulgarians, blocked the supply of food to Kavalla, a famine broke out there. On August 27th, the Germans interfered and saw to it that the food supply was resumed. At the end of August, the troops of the 5th Division were surrounded by the Bulgarians. The 6th Division withdrew to Kavalla. The forts north of Kavalla (Lisse and Perithori) were in Bulgarian hands. The commander of the 4th Corps demanded that the equipment be brought from Kavalla to ancient Greece and that the civilian population be supplied and protected. At the same time, officers of the National Defense Committee (*Ethniki Amyna*) appeared, in order to win the 6th Division for the mutineers in Saloniki.[354] The further developments, which led to the internment of the troops of the 4th Corps from Kavalla to Görlitz in Silesia, will be described in later chapters.

Since the Greek troops did not oppose the occupation of East Macedonia by order of Athens, it was assumed in Allied circles that it was a deliberate game between the Central Powers, the King and the General Staff. Venizelos asserted to Elliot that, according to his information, the occupation was based on a plan that the Greek General Staff had worked out to bring all of Greece under Central Power control, with its help. These were to advance via West Macedonia to Larissa and finally to Athens. In London and Paris, it was easy to see the lack of truth in this information. How was such a German-Bulgarian army supposed to march south without first defeating the Allied troops in Saloniki? But to appease the Allies, Constantine dissolved the Greek General Staff on August 25th. This in turn opened the way to the Defense Committee's Veinzelan Revolution in Saloniki.[355]

Above & Below: "Our cemetery in Drama in Greece. Comrades! Rest in peace in foreign soil!" View from the cemetery to the city of Drama. Unfortunately, not all the names which are noted on the photo can be assigned to people who fell. On the bottom photo one can recognize, from left: Griesemann, Lenz, Deetjen and Hermstedt.

Above: Observer KLt.d.R. Ernst Griesemann, born on December 2nd, 1890 in Wolmirstedt, died on December 14th, 1916 near Sofia, but buried in Drama.

Above: Observer Lt. Wilhelm Lenz, born on December 5th, 1895 in Berlin, died on February 5th, 1918 in Topoljani near Üsküb.

Above: Pilot Lt.d.r. Friedrich Hermstedt born March 5th, 1892 in Erfurt, died on February 5th, 1918 in Topoljani near Üsküb. Lt. Hermstedt transferred from IR 64 to the aviators.

Above: Lt. Hans Deetjen, born on August 19th, 1894 in Allenstein, died on March 29th 1918 in Drama. No further details.

Romania's Entry into the War

Above: On February 7th, 1918, Vzf. Keller returned from a long-distance reconnaisance flight with Lt. Fröhner, from the Marmara Sea. During the air fighting the German aircraft was followed to the front. The aircraft is a Rumpler C.I.

As already mentioned, Romania was also heavily courted by the Central Powers as well as the Allies, but remained neutral despite attractive territorial offers from both sides. But in the summer of 1916 this changed. After the successful first Brusilov offensive in Galicia and the Italian Isonzo offensive, the effect of which was however completely overestimated, Romania concluded an official alliance with the Entente on August 17th, 1916. This was preceded by months of bargaining over territorial gains. The Entente guaranteed the new ally the possession of Transylvania (Siebenbürgen and Banat) and parts of the Bukovina, at the expense of Austria-Hungary. What Romania did not know, of course, was the fact that France and Russia had secretly agreed not to honor this part of the agreement when the time came.[356]

Thereupon Romania declared war on Austria-Hungary on August 27th, 1916, followed a day later by the German declaration of war on Romania. Russia promised to send three divisions to Dobruja and to protect the Romanian coast against attacks by sea. At the same time, the Allies promised to launch an attack on Bulgaria from Macedonia at the beginning of August with the aim of furnishing Romania with military supplies. Romania's entry into the war on August 27th was actually a month late. The Russian attack in the south had been stopped and the offensive in the north near the Pripet swamps had failed.[357]

Basically, Romania's entry into the war on the Allied side was a capital mistake, because where should Romania actually get help – with hostile Bulgaria in the south and Austria-Hungary in the west? After the end of the Brusilov offensive, nothing more could come from Russia either.

The Allies' promises to help Romania could not be kept, which was actually clear from the get-go. Even before the offensive began, Milne had determined that little could be done against the Bulgarian mountain positions. At best, supplies could come via a single-track railway line, and there was no heavy artillery. While Sarrail directed his attack in the west against Monastiri (Bitola), the English troops secured the right flank against the Bulgarian positions in eastern Macedonia. The attack was carried out by 14 divisions altogether, which faced 23 Bulgarian divisions. The locally successful attack against Monastiri in September had no effect on the fighting in Romania.[358]

The Romanian attack on Austria-Hungary was in itself wrong. If Romania had attacked Bulgaria, there might have been a vague chance that it could have led to contact with the Saloniki Front. After the Romanians' initial successes, the Austrians reorganized and the Germans sent troops.[359]

Falkenhayn had not expected the Romanian offensive until September, so when Romania attacked, the Central Powers were too weak to repel the attack. Therefore, the Romanians managed a deep cut in Transylvania. This shook Falkenhayn, already battered by the failures in France, to such an extent that he was forced to resign. The new commander-in-chief was Marshal Paul von Hindenburg, who together with Erich von Ludendorff, comprised the new Supreme Army Command. As compensation, Falkenhayn received on September 6th, 1916 the supreme command of the 9th Army against Romania. He forced the invasion of Siebenbürgen, defeated two Romanian

Right: LZ85 was shot down during the night of May 4th to May 5th, 1916 over the Vardar estuary. This marks the end of the military use of all airships. Here on the colorized photo LZ85 is presented to the Greek population as a spoil of war.

Below Right: "The good old comrades – in fighting and suffering – on the Balkan." Left Vzfw. Johannes Keller, Vzfw. Erich Dürre, born on October 15th, 1892 in Berlin, died on June 18th, 1918 in Stojakovo (Macedonia), Walter Gnädig, Meisner. According to a note on the back, the photo was taken in 1917. The stylized iron crosses meaning "Fallen" are probably written on the back after the fact. In a dedication card, Walter Gnädig was referred to by his given name "Hans Reinhold".

armies near Sibiu (Hermannstadt) and Kronstadt, and forced an exit from the mountains into Wallachia. On December 6th, 1916, in cooperation with the Danube Army under August von Mackensen, he succeeded in conquering Bucharest.[360]

The Romanian entry into the war was a kind of catalyst for Greek developments. Without it, Sarrail would hardly have driven his malaria-stricken troops into an offensive. The preparations for this attack triggered the occupation of the *Roupel-Strait* and eastern Macedonia by Bulgarians and Germans. Domestically, these in turn caused the Venizelist uprising and the division of Greece, and triggered its entry into the war with all of its catastrophic consequences.

Endnotes

1. Keegan, op. cit., p. 306.
2. Falls, op. cit., p. 139f.
3. Winston Churchill, The World Crisis 1911-1918 Vol. I (London: Odhams Press, 1938), p. 1098f.
4. Keegan, op. cit., p. 307f.
5. http://de.wikipedia.org/wiki/Erich_von_Falkenhayn#F.C3.BChrungsschw.C3.A4chen_an_der_Ostfront.

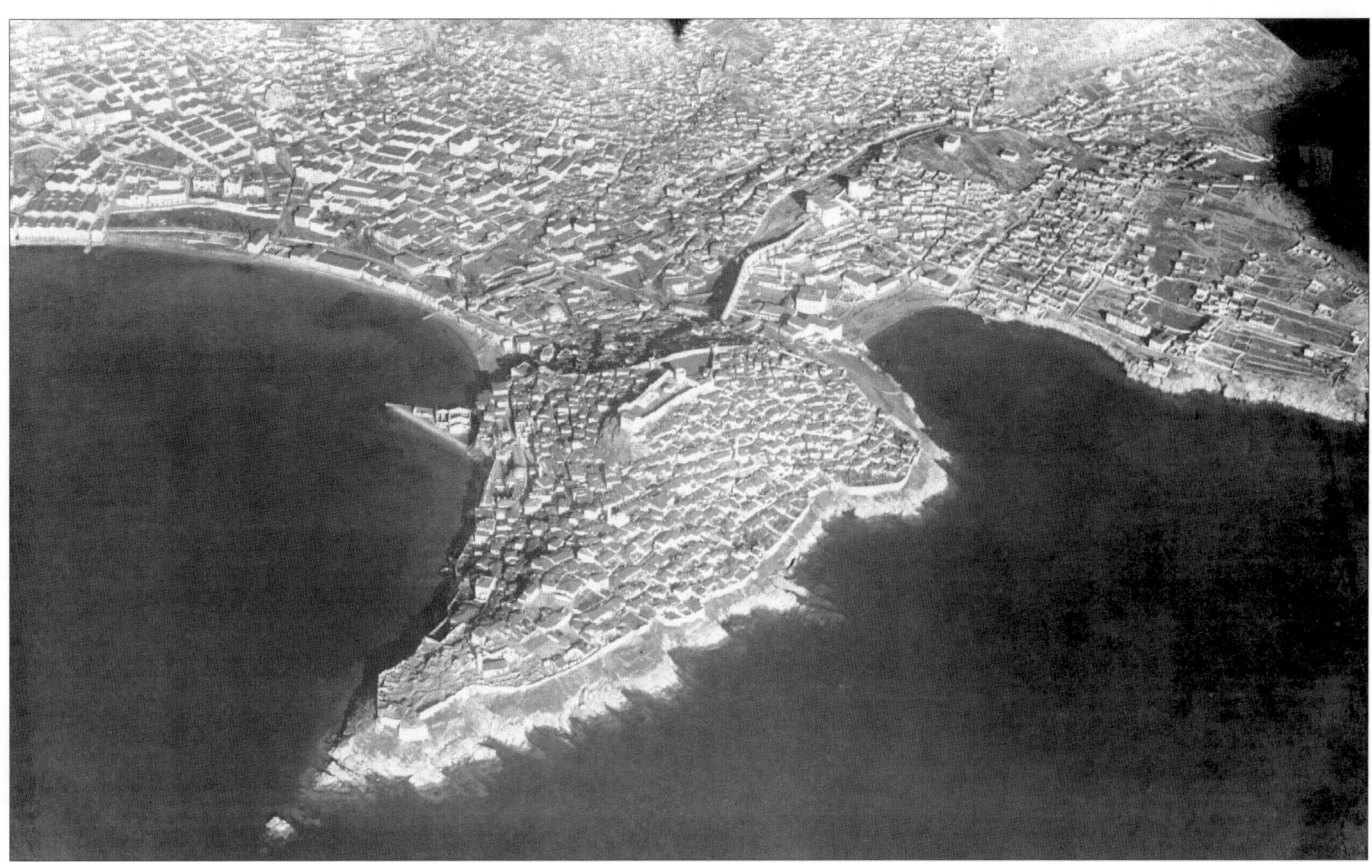

Above: Johannes Keller noted on the back: "The wonderful Kavalla."

Facing Page: The observer in a tethered balloon was subject to many dangers. In the first world war only one balloon observer received the Pour le Merite medal for bravery: Peter Rieper. The official designations changed from 1914–1918 from airship troop, field airship unit to balloon unit.

Split of the Country – Ethnikos Dichasmos, August–December 1916

The Evacuation of the 4th Army Corps to Görlitz

At the beginning of September 1916 the following situation existed in and near Kavalla: The 5th Division in Drama, as well as the 16th Infantry Regiment of the 6th Division and the 20th Infantry Regiment of the 7th Division in Eleftheroupolis, were surrounded by Bulgarian troops. The remnants of the 6th Division moved towards Kavalla, which they reached on September 4th. The forts north of Kavalla were now under Bulgarian rule. The commander of the 4th Corps had repeatedly asked Athens to transport the military equipment to ancient Greece and to take care of the starving population, a result of the British naval blockade. The 15 Venizelian officers of the corps went from Saloniki to Thasos to pull the members of the 6th Division onto the side of the mutineers. The order came from Athens that the 5th and 6th Divisions should retreat to Kavalla, but this was not possible because both units were surrounded by Bulgarian troops.[361]

On September 6th, the commander of the 10th Bulgarian Division, accompanied by a German lieutenant, appeared at the commander of the 4th Corps and demanded that the heights north of Kavalla be occupied in order to repel an Allied attack. Refusal would be seen as an unfriendly act. The corps commander knew that no attack was imminent, but since he had instructions from Athens to avoid problems, he gave in and cleared the heights. The garrison of Kavalla was thus concentrated in the city's urban area, with no possibility of defending itself.

The day before, the commander of the 6th Division, Colonel Ioannis Chatzopoulos, on behalf of the sick commanding general, and a few other officers, had met in secret with the British Vice Consul of Kavalla outside of the city. They agreed that the troops of the 6th Division should be brought to Saloniki on Allied ships and join the mutineers there. This way they would not end up in Bulgarian-German captivity.[362]

On September 9th, Chatzopoulos met Major Wolfgang von Schweinitz, the liaison officer with the Bulgarians. He wanted to prevent armed clashes between Greeks and Bulgarians. They agreed that they would meet again the next day, and then von Schweinitz would convey to him the German high command's proposals. In the afternoon of the same day, a British naval unit landed and destroyed the radio of the 4th Corps. The crew was completely isolated. During the night some English ships appeared in the port of Kavalla to pick up the troops of the 6th Division and bring them to Saloniki. When Chatzopoulos learned, however, that that the transport was reserved for only those troops who were prepared to join the mutineers in Saloniki, he forcibly prevented the embarkation. Only 15 officers and less than a hundred soldiers managed to leave for the island of Thasos in small boats.[363]

On the morning of September 10th, von Schweinitz and Chatzopoulos met again. The former said that Hindenburg was demanding that the 4th Corps gather at Drama. Dispersed across East Macedonia, it was hindering Bulgarian operations. If the demand was not accepted, there would be use of force. Chatzopoulos wanted to consult with Athens, which the Major refused because it would take too long. Chatzopoulos should decide *ad hoc*. But he wanted to consult with his commanders. He suggested to von Schweinitz that he should ask Hindenburg what guarantees the German side could offer in the event of a transfer of the corps and its weapons to Germany. The major promised to inquire and it was agreed that they would meet again the following day.[364]

After his return to headquarters, Chatzopoulos convened a council of war in which the division commanders of the 6th and 7th Divisions participated. They decided to try to bring the troops to ancient Greece, in cooperation with the Allies. Then the Chief of Staff of the Corps contacted the command of the British fleet in the Mediterranean via the radio of a British warship docked in the port of Kavalla, and asked them to transport the troops to Volos, Chalkis or Piraeus, so that they could be available to the Greek government. The admiral promised to answer. In the meantime, Chatzopoulos had the troops line up with their equipment for embarkation on the night of September 9th. At 9:30p.m. Chatzopoulos and his chief of staff drove to a British transport ship to initiate the evacuation. But on board this ship was the commander of the Serres division, a Venizelian officer who said that only supporters of the National Defense (*Ethniki*

Amyna) would be permitted on board. Chatzopoulos gave up. Panic ensued among the troops, who were now threatened with Bulgarian captivity. Riots broke out, prisoners were freed and many tried to escape captivity on fishing boats.[365]

Desperate, Chatzopoulos asked the British to inquire for him in Athens whether he could surrender to the British and be interned on the island of Thasos. Athens told him to embark on Greek or British ships and go to Volos. But this order does not seem to have reached Chatzopoulos. The British, on the other hand, were only willing to transport supporters of Venizelos to Saloniki.[366] On the morning of September 11th, Chatzopoulos met Major von Schweinitz again. The latter informed him that Hindenburg was ready to have the 4th Corps brought to Germany with its weapons. The men of the corps would not be treated as prisoners of war but as guests of Germany. Details should be determined in cooperation with the Greek ambassador in Berlin. They could take their time with the transport.[367]

In the meantime, Athens had learned of the events in Kavalla and turned to the British. They were now ready to evacuate the troops from Kavalla by sea. The Ministry of the Army briefed Chatzopoulos about this through the British Vice Consul in Thasos. The corresponding message reached Chatzopoulos around 9 p.m. But this order came too late, because as early as 6 p.m. von Schweinitz, to whom information about a possible evacuation had probably been leaked, ordered that the troops of the 4th Corps should start moving towards Drama that evening, which then also happened.[368]

On the evening of September 11th, 1916, a column of 400 officers and 6,000 men set out, heading north. 2,000 men of the 6th Division including their commander, Colonel Nikolaos Christodoulou, managed to leave for Thasos. There the troops split up: some would be brought to ancient Greece and another part to Saloniki. Among the troops on Thasos was also an artillery regiment. With the consent of the Allies, its equipment was loaded onto a ship to be transported to Volos. On the way, the ship was stopped by a French destroyer and redirected to Saloniki. The weapons and vehicles were handed over to the Venizelist armed forces there.

The Kavalla garrison reached Drama on September 12th. Two days later the soldiers were loaded onto the train. On September 13th, the Serres division, which had also been transferred to Drama, followed. The last transport left on September 27th. The destination was Görlitz in Silesia.[369] Altogether

Above: Comrade of Johann Keller: Portrait of Paul Wedekämpfer.

there were 10 transport trains.[370]

Abbott writes about the evacuation: *"Nothing that has taken place so far has been so suitable to make the commanders of Greece appear in a bad light before the eyes of people of the Entente. It was claimed that there were 25,000 surrendering troops, even 40,000 men. Numbers that have now been reduced to about 8,000: 3 divisions, each consisting of 3 regiments of 800 men each. The surrender was presented as if ordered by the Athens government. Out of affection for Germany and Bulgaria, as well as out of hatred for France and England, King Constantine allegedly gave up not only rich territories which he himself had conquered, but also the soldiers whom he had twice led to victory."*[371]

The Allied and Venizelos propaganda machines went to great lengths to portray the evacuations as a perfidious interplay between the governments of the Central Powers and Greece. But the Greek government knew nothing of the events in connection with the capitulation of the IV Corps, because there was no longer any direct connection between Athens and Kavalla. Chatzopoulos had chosen this solution himself, because he did not want his troops to become prisoners of war of the

Above: "To my dear friend and fighting comrade as a remembrance – Hans Reinhold Gnädig, Macedonia, 22.IV.18." Reinhold Gnädig crashed on December 7th, 1924 as the workshop aviator in Liegnitz. Birthdate is unknown. He was also awarded the "iron half-moon" medal.

Bulgarians. Abbott reports that Athens learned of the whole affair only after the first Greek units had arrived in Germany. There were violent protests: the Greek troops were neither prisoners of war nor internees, because only neutral countries could intern troops from warring states and Greece was not a belligerent state. They asked Germany to send the troops back via Switzerland. Berlin replied that they would be happy to do so, but wanted guarantees that the Greek troops would not be punished by the Allies for their loyalty to their King. This was a reasonable request, given that the Allied press spoke of *"shameful desertion, mutiny, perjury and treason"* which should be severely punished.[372]
It was grotesque, however, that the Allied press spoke of treason, etc., because at most only King Constantine could have complained about that.[373]

But the Greek government also complained to the Germans for another reason. Originally they had assured it that the Bulgarians would behave correctly. But the Bulgarian troops were accompanied by *Komitatschis* who did not keep their promises and wanted nothing more than to drive out the Greek population and annex Eastern Macedonia for itself. The result was rioting. Athens complained to Germany, which on August 23rd urged the Bulgarians to behave better. But in the meantime the *Komitatschis* had also brought the army under their control and so the clashes continued in September. After the war, a Greek university commission produced a detailed report that contained hair-raising details.[374]

While on the side of the Entente and the neutral states, the transport of part of the armed forces of a neutral country was received with indignation, the German side was enthusiastic. The *Görlitzer Nachrichten* wrote on September 28th, 1916: *"The first transport of the Greek guests arrived at the local train station yesterday afternoon at 3:27 pm. It consisted of 22 officers, 427 men and 15 mountain cannons, the latter being unloaded at the so-called blockhouse ramp and transferred directly to the shed intended for this purpose. Accompanied by the officers were some women and children [...] At the reception: [the] wing adjutant of the emperor, [...] the representative of the garrison command, [the] mayor, [...] the station commandant [...] and the officers of the garrison. When the train arrived, the band of the reserve battalion [...] played the Greek national anthem. The Greek soldiers marched to their new location, the entrance of which was decorated with the lettering* Χαίρετε, *and garlands. The Lord Mayor of Görlitz said: "We'll shake hands with you in German loyalty and comfort, and promise to you we will do everything we can to make your stay with us as pleasant as possible. You are very welcome to all of us [...] and when you one day return to your fatherland in good spirits, please remember kindly the times that you stayed with us, just as we will."*[375]

The Greeks integrated well into the life of the city of then 90,000 inhabitants. The daily newspaper *NEA TOY GÖRLITZ* informed them about local and international events. The shops in the city advertised to Greek officers, who received their full peacetime salary, paid by Germany. There was even a bar called *Drei Raben* (Three Ravens) that poured Greek wine. The officers lived in furnished apartments in the city. The younger women of Görlitz, whose husbands were at war, were, of course taken with the spirited southerners.[376]

There were many weddings, and after the end of

the war, when the Greek husbands were repatriated, a large number of their German wives went along. But in Greece there were soon tensions, because in the Greek world of that time women played a completely subordinate role, which understandably did not suit the women who arrived from Germany. There were separations and divorces, and many of the women later returned to Germany. Elisabeth Logothetopoulou, née Hell, who was married to the medical professor Konstantinos Logothetopoulos, took care of the women in Greece. He later became the second prime minister during the German occupation in World War II.[377]

At the beginning, the Greek soldiers were not permitted to work, but when idleness diminished discipline, Greek leadership allowed the soldiers to take up work. At the beginning of 1918 there were around 4,000 men working in agriculture and industry, not only in Görlitz, but throughout the entire Reich. Their wages were the same as those of their German colleagues.[378]

There were also language courses and vocational training. The instruction was led by Prof. August Heisenberg, who at that time held the only chair for Central and Modern Greek Philology in Munich. From December 1916, Heisenberg was assigned to the 4th Corps as a liaison officer with the rank of captain. He pursued studies in Greek dialects. With the support of the *"Royal Prussian Phonographic Commission"* he made the first sound recordings of dialects and music. In July 1917, in addition to recordings of poems, stories and songs, the first recordings of a Rebetiko song with *bouzouki* accompaniment were made, years before something like this was done in Greece. In recent years these *"treasures"* have been rediscovered and analyzed.[379]

With the abdication of King Constantine in June 1917 and Greece's entry into the war, the situation changed. The few Venizelists among the Görlitz Greeks now had the say. Riots broke out. Finally, the German authorities intervened and arrested 36 officers who were branded as *"Venizelist agitators"* and put in a prisoner-of-war camp near Werl in Westphalia. The common soldiers were now committed to labor service.[380]

When the German revolution broke out in November 1918, workers' and soldiers' councils were also formed in Görlitz. The movement also spread to the Greeks, the officers were chased away, and the royalist-minded commander was deposed. Leading the group was a *"Görlitz Soviet"*, as it was later called in Greece. When talk of repatriating the Greeks began, many took flight towards the Bohemian border. On February 21st, 1919, the first who were willing to return home left Görlitz for

Above: "To my dear friend and fighting comrade as a remembrance – Hans Reinhold Gnädig, Macedonia, 22.IV.18." Reinhold Gnädig crashed on December 7th, 1924 as the workshop aviator in Liegnitz. Birthdate is unknown. He was also awarded the "iron half-moon" medal.

Italy. A day later a hospital train with the sick left for Greece.[381]

In Greece, the returnees were not received well. The ruling Venizelists suspected them of treason and locked them into camps. From there they were brought before examining magistrates. From May 22nd to July 13th, 1920, many were tried. Officers were dismissed, deported and exiled. Eight officers were sentenced to death but not executed. The trials deepened the national division and the legend of the *"Görlitz betrayal"* continued for decades.

About 200 Greeks remained in Görlitz, and many of them started families. The seven restored grave sites of the officers who died in Görlitz, including Chatzopoulos, can still be seen in the city's *"New Cemetery"*. The Greeks' stay in Görlitz cost the German army command and the government exactly

Above: The medal "Iron half-moon".

Above: Two officers in a building in the bay of Kavalla. "Oberstleutnant Geissler and Leutnant Baron von der Ropp, Kavalla." Hptm. Haupt-Heydemarck takes over the Squadron 30 in Drama (Greece) from Oblt. Geisler (February 1917).

10,869,400.99 marks from the specially created *Greek fund*. Greeks returned to Görlitz thirty years later. They were left-wing political refugees who were distributed to the people's democracies after the end of the civil war in 1949. About 14,000 went to Poland, some of whom were housed in the eastern part of Görlitz, which was now called Zgorzelec. In the beginning it was said there were more Greeks there than Poles. These Greeks were only allowed to return to their homeland after 1974.[382]

Endnotes

1. Hellenic Army General Staff, A Concise History, p. 86.
2. Ibidem, p. 87.
3. Ibidem; Leon, op. cit., p. 399.
4. Ibidem, p. 87.
5. Ibidem, 88; Abbott, op. cit., p. 130f.
6. Abbott, op. cit., p. 131.
7. Hellenic Army General Staff, A Concise History, p. 88; Abbott, op. cit., p. 131.
8. Hellenic Army General Staff, A Concise History, p. 89; Abbott, op. cit., p. 131.
9. Hellenic Army General Staff, A Concise History, p. 89f.
10. Klaus-Dieter Tietz, "Griechen in Görlitz" Hellenika N. F. 5 (2010), p. 60.
11. Abbott, op. cit., p. 131f; Cosmetatos, op. cit., p. 204 explains that the French press published the number 40,000 and the transport to Görlitz represented a new treason of Greece.
12. Ibidem, p. 132.
13. Leon, op. cit., p. 401.
14. American Hellenic Society, Report of the Greek University Commission upon the Atrocities and Devastations Committed by the Bulgarians in Eastern Macedonia (New York: Oxford UP, 1919)
15. Tietz, op. cit., p. 63f.
16. Ibidem, p. 64f
17. Elisabeth Logothetopoulou said this in discussions with the author. The author got to know some of these women, who stayed in Greece, in 1967/68
18. Ibidem, p. 65f.
19. Ibidem, p. 66.
20. Gerassimos Alexatos, "Xairete: Ein griechisches Armeekorps in Görlitz." in: Wolfgang Schultheiß (ed.), Meilensteine deutsch-griechischer Beziehungen (Athen: Stiftung für Parlamentarismus und Demokratie des Hellenischen Parlamentes, 2010), p. 192.
21. Tietz, op. cit., p. 67.
22. Ibidem, p. 68f.

Above: The Halberstadt fighter at left with the initial 'M' (for Meier?) was Halberstadt D.II(Han) 820/16 serving with *Jasta* 25 in Macedonia. The serial is not visible on the Halberstadt D.V at right with initial 'L'.

Above: A full view of *Kampfstaffel* 2 of *Kagohl* I in Macedonia with Halberstadt D.II(Han) 810/16 and 813/16 heading the line up.

The Governments of Zaïmis and Kalogeropoulos

On June 23rd, Alexandros Zaïmis became Prime Minister. He had accepted the demands of the Allies of June 21st, demobilized the army, dissolved parliament and scheduled elections, and dismissed some police officers who had disapproved of the Entente. But then came the German-Bulgarian occupation of East Macedonia, which he was forced to watch helplessly. The Allies were also unsettled and wanted to know what he was going to do against a further advance to Saloniki and Larissa further south. He replied that there was nothing he could do about the demobilization desired by the Allies.[383]

The alleged advance of the Bulgarians southwards was nothing more than a rumor, which Venizelos picked up and presented as truth in order to discredit the government among the Allies.[384] How should the Bulgarian-German troops have pushed past the Allies in Saloniki?

In mid-August Zaïmis learned of Romania's impending entry into the war and of Italy's intention to occupy northern Epirus. Indeed, in late August, Italian forces began to occupy the area up to the Delvino-Argyrokastro-Premeti-Leskoviki line. Upon the orders of Athens, the Greek troops stationed there offered no resistance. The uprising in Saloniki and Romania's entry into the war finally prompted Zaïmis to carefully change course. On August 30th, Zaïmis informed the British envoy Elliot: Although the poor financial situation and the lack of war material meant that they could not immediately enter the war, they would like to know whether the Allies would be willing to support Greece financially and materially. Of course, these negotiations were to remain absolutely secret. The Greek ambassadors in Paris and London received the same instructions.[385]

Paris, London and Petersburg responded positively. Only the Italians made excuses, hoping for territorial gains in Northern Epirus. Briand rejected lengthy negotiations, however, asking Athens to make a formal statement about its intentions and the date of the likely entry into the war. He promised financial and material support from the Allies, but all other issues would be discussed by the Allies after Greece entered the war. London reacted similarly. But before the discussions got going, the Bulgarians occupied eastern Macedonia without a fight, which convinced the French that Athens was cooperating with the Central Powers. On August 29th, Paris proposed that the Allies take control of the Greek postal service, telegraphy, radio stations, railways and ports. The German ambassador Karl von Schenck and all enemy agents were to be expelled from the country and all trading ships of the Central Powers were to be confiscated. In addition, troops should be landed in Piraeus.[386]

In the meantime, Constantine had decided to reshuffle the government and possibly include Venizelos in the new one. His goal now was to enter the war on the side of the Allies. On September 1st, the British envoy, Elliot, visited Constantine. He explained his plan to the ambassador and Elliot was more than pleased. But a little later the news arrived that ruined everything: The French fleet had arrived in the Bay of Salamis.[387]

In order to support the French demands, a fleet of 37 ships had been sent to the Saronic Gulf under the Commander-in-Chief of the Allied naval forces in the Mediterranean, Admiral Dartig du Fournet. They arrived in Salamis Bay on September 1st. On September 2nd, he had the *"interned"* ships of the Central Powers that had been anchored there, occupied and towed to the anchorage of the Allied fleet, which was a blatant breach of international law.[388] The British knew nothing about this.[389]

In Athens, Du Fournet had to play the role that the politically cunning Sarrail played successfully in Salonika. Du Fournet did not like the British, did not care much about the activities of the French diplomats and was an inexperienced negotiator. In the whole time he was unable to visit the Acropolis, but contented himself with looking at it through his binoculars from one of the cannon towers of his flagship. He was also under pressure from the anglophobic Minister of the Navy.[390]

Elliot turned down a fleet demonstration. In his opinion, the Greeks' mood had been tipped to favor the Allies after the Bulgarian invasion. Imposing such harsh conditions would be counterproductive. The British were able to prevent the troops from landing in Piraeus, but not the fleet demonstration. On September 2nd, Guillemin presented a memorandum containing the above requests. If the Greek government did not expel the German agents, Admiral Dartig du Fournet would take matters into his own hands. Zaïmis accepted the conditions without restrictions.[391]

Above: Aerial view of a large British base in 1918 in Greece, with airfield, barracks and supply trains via road and rail.

A little later, the secret services of France and England began to take action against all enemy agents in Athens and Piraeus, often using force. They even arrested the general secretary of the Austrian embassy and the Romanian military attaché for being a Germanophile. On September 8th, the German envoy Schenck and 50 German – alleged – agents were arrested and deported.[392] The British were confused by the brutal French approach; they were particularly irritated by the promotion of an anti-dynastic movement. Foreign Minister Gray advocated that *"the French would be told bluntly that support for a revolutionary movement against the Greek King would be resented by the Russian ruler. This could have an unfavorable impact on Franco-British relations with Russia and become a serious opportunity for the pro-German reactionaries in Petrograd."*[393]

The French did not trust the Greeks and doubted that they would actually keep the commitments they had made. They suspected them of sending encrypted radio messages to the Germans from the cruiser *Averoff*. The French secret service in Athens provided Paris with exaggerated or even falsified information in order to induce Paris to take an even tougher course. The chief of this group even managed on September 9th to hire 16 Venizelists under the leadership of a Cretan who belonged to the French secret police. These broke into the garden of the French embassy, fired a few shots and shouted slogans such as *"Long live the King"* or *"To hell with France and England."*[394] The whole thing was organized by French naval attache de Roquefeuil.[395] Guillemin immediately blamed the reservists for this action, and demanded banning the league.[396] The goal of this action was to disturb the negotiations between Constantine and the British, primarily, about a Greek entry into the war. With the help of their loyal reservists from the Balkan wars, Constantin could have easily and quickly mobilized an army. If the league was dissolved, this would have been much more difficult. If Greece were to enter the war, it would be under Venizelos' leadership. Had Greece entered the war under Constantine,

Above: A German aircraft tent in Hudova, with aircraft, was destroyed by a British bomb.

Above: The difficult transport situation is seen well here. This road is passable during good weather, but when it rained, only camel caravans were able to pass.

Elliot would have stolen the glory, which Guillemin, of course, did not want. So it was an intrigue.³⁹⁷

Zaïmis, who of course had no idea what was actually going on, formally apologized. Guillemin played innocent and replied that he was awaiting instructions from Paris. Admiral Dartig du Fournet sent 25 armed seamen to protect the legation in the city center. On September 10th, the French envoy demanded the immediate arrest of the perpetrators and police officers who had failed to protect the legation. He also demanded an official distancing of the King from the people who were behind the attack. French pressure and the events taking place in Macedonia at the same time up to the surrender of the 4th Corps led to Zaïmis declaring his resignation on September 11th – allegedly for health reasons.³⁹⁸ Venizelist circles were satisfied. With every cabinet overthrown, it became more difficult for Constantine to find a prime minister who was acceptable to the Entente. Obviously everything was going Venizelos' way.³⁹⁹

King Constantine commissioned Nikolaos Dimitrakopoulos, the leader of a small party in parliament, to form a new interim government. Dimitrakopoulos was a friend of the Entente. He held talks with their ambassadors and declared that the country was on the side of the Entente. He also wanted to hold elections. Since it was not certain that Venizelos would have emerged victorious, he intrigued with the Allies against Dimitrakopoulos. The result was that the Allies made it clear that they would not recognize Dimitrakopoulos as prime minister. Thereupon Dimitrakopoulos refused the mandate to form the government because he realized that, given the attitude of the Allies, his government would have little political leeway. At the same time, the Allied envoys let Constantine know that they would not recognize Gounaris, Rallis, Skouloudis, Stratos or Streit as prime minister. Obviously they hoped that the King would call on Venizelos to serve in that role.⁴⁰⁰

Constantine then gave the assignment to form a government to Nikolaos Kalogeropoulos, a well-known Francophile, who had lived in France for quite some time, and had pursued his doctorate degree at the Faculty of Law at the Sorbonne.⁴⁰¹ Constantine described his view of the situation in a letter to his brothers Andreas in London and Nikolaos in Petersburg: *"The situation can no longer be unraveled. The Entente Army and Navy commanders spark and encourage revolution and armed turmoil in the country. In doing so, they favor the Saloniki movement by all means by continuing the oppressive measures and restricting any freedom of thought and action. The Entente*

Above: Front of a photo post card to Johannes Keller, FA 20: "Friend Paul Fröhner" and on the back [to remember] "the Balkan, from his observer, Leutnant Paul Fröhner, Drama, 25.II.[1918]."

ministers are crippling the entire government. This is how a country is driven into anarchy. This approach not only contradicts the assurances given to us, but also excludes any practical possibility of rethinking our policy in order to ultimately make a free, beneficial decision."

Then came statements that showed that Constantine was not a friend of the Central Powers:

"In the circumstances, a decision to take part in the war would involve the risk of violence and suspicion. Even if you make the decision without the risk of harm, you still need caution and discretion. This is necessary in order not to provoke a German-Bulgarian attack. They are in our land before we are able to make any real contribution to the Entente. […] Under certain circumstances which make Greece's participation in the war useful and in line with our interests, I have already declared

Above: The honor guard of Lt. Fröhner and Pilot Hardt. The solder left has the sleeve insignia of the FA 20.

Above & Facing Page, Below: The funeral of BO Lt.d.R. Paul Fröhner, born on April 22nd, 1893 in Berlin, died on September 2nd, 1918, Tefik Bey and BO pilot Johannes Hardt born on April 25th, 1896 in Kelsterbach, died on September 2nd 1918, Tefik Bey. On the back Johannes Keller wrote: "My best comrade, Leutnant Paul Fröhner! Forever in our memory, rest peacefully! A brave officer!" Bulgarian allies also came.

that I am ready to go to war on the side of the Entente. With that in mind, I am ready to consider negotiations." Kalogeropoulos saw things the same way.[402]

AP correspondent Hibben added: Constantine *"telegraphed to his brothers [...] to give the governments of the three powers his personal assurance that his intention would be to enter into active military cooperation with the Allies in a battle against Bulgaria. He expressly offered the full support of Greece, but on the basis of the conditions that have already been outlined to Sir Francis Elliot. That is, that the full integrity of Hellas is guaranteed. I saw the original message myself; there is no doubt that King Constantine intended to bind himself to the Entente. There was no going back. The governments of Great Britain, France and Russia only have to accept the offer in order to conclude the agreement for Greece to enter the war on the side of the Entente."*[403]

Kalogeropoulos was sworn in on September 16th. His cabinet's first act was the formulation of a formal alliance agreement between Greece and the Entente, which was forwarded to London, Paris and Petersburg on September 18th. On the same day, Constantine asked the editor of the independent daily *Empros* to start a campaign for the entry into the war in order to get the majority of the MPs behind him to approve the necessary war credits that would be required.[404]

Kalogeropoulos was held in high regard in Parliament. He was a supporter of Theotokis and thus an anti-Venizelist. He openly admitted to the *London Times* correspondent that he sympathized with the Entente. He was a member of the Franco-Greek League. He said that he intended to continue the policy pursued by Zaïmis *"in order to strengthen ties between Greece and the powers of the Entente."*[405]

Foreign Minister Alexandros Karapanos was a

Above: Portrait of Vzfw. Wilhelm Grasmeher, born on May 31st, 1891 in Steedem, Oberlahn district. He also flew in the FA 30 in Drama. In March 1919 he was a flight instructor in Bromberg. He was expelled as a German from Piotrowska (Poland) on October 1st 1920.

well-known Entente sympathizer. It's not entirely understandable as to how one can characterize the cabinet as pro-German because three ministers wrote articles about Germany, as Leon does. It must also be noted at this point that no one had so far provided verifiable *hard evidence* of the government's alleged pro-German behavior. The unsubstantiated allegations can be explained with the following reasons.[406]

As soon as Kalogeropoulos was in office and had spoken about his intentions, Venizelos' propaganda began: he said he could not believe that the composition of the cabinet would change course in the direction of a *"national"* policy, since it was a traditionally pro-German party. The British envoy promptly said the following in an interview with a Venizelist newspaper: *"One can certainly not agree with this situation. I read the announcement of the new prime minister in the newspaper. What surprised me is that Mr Kalogeropoulos describes his cabinet as a political [business cabinet], whereas the last Allied announcement called for a transitional cabinet [gouvernement de service] for Greece.[407]*

By "business cabinet", Elliot meant a

Above: Vzfw. Theodor van Ahlen of the Fl.-Abt. 30. A very successful aviator in the southeastern war theater. [I was unable to find any further life dates.]

"*gouvernement de service*" (interim government) as is usual in Greece prior to elections. The only role of such a government is to carry through the next elections, and it may not make policy. And of course such a government would not need to be taken seriously by the Entente.

For the Entente press, the Kalogeropoulos government was, of course, pro-German. Its representatives adopted the untrue claims of the Venizelist press that the government was friendly to Germany. The representative of the "national will" was Venizelos, King Constantine was threatened with the same fate as King Otto, who had been deposed, and the country was facing *"terrible and hopeless things"*.[408]

Shortly after taking office, Kalogeropoulos let London know that they were ready to side politically

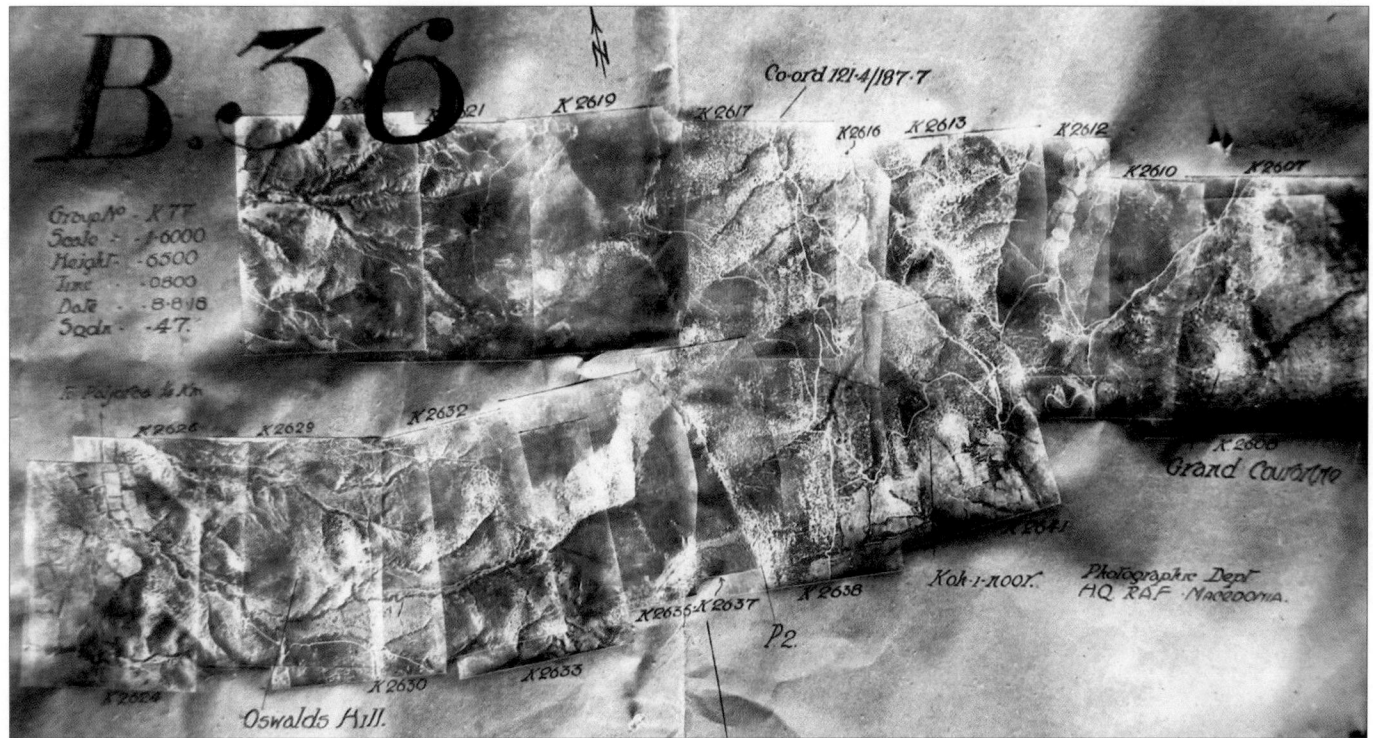

Above: A British aerial photograph of the Macedonian region.

with the Allies. From a military point of view, this was currently not possible due to (lack of) finances and equipment. They wanted to know from the Allies whether they would be willing to support an eventual Greek entry into the war with money and weapons. When the answer was delayed, the Greek envoy handed over a note on September 20th which said: *"The Government of Greece declares that it is refraining from its neutrality policy in favor of the Entente powers. As soon as it has re-equipped its military forces thanks to financial and military support, Greece will proceed within a mutually agreed timeframe towards the general mobilization of its army."*[409]

Paris and London viewed the statesmen there and the Greek offer as a ploy to buy time. There were three ministers in the Greek cabinet who were opposed to the Entente. Only Venizelos would pursue a pro-Entente policy. Elliot suspected Constantine was double-crossing him. His offer to work together was never meant seriously. That is why Zaïmis resigned when he found this out.[410]

The Foreign Office spoke of the *"emptiness and duplicity"* of the offer. Only if Greece declares war on the Bulgarians could it be believed. The argument that Greece was not prepared was allegedly only an excuse to buy time. The Greek ambassador was told that Greece was asking for a lot without promising anything concrete in return. The conditions presented by Greece only served to release the country from all obligations. The request for secrecy meant that negotiations were doomed to failure from the outset, because how should the Greek army be reorganized under the mantle of secrecy?[411] It was obvious that every flimsy argument was being sought in London to avoid an alliance with Constantine.

The King's declaration that he would only enter the war if Sarrail's offensive was successful, that he refused to cooperate with Venizelos and expected a binding promise, that Thrace as far as the Maritza, the Dodecanese and Albania as far as the Skumbi River would become Greek, was understood as an excuse and even sabotage of the policy of the Greek government.[412] Constantine was not permitted to make demands. Postponing the entry into the war until a favorable point in time was interpreted by the Allies as a rejection.

The French reaction was even stranger. Briand let Constantine know that the Allied arrangements made it impossible to enter into official alliance negotiations with Greece, but if Constantine took sole responsibility for an attack on Bulgaria, France and England would be ready to accept Greece as an equal ally. Briand's offer contained no guarantee about Greece's legitimate territorial claims. Greece would receive support at the peace conference. Briand did not even mention the Greek demand for a

Above & Above Right: Immelmann as a young person

guarantee of its territorial integrity. *"Premier Briand stated that he had delayed a formal response to the proposals of the Kalogeropoulos Cabinet in the hope that King Constantine himself would take the initiative. By declaring war on Bulgaria without an agreement with the Entente, the king would present the Allies with a fait accompli."*[413]

Constantine had expected a clear answer to the clear offer, and was more than disappointed. He considered Briand's proposal *"childish."* *"If you look at the situation through a military lens, as he said, why should I declare war before I am able to make war? Let us be able to go to war with Allied support, and I will declare war as soon as we are able to carry it out. A useless gesture of hostility without adequate response is ridiculous. However, the Allies will not give such an answer because it corresponds to our proposals. All we have to do is get our house in order and wait for the official response."*[414]

Constantine was absolutely correct. To begin a war with a widely demobilized army would have been irresponsible. But it wasn't about that, but rather about dismantling Constantine as King and politician. The Allies wanted Venizelos.

Kalogeropoulos said about the Entente press campaign against Constantine: *"I can't understand it! Nothing could be more unjust to the King than the stubborn claim that he was pro-German. He is pro-Greece and only pro-Greece. He is the last man in Greece who allows himself to be influenced by prejudices and preconceived notions."*[415]

Foreign Minister Gray suggested to Briand that Constantine should be asked to form a *"national government"* composed only of Entente-friendly personalities, and named October 1st as the deadline for unconditional entry into war against Bulgaria. Only in this way could confidence in the Greek government be restored. They had nothing against

Above: Crashed German aircraft with the pilot Lt. van Ahlen.

Above: Our airfield from 500m: 11 hangars as well as two aircraft tents are recognizable. The name of the airfield is unknown. This is the first photo from the "Fieseler Archive" of the Jasta 25. A heartfelt thank you here! Slg. Rolf Nagel.

Above: Panorama photo I, with Roland D IIa, about 1917: Jasta 25 in a parade lineup. The name of the airfield is probably Kanatlaci, today Budakovo. One can see 6 aircraft alto-gether. The guard of ground personnel stand next to their aircraft. Slg. Rolf Nagel.

Above: Panorama photo, part II, with Roland D IIa, spring 1917: The Jasta 25 in a parade lineup. One can see 5 aircraft with 8 pilots. The guard of ground personnel stand next to their aircraft. Slg. Rolf Nagel.

the dynasty or the monarchical principle, but only if Greece immediately declared war on Bulgaria would there be a chance of preserving the country's unity and stopping the revolutionary movement. Briand agreed, but suggested that Greece should immediately declare that there was a state of war with Bulgaria. If the King refused to enter the war immediately, then the current government would have to be replaced. The Greek troops would have to withdraw completely to ancient Greece, the royalist reservist league must be disbanded and the troops and gendarmes would have to be withdrawn from all the islands occupied by the Allies. London and Petersburg rejected these proposals and Briand gave in.[416]

On September 30th, Briand expressed his personal opinion to the Greek ambassador in Paris, Athos Romanos: If Greece immediately declared war on Bulgaria and the King formed a new government that would give the Allies the necessary guarantees, the Allies would finance and materially support Greece, guarantee its territorial integrity and support the Greek territorial claims at the future peace conference. These were vague promises from the Allies, whereas Greece would be forced to take a leap into the dark. Kalogeropoulos and Karapanos and two other ministers agreed to accept this, but the King found these promises too vague and suggested an answer based on the September 20th note. At a meeting of the Council of Ministers on October 4th, the differences between the government and the King came to light. Kalogeropoulos then announced his resignation.[417]

The question arises as to why the Entente rejected

the offers of the Greek government and the King. AP correspondent Hibben came to the following verdict: *"The Kalogeropoulos cabinet supported an intervention. The Hellenic government [...] submitted proposals to the Entente embassies for immediate participation in the war and King Constantine approved the course of the government."*

But nothing could change the minds of the Entente envoys. *"After the they labelled the entire cabinet as pro-German, the Entente ministers did not dare to admit their mistake."*[418] This much narrow-minded stubbornness is amazing.

Right: Lt. Rudolf von Eschwege (born on February 27th, 1885 in Homburg v.d.H., died on November 21st, 1917 in Orljak). Slg. Lance Bronnenkant

Above: Again the lineup of the 11 Roland D IIa aircraft of Jasta 25 at Hudova, 1917. Here one can also see the personal insignia of each pilot.

Above: Jasta 25: Aircraft Roland D IIa (late spring/early summer 1917) with mechanics in front of the aircraft tents. Slg. Rolf Nagel.

Facing Page: Jasta 25: Fw. Toni Bauhofer in front of a Roland D IIa (about 1917). "To remember little Toni". Behind the first strut is a small coat of arms. (Vzfw. Anton Bauhofer, born on January 25th, 1892 in Munich) Both photos: Slg. Rolf Nagel.

Endnotes

1. Leon, op. cit., p. 383.
2. Falls, op. cit., p. 211.
3. Leon, op. cit., p. 390f; Falls, op. cit., p. 213.
4. Ibidem, p. 391f.
5. Hibben, op. cit., p. 53.
6. Hellenic Army General Staff, A Concise History, p. 106. Four German and three Austrian merchant ships. The crews were taken prisoner and locked up on French ships. Hibben, op. cit., p. 54.
7. Hibben, op. cit., p. 53.
8. Palmer, op. cit., p. 101.
9. Leon, op. cit, p. 392f.
10. s Hibben, op. cit., p. 54: "A campaign in France and England had long been in progress against [...] Schenck, the head of the German propaganda in Greece. Rather an insignificant figure in fact, the baron had been raised to a pinnacle of diabolical cunning and almost superhuman influence by the more sensational British and French newspapers."
11. Leon, op. cit., p. 393f.
12. Ibidem, p. 394; Hibben, op. cit., p. 57.
13. Hellenic Army General Staff, A Concise History, p. 106. This was even confirmed by a court. Hibben, op. cit., p. 57; Cosmetatos, op. cit., p. 210.
14. Falkenhausen, op. cit., p. 110.
15. Hibben, op. cit., p. 57.
16. Leon, op. cit, p. 395.
17. Hibben, op. cit., p. 58.
18. Ibidem, p. 59.

Above: Jasta 25: "Fw. Eggebrecht, named Shiekh". He had the Turkish as well as the German pilot badge as well as the "red half-moon" medal.

Above: Jasta 25: Vzfw.. Ernst Meyer stands in front of an Albatros D III with the letter "M" painted on the fuselage (late 1917).

Jasta 25: Fw. Golinski stands in front of an Albatros D III (OAW) with the letter "G" painted on the fuselage. Both photos Slg. Rolf Nagel.

Jasta 25: "Fw. Hermann Pinkert" Albatros D III (OAW), late 1917. (Vzfw. Hermann Pinkert, born 1894)

Jasta 25: Lt. Carl Dunkel, photo dedication: "Carl Dunkel, February 1st, 1918", the aircraft has a "D" painted on.

Above: Jasta 25: Fw. Schott stands in front of an Albatros D III (OAW) with the letter "S" painted on its fuselage. The caption says "Schott, April 9th, 1918". (Vzfw. Heinrich Schott, born May 27th, 1891 in Eitra, died April 25th, 1918) Both photos Slg. Rolf Nagel

19. Hellenic Army General Staff, A Concise History, p. 107.
20. Abbott, op. cit., p. 135f.
21. Hibben, op. cit., p. 59f.
22. Ibidem, p. 60.
23. Abbott, op. cit., p. 136.
24. Leon, op. cit., p. 404. Es handelte sich um Lysander Kaftanzoglou, Dimitrios Voktopoulos und Loukas Roufos. Sie waren gegen einen Kriegseintritt. Hibben, op. cit., p. 59.
25. Abbott, op. cit., p. 137.
26. Ibidem.
27. Falls, op. cit., p. 214f; Abbott, op. cit., p. 137; Constantine fully supported this suggestion. Hibben, op. cit, p. 64.
28. Falls, op. cit., p. 216; Hibben, op. cit., p. 65 Politis "had striven without avail to persuade Sir Francis Elliot that the Government was honestly in favor of leaving neutrality on the side of the Entente. [...] He had done all he could to

Above: Jasta 25: Photo dedication: "Burckhardt, Hauptmann, (signed) October 19th, 1918". His Albatros D III (OAW) does not have a letter, but rather a black band painted around the fuselage in front of the pre-1918 iron cross. (Hptm. Friedrich-Wilhelm Burckhardt, Born December 24th, 1889 in Köslin, died June 18th, 1962)

Above: EK II [Iron cross] with award ribbon.

convince the Allied diplomatists that they were committing the greatest political blunder of the war in boycotting the Calogueropoulos cabinet. When Sir Francis and M. Guilemin remained obdurate, he finally gave up his attempt and went to Saloniki, where he joined Venizelos as minister for foreign affairs."

29. Leon, op. cit., p. 403f.
30. Ibidem, p. 404.
31. Hibben, op. cit., p. 64.
32. Ibidem.
33. Ibidem, p. 61.
34. Leon, op. cit., p. 405f.
35. Ibidem, p. 407f.
36. Hibben, op. cit., p. 66.

Above Right: Jasta 25: Fw. Reinhard Treptow (Officers' deputy), born on November 28th, 1892 in Leikow.

Right: Lt.d.R. Gustav Rose.

Above: Jasta 25 at the celebration of the 50th aircraft shot down, middle of August, 1918. Squadron leader: Leutnant d.R. Renatus Heydacker (front middle), to his right probably Lt.d.R. Fritz Thiede (leader Jasta 38). Slg. Rolf Nagel.

Above & Facing Page, Above: Lt. Gerhard Fieseler achieves his fifth kill: a French Nieuport 24, piloted by Marie Eugen, is forced to land. The aircraft was repainted with 1918 German national insignia. Both photos Slg. Rolf Nagel

Facing Page, Below: Vzfw. Keller gets out of the aircraft after a flight Note the 1918-style national insignia.

243

Above: Handwritten note on the last issue of the "Nachrichten der Luftstreitkräfte Nr. 36.": "In light of the ceasefire / revolution Nov. 18, this issue is no longer registered, and was not distributed. Some pages were not written."

Increasing French Pressure

Above: In June 1918, "tank tracks" were photographed for the first time, in Greece, north of Saloniki. These "tanks" were not used in the fighting but through these experiences on the western front (Cambrai), this announcement caused consternation among German units. This photo is a sample from the collection of Falk Breuer, which was colorized by Mr. Matthias Hundt (Dorsten). The photo is from the region of Bapaume in 1917, and was originally black and white. It was important to me to show the caricature of the "surrendering German solder". The Tank was shot down.

The opinion that has been suggested over and over again that Constantine was so stubborn because he was impressed by the German successes against Rumania and had bet on the German victory, cannot be proven and is also so improbable. Constantine was a professional soldier and was able to correctly assess the importance of this local victory. The defeat of Romania did not change the overall military situation. His insistence on a controlled and not a hasty entry into the war, and on written guarantees to support the territorial demands, were sensible. He knew that without them, the Allies would not have been bound to anything. Constantine's behavior was rational, shaped by a healthy distrust, and appropriate for the situation. Venizelos and his supporters, on the other hand, relied on the vague Allied promises.

It seemed like it would take some time until a solution could be found, but then something completely unexpected happened: Venizelos shifted his position. It had been clear to him for some time that he would not be able to spark a rebellion in Athens and ancient Greece, but had delayed leaving the capital. In the middle of August, some conspirators from Crete visited Venizelos in Athens, to talk about a revolution. There, they found out that the British were still counting on reconciliation between Venizelos and Constantine. On September

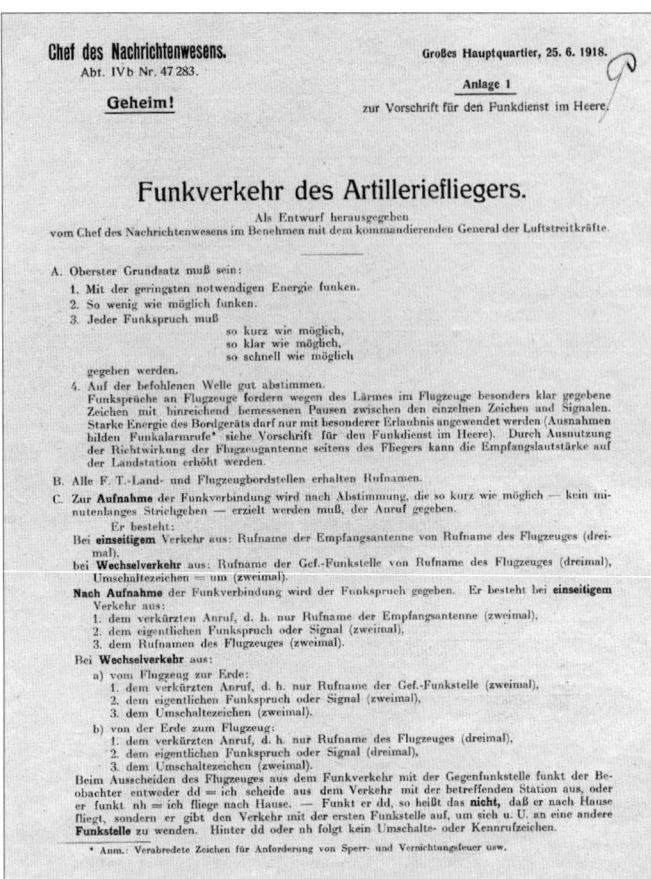

Above & Facing Page: In June 1918 a regulation was issued for the artillery pilots about how to work during a flight. The working methods of the flying personnel, when aiming at targets, should be standardized by the ground organization.

11th, the British consul in Chania showed up in Athens with a few of these conspirators. Elliot told him to stay there. It was essential to avoid any suspicion that they may have been a part of this conspiracy. That hindered the British diplomat from participating, but not the Royal Navy, also present, which supported the revolutionaries from Crete. France went one step further. On September 18th, Guillemin informed Admiral Dartige du Fournet, that Venizelos would go to Crete and a marine escort should be prepared. Venizelos cancelled at the last moment.[419]

On September 13th, Venizelos convinced Admiral Pavlos Koundouriotis to join him in establishing a provisional government and set up an army against Bulgaria. In this way, he hoped to gain the support of the Greek Navy. The popular general Panagiotis Danglis also joined them. On September 20th, AP correspondent Hibben asked Venizelos if it was true that he would go to Saloniki to lead the movement that was splitting the country. Venizelos avoided answering but admitted that the current situation could not continue. It was unbearable.[420]

Since by September 24th there was still no answer to Constantine's offer to enter the war on the side of the Allies, Constantine believed that he could bring movement into the discussion if he swapped out the cabinet. He planned to make Admiral Koundouriotis, a Venizelist and supporter of the Entente, the new prime minister. He invited him into the palace on the morning of September 25th, for this purpose. Constantine was very positive toward AP correspondent Hibben: *"He believed that he had achieved the end of the Entente's hesitation to recognize Greece as an ally. He was very sure that within a few days he would receive a positive answer to his suggestions. It seemed near impossible, that anything could stop the agreement between Greece and the Allies."*[421] But an unexpected event occurred.

On the late evening of September 25th, Venizelos left his house in Patission Street, and headed for the center of the city. He managed to shake off the policeman that was shadowing him. Near the old parliament building in Stadiou Street (today the national historical museum), he was picked up by a French vehicle flying the tricolore. The car drove down Piraeus street to the coast. In Faliron he met with Koundouriotis.[422]

On September 26th, 1916, Venizelos, Koundouriotis and Danglis and about 100 officers left Athens for Chania on Crete under the protection of a French torpedo boat.[423] In the meantime, his supporters had taken power there. The British Navy and the *Secret Service* had prepared everything so perfectly that Venizelos moved into the best hotel and, together with Koundouriotis, proclaimed a provisional government. Three days later Danglis joined the "*government*". Under the protection of Allied warships, Venizelos sailed to Samos on October 5th and from there via Mytilini, Chios, Limnos and Thasos to Saloniki. On October 9th he arrived in Saloniki, where Sarrail met him on the quay.[424]

Ancient Greece and the other islands remained loyal to the King. Koundouriotis' hope that the Greek Navy would join him was deceptive, only two torpedo boats and another small unit actually did. Danglis was also unsuccessful in getting army units to switch sides. The Venizelist movement itself was relatively small. But since England and France supported it, it gained disproportionate weight. For foreign and domestic policy reasons, Venizelos temporarily refrained from all attacks on the monarch or the dynasty. He hoped that Constantine would join him after all.[425]

Venizelos' actions caused considerable confusion on the part of the Allies in Greece. Sarrail suddenly felt that the remnants of the Greek regular army were demoralized and barely capable of action. Instead he wanted to use the Venizelist volunteers of the *Ethniki Amyna*. In his hatred of Constantine, he forgot that even 100,000 Venizelian volunteers were not as valuable as 50,000 professional soldiers in the regular Greek army. The two Allied envoys disliked Briand's unofficial suggestion that Greece should receive certain guarantees if it immediately declared war on Bulgaria. This idea would only harm Venizelos and they were of the opinion that natural developments should take their course. In addition, this could offer the King an opportunity to show a willingness for compromise, which would only harm Venizelos. It would also be better if the official Greek government declared war.[426]

After Kalogeropoulos' resignation, the King commissioned the liberal history professor Spyridon Lambros to form a government. On October 10th, the new interim government was sworn in and recognized by the Allies, but that meant little, because shortly afterwards they began to exert

Above: "British encryption key of the radio stations in 1916 until early 1917." With these keys, the German observation unit in Sofia could read parts of the British radio communications.

pressure again. The royalist reservist league was a particular thorn in the side of the Allies. At a meeting of Allied envoys on October 4th, attended by Admiral Dartig du Fournet, Guillemin suggested that the Allies should take control of the Athens police force and that the Admiral should land 300 sailors if the Greek government did not all meet the conditions imposed upon it. The league must be dissolved. Sarrail had difficulty hiring Greek volunteers because of the league. After the ambassadors met again on October 5th, they gave the government an ultimatum: By October 11th, certain Athens police officers would have to be replaced. The attackers in the garden of the French embassy would have to be arrested and punished. The centers of the reservist league would have to be closed and their meetings banned. Admiral du Fournet was responsible for enforcement. London and Petersburg disapproved of French activities and were concerned about the unlimited power of Admiral du Fournet.[427]

However, that did not prevent the French from getting to work. On October 6th, the French Naval Minister instructed du Fournet to discuss the following demands with the Allied envoys: replacement of army officers who had made negative comments about the Entente and control of the police, the railroad and the port of Piraeus, as well as other ports that would be important to the Allied forces and the Navy. Grey complained in Paris that they had not spoken to London first before making such proposals. The presence of the French fleet did not justify the admiral's arbitrary behavior. But this did not bother the Minister of the Navy, because he ordered du Fournet to carry out all measures that were necessary for the security of the *Armée d'Orient* and the Navy, even without the consent of the British. London gave in again and agreed that du Fournet could land 300 sailors if the Greeks did not comply with his demands.[428]

On October 10th, the day before the old ultimatum of October 5th expired, du Fournet presented a new one: In addition to the control of the police and the railway, he required the decommissioning and disarming (sequestration) of 30 Greek warships. The great warships *Kilkis*, *Limnos*, and *Averoff* were to deliver the breeches of their cannons; the island of Leros near Salamis was to be occupied and all coastal batteries disarmed and the breeches of the cannons removed. If the Greeks did not accept by October 11th, he would use force. This did not prevent du Fournet from immediately occupying the artillery positions, and the Greek fleet was also occupied. The British admiral on site was outraged, after all, the British naval mission he led had modernized the Greek fleet. He protested violently, but that was all. The British ships in du Fournet's fleet remained on site. Foreign Secretary Grey was outraged by these new demands; the Czar ordered that no Russian sailor should take part. An inter-allied crisis loomed, and was narrowly avoided when the Greeks accepted the demands.[429] With that, Greece was effectively an occupied country.

The demand for demobilization of the fleet was ridiculous because the fleet consisted of four old battleships (built in 1891/92) and a more modern battleship (built in 1908), as well as two cruisers (built in 1911 and 1914). Du Fournet himself admitted that this fleet posed no threat. To a certain extent he was ashamed: "*Officially we allowed the Greeks to be neutral, but we still imposed our will on national life, even on the secrets of the private lives of every Greek person. We executed the plan which the admirals assembled in Malta had rejected in March 1916.*" The reasons put forward by the French government were so winding that they were easy to see through. But Greece could only protest against them, and submit to them.[430]

The fleet was demobilized on October 11th. The crews marched on to Athens, which elicited great unrest there. On October 16th, there were protests and demonstrations against the Entente. The league reservists threatened the Venizelists. Du Fournet informed the Greek government that he would land a division of marine infantry to assist the Greek police in maintaining law and order. On October 17th, 300 marines marched from Piraeus into central Athens and occupied the town hall and the city theater. With the consent of Prime Minister Lambros, Du Fournet let mixed Franco-Greek units patrol through the streets of Athens. This provoked the Athenians even more, and in order to avoid clashes, the French troops were transferred to Zappeion. It was obvious that the Allies wanted to drive Constantine from the throne and bring Venizelos to power.[431]

The Anglo-French press portrayed the protest demonstrations as the work of the Germans. The sailors' protests were described as an attack on the Entente and the peaceful demonstrators were described as a destructive mob. This served as a justification for the intervention.[432]

The French actions were highly counterproductive: *"These French activities were very questionable in relation to international law. They were also a heavy blow to Entente sympathizers. Naval officers' move to the side of Venizelos and Koundouriotis was harshly prevented, and the British, for their part, were very upset. The Greek fleet had been formed into a modern combat unit by officers of the Royal Navy, and a rear admiral who had led the naval mission protested to the admiralty. As a result, the British units in the Allied squadron should be withdrawn from Salamis. But the British shrank away from a public break with their French ally with a deplorable weakness."*[433]

On October 17th, Constantine proposed to the British envoy, Elliot, that the size of the Greek army should be halved. The 3rd and 4th Corps, which were now in Thessaly, were to be withdrawn and only one division was to remain there to maintain law and order. In return, the Allied repression should end. He also asked for a guarantee that the rebels would not march on Athens. Elliot tentatively agreed, but wanted to consult with his colleagues. On October 20th, the Allied envoys demanded the immediate dismissal of the reservists who had served more than two years. The units should return to their peacetime locations. The units of the 3rd and 4th Corps were to be relocated from Thessaly and those of the 5th Corps from Epirus to the Peloponnese. From then on, the Allies would have to approve all transfers beforehand. The King accepted the first two demands, but refused to transfer the troops to the Peloponnese, which he called a mousetrap.[434] Apparently the units that remained loyal to the crown were to be concentrated and isolated on the Peloponnese, so that Venizelos could have free hand in the north of Greece.

Above: Bulgarian stamp from Sofia.

In the meantime, the major political and strategic differences between the Allies emerged. Falkenhayn drove the Romanians out of Transylvania and threatened Romania itself. In Paris, the idea of reinforcing the troops in Saloniki and starting a relief offensive for Romania was developed. In London, opinions were still divided. Lloyd George, who had since become Secretary of War, agreed to send eight divisions to Saloniki, but Chief of Staff Robertson refused. In his opinion, the best way to help the Romanians was to continue the Somme Offensive. At the Boulogne conference on October 20th, participants agreed that reinforcements should be sent to Saloniki. The British reserved the right to withdraw the troops at any time if they were needed more urgently elsewhere.[435]

The overall military situation at the time was not particularly good for the Allies. The Battle of the Somme brought great losses and no successes. The successes of the Brusilov offensive fizzled out and the Russians were driven further and further east. The offensive against Monastir was also not making any progress. The Romanians were on the run. Then there was the politically and militarily confusing situation in Greece, where there were now two governments. The legitimate government could hardly be completely disempowered. Recognition of the Venizelos government could not happen without the abolition of the legal government, which was rejected by the Russians and the British. In other words, the situation in Greece was completely chaotic.

In Boulogne they had also agreed on an answer to

Above: Fritz Kempf.

Above: Fritz Kempf, died on September 2nd, 1966 in Freilassing. "Could still see grandchild Per Fredrik, who was born on January 3rd, 1966 in Oslo."

the Greek proposal of September 20th to go to war. On October 23rd, the text was handed over to the Greek envoy in London. The decisive passage read: *"The royal government has offered us several times since the beginning of the war to join the conflict on our side. These offers, especially the last one, were always linked to conditions that made acceptance unacceptable to us. In either case, they did not contain any guarantees. Without indispensable guarantees, the powers will not approve of Greece's consent. Unless the royal government recognizes that, as a fact and consequence of the Bulgarian actions in Kavalla and Florina, a state of war exists between Greece and Bulgaria. This is the only way to achieve unity in Greece and win the trust and support of the Allies."*[436]

In truth, the Allies should have realized that the demand for an immediate entry into war against Bulgaria, given the military situation of Greece, could not be accepted if the King was to act responsibly. They were probably aware of it too. Had Constantine actually entered the war on the side of the Allies, they could have hardly supported Venizelos as they did. However, the Allies knew that Venizelos was ready to immediately enter the war without conditions, but Constantine was not.

By mid-October Briand had realized that the tactics used by the French diplomatic and military representatives were inadequate, and that the situation was deteriorating further. He therefore sent a person he trusted, MP Paul Bénazet, to Greece to meet Venizelos and Constantine to find a way out of the crisis. Bénazet met the King twice. At the second meeting on October 23rd, they reached an agreement on the demobilization and transfer of the remaining army to the Peloponnese. Equipping the army and the fleet should be left to the Allies for a fee. In return, the Allies would guarantee neutrality, no longer mistrust the King, loosen the measures of repression against Greece, adopt a friendlier tone in future diplomatic notes, and finally keep the whole agreement a secret. Bénazet had the impression that the King was very cooperative and willing to pay a heavy price for Greek neutrality to be respected.[437]

At another meeting with the King and the premier, the latter made the following demands in return for the surrender of all war material: repeal of the measures of June 21st, guarantees of Greek territorial integrity, a promise that Venizelos' controlled area would not be allowed to expand south, compensation for the war material surrendered to the Allies, support for the Greek territorial demands at the future peace conference, support of the King and the dynasty, refusal to recognize the Provisional Government, maintenance of the fleet left to the Allies. Bénazet largely agreed, Paris of course only partially.[438]

When Elliot and Guillemin agreed, du Fournet promised to withdraw the French troops from Athens as soon as the Greek troops were in the Peloponnese. But when a German submarine sank two Greek merchant ships, Admiral du Fournet no longer felt bound by this agreement and on November 2nd demanded the surrender of the light units of the Greek fleet and that the French forces occupy the Salamis naval station and its annexes. Before making this demand, he went to see the King, who had assured him that he would meet all demands once they were agreed with Bénazet.[439]

But what happened next proved that the French representatives did what they wanted, and that Bénazet's presence did not bother them in the least. On November 4th, Venizelos, with Sarrail's consent, ordered a battalion of the *Ethniki Amyna* to attack and occupy the city of Katerini. The battalion crossed the Aliakmon and marched on Katerini. When the commander of the rather small crew at Katerini heard of this, he sent a telegram to his superior in Larissa and asked for reinforcements. But the French officer who controlled the Greek telegraph system delayed sending the telegram until the Venizelist troops showed up. But the information about the advance had reached the command in Larissa even without the telegram, and it immediately set a battalion with machine guns and mountain artillery on rugged paths across the Olympus massif behind the attackers. When a second battalion was to be transported to Katerini by rail, the Allies refused to transport it. It then set off on foot. Since Constantine had ordered his commanders not to shed Greek blood, the national troops refrained from attacking, but let the attackers know that they were surrounded and could be put under artillery fire at any moment. The commander of the Venizelist battalion telegraphed to Sarrail for help. On November 5th, a French unit arrived and occupied the city, the train station and the roads leading south. The soldiers of the national army withdrew and the Venizelists moved in. Their march south to conquer ancient Greece was stopped.[440] But the threat of civil war was still very real.

When Bénazet found out about this, he immediately asked Venizelos to recall the battalion and Sarrail to push French troops between the adversaries in order to separate them and put an end to the fighting. Public opinion in Athens was violent. The government and the King felt that the agreement with Bénazet was being undermined. They recognized that disarmament gave the

Above: With 19 victories Lt. Gerhard Fieseler was the leading ace of Jasta 25.

Venizelists the upper hand while the Allies looked the other way. The order was given to stop the transports to the Peloponnese. Comments were made to Bénazet that the events in Katerini clearly showed that the Entente and Venizelos were collaborating. Indeed, there was such an interplay, as the following telegram from Guillemin to Briand of November 1st shows: *"If, under these circumstances, the Venizelists' attack on Katerini takes place, as declared, we should let the Greeks settle their differences themselves, without our*

Above: Lt. Gerhard Fieseler in a captured Nieuport fighter (born on April 15th, 1896 in Bergheim, died on September 1st, 1987 in Kassel).

assuming any responsibility. We must certainly not prevent the Venizelists from expanding to the borders of Thessaly. If you so wish and you are able to. In my opinion, they should wait until Army Groups C (3) and D (4) have been transferred to the Peloponnese. I would be happy to receive your approval. The Venizelists were angry about the Bénazet agreement."[441]

This meant that Guillemin knew of the impending attack on Katerini and did nothing about it. On the contrary - he supported it and tried to convince Briand of his own view of things. He knew that in doing so he was torpedoing the Franco-Greek agreement that Bénazet had concluded. But he didn't care. All he wanted was for Venizelos to gain control over large parts of Greece.

On November 6th, the Greek government rejected Admiral du Fournet's demands of November 2nd. The delivery of the light fleet units and the naval station violated Greek neutrality. Du Fournet didn't care. On November 7th he announced that he would hoist the French flag on the light units and have them manned by French sailors. In addition, he would have French troops occupy the naval station. This happened on the same day.[442]

It was clear that the King was livid. Bénazet tried to appease him. A new agreement was signed on November 7th. The disarmament of the army and its relocation to the Peloponnese remained, but every officer and soldier was free to choose whether to join the armed forces in Saloniki or not. They wouldn't be prevented from doing so. The Allies guaranteed to the King that the weapons handed over would not fall into the hands of the Venizelists. The Allies also promised to lift all repressive measures they had ordered and to put an end to acts of rebellion. After that, Bénazet left for Paris on November 7th.[443]

Shortly thereafter, the French army minister

Johannes Keller with a with Fokker D.VIIF wearing either ex-Jasta 6 or 73 striped nose markings, possibly at FEA 9.

Above: Johannes Keller as a "silhouette paper cutout" from his album and below as one of his last photos at the FEA 9 in Darmstadt, shortly before the end of the war.

arrived in Athens. On November 15th, he announced to the Greek government that he had ordered the establishment of a 5km wide neutral zone across northern Greece from the Albanian-Greek border at Leskovik to the Aegean Sea near Litochoro on Olympus. This zone will be under French sovereignty. French law would apply there and Sarrail would monitor compliance with it. The Greek government also submitted to this condition. On December 2nd, Sarrail issued the relevant orders.[444]

On November 16th, in a rather rude note to the Greek government, du Fournet demanded the surrender of a large part of the artillery and its ammunition, 40,000 *"Mannlicher rifles"* and 140 machine guns.[445] The note contained no promise that the weapons would not end up in the hands of the Venizelist troops. This was in no way in accordance with the agreement reached with Bénazet.

Apparently that did not matter to the French military. Du Fournet justified the claim by saying that Greece was by no means benevolently neutral towards the Entente, as the surrender of *Fort Roupel* and Kavalla had shown.[446]

The Greek population was indignant about this, as well as about the seizure of the fleet, after all the weapons and ammunition had been paid for with their taxes. They rightly had the impression that Greece was being disarmed. The Entente censorship prevented any information about it from being leaked outside of Greece. Hibben wrote 20 articles about it for the AP, all of which were censored. Prince Nikolaos stated bitterly: *"We are faced with the following alternatives: Handing over the weapons and ammunition of our army to one of the war participants, with which the other party to the war is then fought. This will force us into war whether we like it or not. Or we suffer from pressure that includes a delivery stop of goods from neutral states like the United States. This would lead to hunger among our people in a short time. We only have this choice: declare war or starve. There is no other option."* The Venizelists were satisfied, as the development resulted in the removal of the King from the throne, which had been their goal this whole time.[447]

In view of the events in Katerini, the demand was rejected at a cabinet meeting chaired by the King on November 19th. On the same day, Constantine met the admiral. The King pointed out that the note came from France only and contained no guarantees. He was still ready to join the war on the Entente side, but he wanted a guarantee of Greece's territorial integrity.[448] They parted ways with no result. On November 22nd, the prime minister replied officially: He stressed that the handover of weapons was a violation of neutrality and could not be accepted. On the same day the Allied Railway Control Commission announced that it would not allow arms to be transported to the Peloponnese. Briand said in a statement that the only solution for Greece was, not to enter the war on the side of the Entente, as the King had suggested, but that Constantine should transfer power to Venizelos. Constantine refused this to Guillemin: He and the Greek people believed Venizelos to be a traitor. He would not call him back, under pressure from outside of the country, but rather only if the Greek people demanded it of him. The Privy Council, the assembly of all former heads of government - naturally without Venizelos - unanimously rejected the recall of Venizelos at a meeting on November 20th.[449]

Du Fournet now obviously felt like a French

Above: Josef Jacobs with Flieger-Abteilung 11 in the front cockpit of an LVG C.II and Schwabedissen in the observer seat. Jacobs did not serve in Macedonia.

governor of Greece with unrestricted powers: In disregard of international law, he gave the envoys of Germany, Austria-Hungary, Turkey and Bulgaria, with all members of their respective embassies, until 9 a.m. on November 22nd to leave the country in which they were accredited. The diplomats protested to the Greek government, but there was nothing it could do about it. The diplomats were repatriated via Kavalla.[450] In the press of the Entente states, this was celebrated as a great victory. Now everything was headed towards a takeover by Venizelos. It was simply not understood that the Greeks' behavior was based on their own patriotism and not controlled by German propaganda.

On November 24th du Fournet issued an ultimatum: 10 mountain batteries with all accessories would have to be delivered by December 1st and the rest by December 15th. These batteries would be used to liberate occupied Greek territory. If the deadlines were not met, he would have to use

Above & Below: Lt. Walter Schwabedissen crashed Aviatik C.I upon landing on May 28th, 1916, with C. 823/16.

force. This demand meant, *de facto*, disarming the Greek government. After that, it would be easy for Venizelos to take power in Greece. And this was exactly the aim of the French on the ground, who acted without the knowledge of the government in Paris. In Athens, Du Fournet's demands were understood to mean that after the disarmament of the royal army, Venizelos would march with his troops to Athens and depose Constantine as King. The King's supporters decided to use all means to defend themselves against this. Talks between Constantin and du Fournet and between the chief of the 1st (Athenian) Army Corps, Konstantinos Kallaris, and the French military attaché were unsuccessful. On the contrary, du Fournet let the King know that he was planning landings on December 1st to occupy certain key positions in Athens until the cannons were handed over to him. Kallaris flatly refused to surrender, indicating that if the French insisted, it would lead to armed clashes.

Above & Above Right: Walter Schwabedissen.

On November 30th, the Greek government replied in the same vein.[451]

In other words: Du Fournet had enough evidence that neither the King, nor the government, nor the Greek people were ready to hand over the Greek army's weapons. According to the press telegrams that went through the French censorship every day since November 25th, du Fournet's attempt to acquire the weapons would meet with resistance. A telegram on November 30th said: *"The intention to oppose disarmament cannot be overlooked [...] tomorrow, bloodshed is highly likely."*

Constantine was quoted in another telegram: *"While one does not even think about attacking the Allies, the armament of Greece will not be surrendered, nor will anyone be permitted to acquire these weapons by force."*

The French censorship officer stopped each of these telegrams, which shows that he understood well the message they contained. One can assume that he informed du Fournet, but the latter obviously ignored them.

On November 27th, Constantine had sent a 34-page memorandum through his brother Georgios to Prime Minister Briand, in which he referred to the four times he had declared his willingness to join the Entente. But he had never received an answer. He shredded the lies fabricated by the propaganda that he was pro-German: *"He declared that neither Greece, nor he personally, had any kind of agreement, understanding or contract with the Germans."*

The claim that German submarines were being supplied by Greece was a lie, as was the claim that he broke the Constitution by dismissing Venizelos from the office of prime minister. He would also have been willing to reappoint Venizelos if he had won a majority in the December 1915 elections. If Briand had made the text public, the image of Constantin as an agent of the Germans, as, for example, the Paris newspaper Le Temps concluded on April 11th, 1917, would have vanished into thin air. [452] If this had happened, it would hardly have led to the events of December 1st, 1916.

The Noemvriana (November Events)

The following events are referred to in Greek historiography as the *Noemvriana*, the November events, even though they took place in December 1916. The reason: The Julian calendar was still being used in Greece at that time.

Robert David recalls the events on this day. At the time he served as an officer in the *Armée d'Orient*, that was stationed on the Macedonian front. David, who belonged to the PRDS (*Parti républicain démocratique et social*) later became *Sous-secrétaire de guerre* (undersecretary of war). In his memoir, which was published in 1927, he wrote: *"As I have heard, contingents of French soldiers marched out on December 1st at around 4 a.m. to Athens from Piraeus and Phaleron. They took enough provisions with them to make their soup, but only a few had wagons. A simple show of force, they were told, a stroll to see the Acropolis up close, and they'd be back by evening. Before 10 o'clock the squares were occupied, the bars put up, we went to lunch. Machine gunfire burst suddenly from the Pnyx and State Hill. Batteries had been installed on the Acropolis itself in order to attract the fire of the attacked troops, and later to raise the indignation of the civilized world against the barbarism of the attackers (!).* [453] This version of events is a typical piece of propaganda, as was the whole book.

In reality, du Fournet and the Allied governments did their best to create the impression that the landing force had wanted to conduct a peaceful demonstration and that the Greeks treacherously attacked it. The actual course of events was very different.

After Du Fournet received the answer from the Greek government, he prepared a military show of power. He put together a 3,000-man landing detachment of French, English and Italian troops under the leadership of the French captain Henri Pugliese-Conti. During the fighting that followed, a copy of the Pugliese-Conti order of the day fell into the hands of the Greek government troops. AP Correspondent Hibben reported: *"Order No. 12 was for general arrangements for a serious land campaign, with equipment for two days of combat. The ammunition issued consisted of 96 cartridges and 8 empty cartridges."*

The wounded would be sent back to Piraeus and taken aboard the *Provence*. Instructions followed as to which maps were to be used. The order also provided for bombardment by ship artillery. Order No. 13 defined the objectives of the operation: *"The landing forces must arrest themselves by force, if necessary in positions which our troops would pose as a threat to Athens. The places to be occupied are: (1) the hills dominating Athens: Nymph, Pnyx and Philopappos. (2) The Zappeion and its surroundings. Buildings to be captured are (a) a powder storage facility [...], (b) the buildings belonging to Greece's engineering corps, the Rouf [...]. (c) an ammunition factory [...]. In addition, the landing force must occupy Pirasus [sic]."* [454]

In Athens the outrage, fueled by the press, rose to a boiling point. The league reservists flocked to their weapons. On November 29th, the cabinet met with military leaders. It was decided to strengthen the Athenian garrison with troops from Corinth, Chalkis and Thebes. Lieutenant General Kallaris, the commander of the 1st Corps, took over the command. So many reservists arrived on November 30th that the Athens garrison numbered 20,000 men. In coordination with the government, Kallaris ordered all key positions and public buildings to be occupied. Any attempt by Allied troops to take the military warehouses were to be prevented. But the troops should not be the first to open fire. That afternoon the French military attaché and the King met again. Constantine protested against the humiliation and repression of the Allies, and stated that neither the people nor the army were ready to hand over their weapons to them.[455]

The military attaché informed du Fournet and advised him to reduce his demands somewhat and to demand the delivery of a few fewer cannons, in order to appease the Greeks. Du Fournet agreed and the attaché informed the King. But the King refused any weapons delivery. Du Fournet then announced that he would carry out the military show of strength the next day.[456]

The military show of strength, according to Du Fournet's orders was as follows: *"The landing force will establish itself, if necessary by force. It will take positions which our troops will pose a danger to Athens. And they will ensure the possession of military establishments or facilities of military value in the zone of occupation. The positions in question are ..."*[457]

On the same evening Constantine sent a personal representative to Du Fournet in Provence. He sent word to the admiral that any attempt to acquire weapons by force would lead to disaster. Du Fournet couldn't and didn't want to see the danger. He told some newspaper journalists: *"I will not give up any*

Right: Flight log book cover.

Below: Flight log book interior.

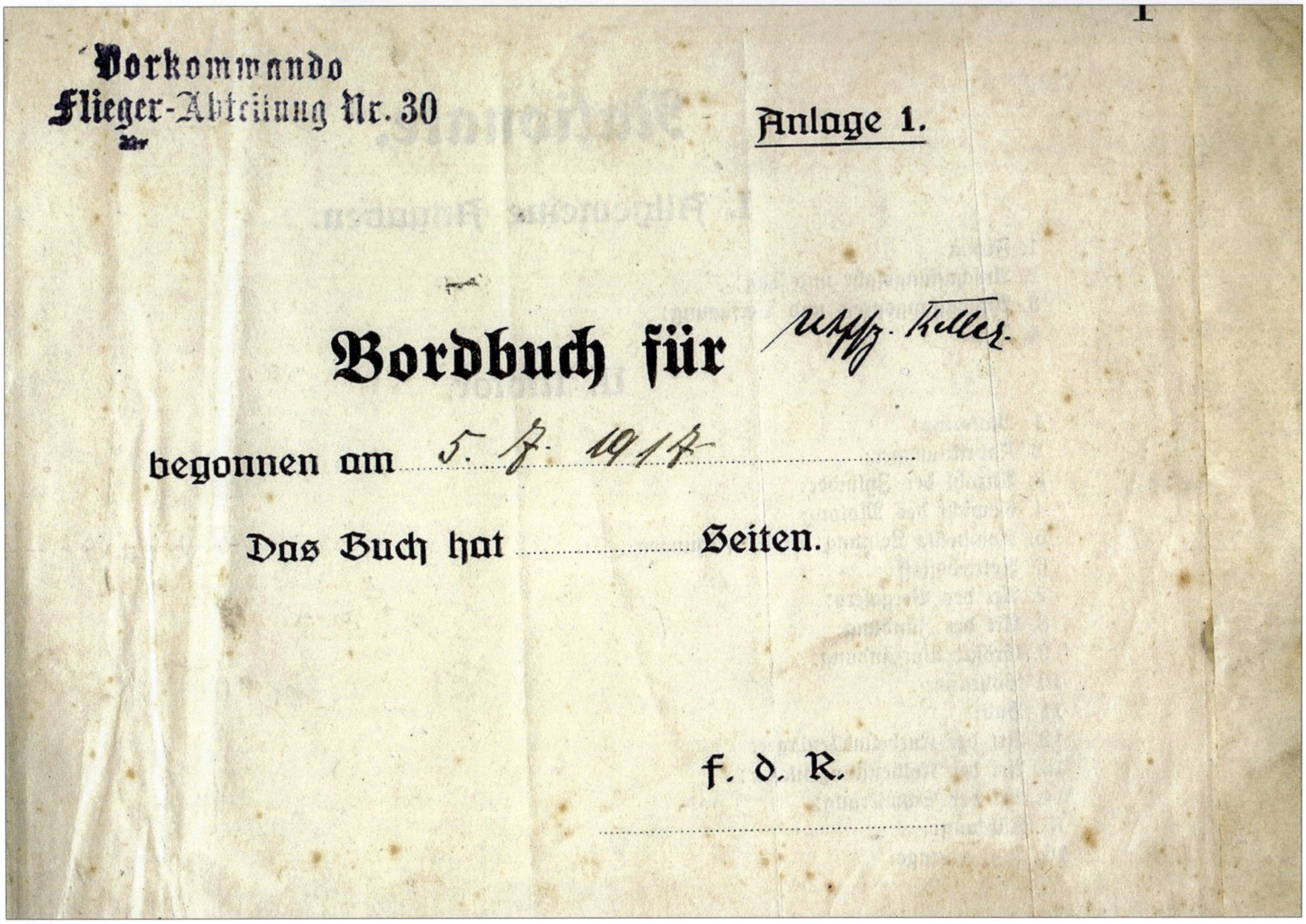

Vorkommando Flieger-Abteilung Nr. 30

Anlage 1.

Bordbuch für *Uffz. Keller*

begonnen am *5.7.1917*

Das Buch hat _____ Seiten.

f. d. R.

Nr.	Führer	Beobachter	Zweck	Flug								1. Auftrag 2. Ergebnis des Fluges 3. Besondere Vorkommnisse (besondere Bemerkungen und Erfahrungen am Flugzeug und Motor, bei Start und Landung und während des Fluges, Unglücksfälle,	Einheit
				Abflug					Landung				
				Ort	Datum	Tages-zeit	Höhe	Weg	Ort	Datum	Tages-zeit		
1	2	3	4	5	6	7	9	10	11	12	13	16	
1	Uffz. Keller	Lt. Rottka	Feindflug	Drama	7/5/17	6:30	4,500		Drama	7/5/17	8:20	Ru C Ia 4713/16	Vkdo. FA 30
2	Uffz. Keller	Lt. Rottka	Feindflug	Drama	7/8/17	10:05	5,000		Drama	7/8/17	12:40	DFW C V 5272/16	Vkdo. FA 30
3	Uffz. Keller	Lt. Rottka	Feindflug	Drama	7/10/17	4:20	5,000		Drama	7/10/17	7:35	DFW C V 2165/17, Luftkampf	Vkdo. FA 30
4	Uffz. Keller	Lt. Rottka	Feindflug	Drama	7/13/17	5:20	4,500		Drama	7/13/17	7:20	DFW C V 2165/17, Luftkampf	Vkdo. FA 30
5	Uffz. Keller	Lt. Rottka	Feindflug	Drama	7/17/17	6:00	4,500		Drama	7/17/17	8:15	DFW C V 5272/16, Luftkampf	Vkdo. FA 30
6	Uffz. Keller	Lt. Rottka	Feindflug	Drama	7/21/17	5:55	4,600		Drama	7/21/17	8:25	Ru C Ia 4713/16	Vkdo. FA 30
7	Uffz. Keller	Lt. Lenz	Feindflug	Drama	7/22/17	7:40	4,900		Drama	7/22/17	10:40	Ru C Ia 4713/16	Vkdo. FA 30
8	Uffz. Keller	Lt. König	Feindflug	Drama	7/23/17	8:00	2,000		Drama	7/23/17	9:45	Ru C Ia 4713/16, Bombenflug Thasos	Vkdo. FA 30
9	Uffz. Keller	Lt. König	Feindflug	Drama	7/29/17	9:00	4,000		Drama	7/29/17	12:20	Ru C Ia 4713/16	Vkdo. FA 30
10	Uffz. Keller	Lt. König	Feindflug	Drama	7/30/17	8:35	4,000		Drama	7/30/17	10:45	Ru C Ia 4728/16	Vkdo. FA 30
11	Uffz. Keller	Lt. König	Feindflug	Drama	8/1/17	4:05	4,400		Drama	8/1/17	7:20	Ru C Ia 4713/16, Fernaufklärung	Vkdo. FA 30
12	Uffz. Keller	Lt. König	Feindflug	Drama	8/4/17	7:25	4,000		Drama	8/4/17	10:35	Ru C Ia 4713/16, Bombenflug Thasos	Vkdo. FA 30
13	Uffz. Keller	Lt. König	Feindflug	Drama	8/11/17	5:30	3,000		Drama	8/11/17	7:25	Ru C Ia 4738/16	Vkdo. FA 30
14	Uffz. Keller	Lt. Rottka	Feindflug	Drama	8/15/17	7:05	3,200		Drama	8/15/17	9:10	Ru C Ia 4738/16	Vkdo. FA 30
15	Uffz. Keller	Lt. Rottka	Feindflug	Drama	8/22/17	12:15	4,500		Drama	8/22/17	14:35	DFW C V 2194/17	Vkdo. FA 30
16	Uffz. Keller	Lt. Rottka	Feindflug	Drama	8/23/17	7:10	4,500		Drama	8/23/17	9:10	DFW C V 2194/17	Vkdo. FA 30
17	Uffz. Keller	Lt. Lenz	Feindflug	Drama	8/24/17	7:15	4,800		Drama	8/24/17	10:00	DFW C V 2194/17, von Kampfeinsitzer-flugzeug verfolgt, [er ist in] Thasos gestartet	Vkdo. FA 30
18	Uffz. Keller	Lt. König	Feindflug	Drama	8/30/17	9:20	2,500		Drama	8/30/17	11:45	Ru C Ia 4738/16, Luftkampf	Vkdo. FA 30
19	Uffz. Keller	Lt. König	Feindflug	Drama	9/1/17	8:05	3,000		Drama	9/1/17	10:45	Ru C Ia 4738/16, im Luftkampf feindliches Flugzeug zum Landen gezwungen	Vkdo. FA 30
20	Uffz. Keller	Lt. Rottka	Feindflug	Drama	9/6/17	6:50	3,500		Drama	9/6/17	8:15	Ru C Ia 4728/16, Bombenflug Stavros	Vkdo. FA 30
21	Uffz. Keller	Lt. König	Feindflug	Drama	9/12/17	8:50	#		Drama	9/12/17	10:50	Ru C Ia 4738/16	Vkdo. FA 30
22	Uffz. Keller	Lt. König	Feindflug	Drama	9/13/17	9:00	2,800		Drama	9/13/17	11:30	Ru C Ia 4738/16	Vkdo. FA 30
23	Uffz. Keller	Lt. König	Feindflug	Drama	9/14/17	7:10	3,400		Drama	9/14/17	8:40	DFW C V 2165/17	Vkdo. FA 30
24	Uffz. Keller	Lt. König	Feindflug	Drama	9/25/17	8:10	2,500		Drama	9/25/17	9:35	Ru C Ia 4738/16, Bombenflug Thasos	Vkdo. FA 30
25	Uffz. Keller	Lt. König	Feindflug	Drama	9/30/17	7:55	4,500		Drama	9/30/17	10:30	DFW C V 2165/17	Vkdo. FA 30
26	Uffz. Keller	Lt. Lenz	Feindflug	Drama	10/1/17	7:15	2,800		Drama	10/1/17	8:50	DFW C V 2165/17	Vkdo. FA 30
27	Uffz. Keller	Lt. Lenz	Feindflug	Drama	10/3/17	7:45	3,400		Drama	10/3/17	9:45	DFW C V 2165/17	Vkdo. FA 30
28	Uffz. Keller	Lt. Lenz	Feindflug	Drama	10/6/17	8:30	3,700		Drama	10/6/17	10:25	DFW C V 2165/17. Die Richtigkeit der Eintragungen bescheinigt. Flughafen Drama, den 13.10.1917. Vorkommando Flieger-Abteilung 30. A. B. König, Leutnant	Vkdo. FA 30
29	Uffz. Keller	Lt. Asemissen	Artillerie ein-schießen	Hudova	10/17/17	11:15	5,000		Hudova	10/17/17	13:05	DFW C V 5279/16	FA 30
30	Uffz. Keller	Lt. von der Ropp	Aufklärung und Lichtbild	Hudova	10/27/17	11:15	5,000		Hudova	10/27/17	13:15	DFW C V 5279/16	FA 30
31	Uffz. Keller	Lt. von der Ropp	Auf-klärungs-flug	Hudova	11/1/17	4:45	1,400		Hudova	11/1/17	5:40	DFW C V 5279/16. Feindliche Stellung in 1 - 200 m mit MG beschossen	FA 30
32	Uffz. Keller	Lt. von der Ropp	Auf-klärungs-flug	Hudova	11/9/17	11:00	4,200		Hudova	11/9/17	12:30	DFW C V 5279/16	FA 30
33	Uffz. Keller	Lt. von der Ropp	Lichtbild	Hudova	11/14/17	10:35	4,400		Hudova	11/14/17	11:50	DFW C V 5279/16. Luftkampf mit englischen Einsitzern. Beim zweiten Angriff erhielt mein Motor einen Treffer durch den Vergaser und Gehäuse. Im Flughafen glatt gelandet.	FA 30
34	Uffz. Keller	Lt. von der Ropp	Auf-klärungs-flug	Hudova	11/19/17	10:20	4,900		Hudova	11/19/17	12:00	DFW C V 2199/17	FA 30
35	Uffz. Keller	Lt. von der Ropp	Auf-klärungs-flug	Hudova	11/24/17	11:30	5,000		Hudova	11/24/17	13:15	DFW C V 5279/16	FA 30
36	Uffz. Keller	Lt. von der Ropp	Auf-klärungs-flug	Hudova	11/26/17	11:05	4,000		Hudova	11/26/17	12:20	DFW C V 5279/16	FA 30

#												
37	Uffz. Keller	Lt. von der Ropp	Bombenflug	Hudova	11/28/17	21:40	2,500		Hudova	11/28/17	22:28	DFW C V 5279/16. Nachtflug. Bomben auf Flugplatz.	FA 30
38	Uffz. Keller	Lt. von der Ropp	Artillerie einschießen	Hudova	11/30/17	11:05	4,500		Hudova	11/30/17	14:10	DFW C V 5279/16	FA 30
39	Vzfw. Keller	Lt. von der Ropp	Aufklärungsflug	Hudova	12/6/17	#	4,200		Hudova	12/6/17	#	DFW C V 5279/16. *Die Richtigkeit der Eintragungen bescheinigt. Flughafen Hudova, den 09.12.1917. Flieger-Abteilung 30. A. B. [A.B. = Angaben bestätigt, Unterschrift unleserlich].*	FA 30
40	Vzfw. Keller	Lt. Lenz	Feindflug	Drama	1/20/18	3:21	4,300	Saloniki	Drama	1/20/18	5:25	DFW C V 3997/17, Fernaufklärung	FA 20
41	Vzfw. Keller	Lt. Wolff	Feindflug	Drama	1/23/18	2:19	3,000		Drama	1/23/18	3:25	DFW C V 3997/17	FA 20
42	Vzfw. Keller	Lt. Grube	Feindflug	Drama	1/25/18	11:35	3,500		Hudova	1/25/18	12:35	DFW C V 3997/17. Luftkampf mit B.E.-Einsitzer. Am 25.01. von Hudova zurück.	FA 20
43	Vzfw. Keller	Lt. Lenz	Feindflug	Drama	1/30/18	10:00	4,300		Drama	1/30/18	11:40	DFW C V 3997/17	FA 20
44	Vzfw. Keller	Lt. Lenz	Feindflug	Drama	2/1/18	#	4,500	Kukus	Hudova	2/1/18	#	DFW C V 3997/17. Am 03.02. von Hudova zurück.	FA 20
45	Vzfw. Keller	Lt. Fröhner	Feindflug	Drama	2/7/18	9:45	4,300	Hadzi Badimal-Laharna	Drama	2/7/18	11:45	DFW C V 3997/17. Fernaufklärung. Von feindlichen Kampfeinsitzer bis zur Front verfolgt worden.	FA 20
46	Vzfw. Keller	Lt. Fröhner	Feindflug	Drama	2/11/18	2:57	4,000	Orljak	Drama	2/11/18	4:28	DFW C V 3997/17	FA 20
47	Vzfw. Keller	Lt. Fröhner	Feindflug	Drama	2/12/18	12:16	4,300	Ofrano	Drama	2/12/18	1:45	DFW C V 3997/17	FA 20
48	Vzfw. Keller	Lt. Fröhner	Feindflug	Drama	2/15/18	2:28	3,000	Badima	Drama	2/15/18	3:57	DFW C V 3997/17. Bombenflug, Luftkampf mit Einsitzer. In den englischen Linien abgeschossen.	FA 20
49	Vzfw. Keller	Lt. von Morgen	Feindflug	Drama	2/20/18	21:41	2,000	Flugplatz Vrasla	Drama	2/20/18	23:15	DFW C V 3997/17. Bombenflug: 2 Spreng- und 2 Brandbomben, Nachtflug.	FA 20
50	Vzfw. Keller	Lt. von Morgen	Feindflug	Drama	2/21/18	#	4,500		Drama	2/21/18	#	DFW C V 3997/17. Fernaufklärung. Luftkampf mit zwei B.E. Einsitzer. Ein B.E. über englische Linien zum Absturz gebracht. Beim 2. Angriff Verwindungsseil durchschossen.	FA 20
51	Vzfw. Keller	Lt. Fröhner	Feindflug	Drama	2/25/18	#	4,500		Drama	2/25/18	#	DFW C V 3997/17, Fernaufklärung.	FA 20
52	Vzfw. Keller	Lt. Fröhner	Feindflug	Drama	2/26/18	#	3,200	Orljak	Drama	2/26/18	#	DFW C V 3997/17	FA 20
53	Vzfw. Keller	Lt. Fröhner	Feindflug	Drama	2/27/18	#	4,400	Lahana	Drama	2/27/18	12:05	Im Luftkampf mit zwei feindlichen Einsitzern Steuerung durchschossen. 28 Treffer im Flugzeug. Maschine an den Park abgegeben.	FA 20
54	Vzfw. Keller	Lt. Fröhner	Feindflug	Drama	3/11/18	6:30	4,700	Saloniki	Drama	3/11/18	9:00	Von feindlichen Einsitzern von Badima bis zur Front verfolgt.	FA 20
55	Vzfw. Keller	Lt. Fröhner	Feindflug	Drama	3/12/18	9:42	3,500	Orljak	Drama	3/12/18	11:05	DFW C V	FA 20
56	Vzfw. Keller	Lt. Fröhner	Feindflug	Drama	3/14/18	10:54	2,500	Struma	Drama	3/14/18	11:30	DFW C V. Auftrag konnte wegen Wolken nicht durchgeführt werden.	FA 20
57	Vzfw. Keller	Lt. Fröhner	Feindflug	Drama	3/17/18	12:12	2,800	Badimal	Drama	3/17/18	13:15	Bildaufnahme des Flugplatzes *Seres, ... wurde von den B.E. abgeschossen*	FA 20
58	Vzfw. Keller	Oblt. Wehmaier	Bombenflug	Drama	3/18/18	6:52	2,200	Orljak	Drama	3/18/18	8:05	Bombentreffer in einer Baracke	FA 20
59	Vzfw. Keller	Lt. Fröhner	Fernaufklärung	Drama	3/24/18	12:00	4,500	Kavalla	Drama	3/24/18	14:37	DFW C V, Fernaufklärung Saloniki	FA 20
60	Vzfw. Keller	Lt. Fröhner	Feindflug	Drama	3/26/18	8:05	2,300	Front	Drama	3/26/18	9:00	Auftrag wegen Wolken nicht ausgeführt	FA 20
61	Vzfw. Keller	Lt. Danneberg	Feindflug	Drama	4/1/18	#	3,500	Stavros	Drama	4/1/18	#	Auftrag wegen Wolken nicht ausgeführt	FA 20
62	Vzfw. Keller	Lt. Fröhner	Feindflug	Drama	4/2/18	#	4,100	Badimal	Drama	4/2/18	#	Von drei feindlichen Einsitzer verfolgt	FA 20
63	Vzfw. Keller	Lt. Danneberg	Feindflug	Drama	4/3/18	4:30	3,900	Vrasta	Drama	4/3/18	5:32	Im Luftkampf mit drei feindlichen Einsitzer Propeller zerschossen und in Ofrano notgelandet. Durch Sturzflug entkommen.	FA 20
64	Vzfw. Keller	Lt. Fröhner	Feindflug	Drama	4/11/18	9:32	4,200	Saloniki	Drama	4/11/18	11:37	Fernaufklärung. Von drei feindlichen Einsitzer angegriffen. Durch Sturzflug entkommen.	FA 20
65	Vzfw. Keller	Lt. Schiele	Feindflug	Drama	4/12/18	11:35	3,900	Vrasta	Drama	4/12/18	12:47	DFW C V	FA 20
66	Vzfw. Keller	Lt. Schiele	Feindflug	Drama	4/14/18	#	2,900	Stavros	Drama	4/14/18	#	DFW C V 9103/17, sechs Bomben	FA 20
67	Vzfw. Keller	Lt. Schiele	Feindflug	Drama	4/15/18	9:57	4,500	Orljak	Drama	4/15/18	11:57	DFW C V 9103/17, Flak bei Orljak gut	FA 20
68	Vzfw. Keller	Lt. Schiele	Feindflug	Drama	4/17/18	#	3,000	Struma	Drama	4/17/18	#	DFW C V 9103/17, sechs Bomben auf *Munibase* an der Strumamündung	FA 20
69	Vzfw. Keller	Lt. Fröhner	Feindflug	Drama	4/18/18	9:34	3,500	Struma	Drama	4/18/18	10:50	Auftrag konnte wegen fünf feindlicher Flieger über Badimal nicht ausgeführt werden.	FA 20
70	Vzfw. Keller	Lt. Fröhner	Feindflug	Drama	4/18/18	14:17	3,200	Struma	Drama	4/18/18	15:17	Über *Augusta (?)* drei feindliche Flugzeuge. *[Wir]* verschwanden sofort beim Anfliegen in den Wolken.	FA 20
71	Vzfw. Keller	Lt. Fröhner	Bombenflug	Drama	4/19/18	8:07	2,800	Golf von Orfano	Drama	4/19/18	9:18	Sechs Bomben auf 2.000 t-Dampfer im Golf von Orfano. DFW C V	FA 20
72	Vzfw. Keller	Lt. Fröhner	Feindflug	Drama	4/21/18	9:27	5,000	Orljak, Badimal	Drama	4/21/18	11:16	DFW C V 9103/17, von feindlichen Fliegern verfolgt	FA 20
73	Vzfw. Keller	Lt. Danneberg	Feindflug	Drama	4/28/18	2:14	#	#	Drama	4/28/18	2:43	Stavros, Struma: Von zwei feindlichen Fliegern verfolgt.	FA 20

			Flugbuch Johannes Keller	
	Ordner 12 Flugbuch	01 -	Umschlag	
	Ordner 12 Flugbuch	02 -	Bordbuch für Uffz. Keller	
	Ordner 12 Flugbuch	03 -	Inhaltsverzeichnis	
	Ordner 12 Flugbuch	16 -	FB Keller Seite 16	
	Ordner 12 Flugbuch	17 -	FB Keller Seite 17	
	Ordner 12 Flugbuch	18 -	FB Keller Seite 18	
	Ordner 12 Flugbuch	19 -	FB Keller Seite 19	
	Ordner 12 Flugbuch	20 -	FB Keller Seite 20	
	Ordner 12 Flugbuch	21 -	FB Keller Seite 21	
	Ordner 12 Flugbuch	22 -	FB Keller Seite 22	
	Ordner 12 Flugbuch	23 -	FB Keller Seite 23	
	Ordner 12 Flugbuch	24 -	FB Keller Seite 24	
	Ordner 12 Flugbuch	25 -	FB Keller Seite 25	
	Ordner 12 Flugbuch	26 -	FB Keller Seite 26	
	Ordner 12 Flugbuch	27 -	FB Keller Seite 27	
	Ordner 12 Flugbuch	28 -	FB Keller Seite 28	
	Ordner 12 Flugbuch	29 -	FB Keller Seite 29	
	Ordner 12 Flugbuch	30 -	FB Keller Seite 30	
	Ordner 12 Flugbuch	31 -	FB Keller Seite 31	
	Ordner 12 Flugbuch	32 -	FB Keller Seite 32	
	Ordner 12 Flugbuch	33 -	FB Keller Seite 33	
	Ordner 12 Flugbuch	34 -	FB Keller Seite 34	
	Ordner 12 Flugbuch	35 -	FB Keller Seite 35	
	Ordner 12 Flugbuch	36 -	FB Keller Seite 36	
	Ordner 12 Flugbuch	37 -	FB Keller Seite 37	
	Ordner 12 Flugbuch	38 -	FB Keller Seite 38	
	Ordner 12 Flugbuch	39 -	FB Keller Seite 39	

Nationale.

I. Allgemeine Angaben.

1. Firma:
2. Anschaffungsjahr und Tag:
3. Anschaffungspreis und Verfügung:
4. Typ:

II. Motor.

1. Motortyp:
2. Fabriknummer:
3. Anzahl der Zylinder:
4. Gewicht des Motors:
5. Nominelle Leistung bei Umdrehungen:
6. Betriebsstoff:
7. Art des Vergasers:
8. Art der Zündung:
9. Größte Vorzündung:
10. Bohrung:
11. Hub:
12. Art der Kurbelwellenlager:
13. Art der Kolbenstangenlager:
14. Art der Schmierung:
15. Kühlung:
16. Wassermenge:

of my demands and will not tolerate any resistance. I will take measures against the [Greek] government that are necessary to force compliance with my demands."[458] The admiral and French military attache completely misinterpreted the situation: They were of the opinion that the Greeks were only bluffing, and if push came to shove, they would capitulate, as they had done so often in the past.[459]

When Constantine realized in the early morning hours of December 1st that there could be no good solution, he gave the following clear orders: Under no circumstances should the Greeks open fire first. If attacked, they should use the rifle butts to drive away those who try to break through the lines they occupied. Artillery should not be used unless the attackers tried to capture predetermined locations. Whenever possible, the Greeks should surround the invaders and prevent them from attacking.[460]

AP correspondent Hibben reported on Du Fournet's further planning: Du Fournet wanted to engage the royalist troops on the periphery in skirmishes, during which Venizelist forces in the city of Athens should arm themselves and carry out a coup d'état. They were to occupy the royal palace and capture the King. Then he would be declared deposed and Venizelos summoned to take over government affairs in Athens.[461]

On December 1st, Admiral Du Fournet ordered the military operation to begin. General Kallaris had distributed four battalions in Athens in the meantime, so that the depots and certain key areas were occupied. He concentrated the remaining forces in reserve in the area of Chalandri and Kifissia. The main positions of the Greeks were near a powder factory on the *Iera Odos*, near the Dafni monastery, the botanical gardens, near the Rouf barracks, near the observatory and on the Philopappos and the Ardettos hills.

The three Allied battalions with a total of 3,000 men advanced in three columns towards Athens at around 3 a.m. The Greek security forces backed away. When the Allied columns reached the positions they were supposed to occupy, they discovered that these were already occupied by

Greek troops who refused to vacate them. Pugliesi-Conti then ordered all units to assemble in front of the Rouf barracks, where he and the 2nd French battalion had already arrived.[462]

At around 9 o'clock the situation was as follows: Allied units had driven the Greeks from the positions near the powder factory and were advancing along the *Ieros Odos*. Two French companies had encircled the Rouf barracks and captured 30 volunteers there. The barracks commander, on the other hand, managed to capture the French unit's supply column. The French protested against this, and the Greek commander suggested an exchange of prisoners, which did not occur. The reinforced 2nd French Battalion had meanwhile advanced to the hill of the observatory. The 3rd Battalion marched towards the Zappeion and the powder storage facility near the cemetery.

For two hours the troops faced each other, without a single thing happening. But at 10 o'clock the first shots were fired near the Rouf barracks. There was no specific reason, and to this day it is unclear who fired the first shots. Afterwards, of course, each side blamed the other. It can't even be ruled out that a

20	k.u.k. Flieger Abtl. 30. Drama												
				Abflug					Flug		Landung		
Lfd. Nr. des Fluges	Führer	Beobachter	Zweck	Ort	Datum	Tages- zeit	Temperatur, Wind- richtung, Stärke	Höchste Höhe über dem Abflug- ort	Weg	Ort	Datum	Tages- zeit	km
1	2	3	4	5	6	7	8	9	10	11	12	13	14
13.	Uffz Keller	L. König	Feindflug	Drama	11.8	5:20		3000			11.8	7:25	
14.	" Keller	L. Rotthas	"	"	15.8	7:05		3200			15.8	9:10	
15.	" Keller	"	"	"	22.8	12:15		4500			22.8	2:35	
16.	" Keller	"	"	"	23.8	7:10		4500			23.8	9:10	
17.	" Keller	L. Lenz	"	"	24.8	7:15		4800			24.8	10:00	
18.	" Keller	L. König	"	"	30.8	9:20		2500			30.8	11:45	

group of fanatical Venizelists or royalist reserves shot first. The exchange of fire jumped to the observatory and the powder storage facility. Near the Philopappos hills, the Greeks sustained heavy fire, and many losses. At 2pm a French unit surrendered at the observatory. The French troops were locked in at the Rouf barracks. An Italian company deployed there withdrew on the orders of the Italian envoy,

22	k.u.k. Flieger Abtl. 30. Drama								
				Abflug					Flug
Lfd. Nr. des Fluges	Führer	Beobachter	Zweck	Ort	Datum	Tages- zeit	Temperatur, Wind- richtung, Stärke	Höchste Höhe über dem Abflug- ort	Weg
1	2	3	4	5	6	7	8	9	10
19.	Uffz Keller	L. König	Feindflug	Drama	1.9	8:05		3000	
20.	" Keller	L. Rotthas	"	"	6.9	6:50		3500	
21.	" Keller	L. König	"	"	12.9	8:50			
22.	" Keller	"	"	"	13.9	9:00		2800	
23.	" Keller	"	"	"	14.9	7:10		3400	
24.	" Keller	"	"	"	25.9	8:10		2800	

o. Drama   21

| | Flug | | | | Landung | | | | 1. Auftrag (kann beigeheftet werden). |
Tages- zeit	Tem- peratur, Wind- richtung, Stärke	Höchste Höhe über dem Abflug- ort	Weg	Ort	Datum	Tages- zeit	km	Durch- schnitts- geschwin- digkeit	2. Ergebnis des Fluges (kann auf Meldekarte, Skizze oder Kartenausschnitt beigeheftet werden). 3. Besondere Vorkommnisse (besondere Bemerkungen und Erfahrungen am Flugzeug und Motor, bei Start und Landung und während des Fluges, Unglücksfälle, Instand- setzungen, Erfahrungen mit Abwurfvorrichtung, Waffen).
7	8	9	10	11	12	13	14	15	16
5³⁰		3000			11.8	7²⁵			R.E.I.a 4738/16.
7⁰⁵		3200			15.8	9¹⁰			" " " "
12¹⁵		4500			22.8	2³⁵			D.F.W. C.V. 2194/17.
7¹⁰		4500			23.8	9¹⁰			D.F.W. C.V. 2194/17.
8¹⁵		4800			24.8	10⁰⁰			" " " (Luftk.) v. Kampfeins. verfolgt u. z. Weg. flugzeug gez.
9²⁰		2500			30.8	11⁴⁵			R.E.I.a 4738/16. (Luftkampf.)

23

| Landung | | | | | 1. Auftrag (kann beigeheftet werden). |
Ort	Datum	Tages- zeit	km	Durch- schnitts- geschwin- digkeit	2. Ergebnis des Fluges (kann auf Meldekarte, Skizze oder Kartenausschnitt beigeheftet werden). 3. Besondere Vorkommnisse (besondere Bemerkungen und Erfahrungen am Flugzeug und Motor, bei Start und Landung und während des Fluges, Unglücksfälle, Instand- setzungen, Erfahrungen mit Abwurfvorrichtung, Waffen).
11	12	13	14	15	16
	1.9.	10⁴⁵			R.E.I.a 4738/16.
	6.9.	8¹⁵			R.E.I.a 4738/16. Bombf. Stavros Von Luftkampf herv. u. Flugz. zum landen gezwungen
	12.9.	10⁵⁰			R.E.I.a 4738/16.
	13.9.	11³⁰			" " " "
	14.9.	8⁴⁰			D.F.W. C.V. 2165.
	25.9.	9³⁵			R.E.I.a 4738/16. (Bomben Thasos)

Lfd. Nr. des Fluges	Führer	Beobachter	Zweck	Ort	Abflug Datum	Tageszeit	Temperatur, Windrichtung, Stärke	Höchste Höhe über dem Abflugort	Flug Weg
1	2	3	4	5	6	7	8	9	10
25.	Uffz Keller	L. König	Feindfl	Drama	30.9.	7⁵⁵		4500	
26.	" Keller	L. Lenz	"	"	1.10.	7¹⁵		2800	
27.	" Keller	"	"	"	3.10.	7⁴⁵		3400	
28.	" Keller	L. Lenz	Feindfl	Drama	6.10.	8³⁰		3700	
29.									
30.									

who refused to use force.

In the meantime, a mixed unit of French and British, about 500 men, armed with machine guns, tried to make their way from Zappeion to the powder storage facility near the cemetery. The small Greek unit there was surrounded by the French troops of the 3rd Battalion, and scrappily defended itself. When two Greek companies arrived from the Ardettos hills, there was a violent clash including a bayonet attack, which ended with the escape of most of the Allies in the direction of the Bay of Faliron. Only one British company could hold out under the heavy siege. The crew of the Zappeion itself was surrounded.

As the first shots were fired, King Constantine called the Allied flagship *Provence* and demanded to speak to Admiral Du Fournet, in order to reach a ceasefire. But the Admiral was in Zappeion and was not informed. He was a quasi-prisoner of the Greek army. Toward the evening, Du Fournet met with the Russian envoy, who relayed the King's demand and offered the delivery of six highland battalions under the condition that the Allied landing corps retreat.

At about the same time, Kallaris, who feared that a larger battle could develop in the city center of Athens, sent an officer to De Fournet, who demanded he call back Allied troops to Piraeus. In light of the losses, the admiral was prepared to order a ceasefire. This followed, and both sides initially held to it.

But at around 4:30 p.m. a shootout between Venizelists and the royal troops broke out near Zappeion. The troops which had surrounded Zappeion ended up in the crossfire as the French also began to shoot. There were larger losses. The commander of a mountain battery on the Ardettos Hills, who was observing the situation, fired ten shells at the plaza in front of the entrance to the Zappeion without receiving orders from his superiors. Du Fournet, in coordination with the French military attaché, ordered by telephone that the fleet bombard the area, which happened immediately. 12.5 cm- and 25cm-caliber shells struck the free city over the next few hours. Most of the grenades exploded near the stadium and near the royal palace. A shell hit the courtyard of the palace, forcing the queen and her children to take shelter in the cellar. Around 6 p.m., when the castle was still under fire, the Entente envoys went to see the King to discuss a treaty. At 6.45 p.m., accompanied by the Deputy Chief of Staff of the Greek Army, they met with the Admiral in Zappeion and assured him that the six batteries would be delivered when the fire was stopped. This happened a little later and the retreat of the Allied forces to Piraeus began, observed by a Greek cavalry unit. At 5:30 a.m. on December 2nd, the withdrawal was complete.

Many wounded Allied soldiers were left behind and brought to the hospitals in the city. The French had 60 dead, including 6 officers, and 154 wounded. The Greeks had 30 dead, including 4 officers, and 52 wounded. The numerous prisoners on both sides were exchanged after the completion of the retreat. 11 civilians were killed by the Venizelists, and 12 wounded. The latter had three dead and two wounded. The Allies gave up control of the police, the post and telegraphic services and the railroad. With the Allied soldiers, the agents of the Entente also retreated from the inner city, which they had terrorized for a year.[463]

As mentioned earlier, the Venizelists interfered in the fighting. They shot at Greek troops marching past from apartments, balconies and rooftops. After all, their interference was so great that Admiral Du Fournet spoke in his memoirs that his troops had been involved in a civil war.[464] Even after the Allies withdrew, these disputes continued: *"On December 2nd, various battles took place in many parts of Athens; suspicious homes, offices and shops were attacked with murderous fury and defended. M. Venizelos' house was fittingly at the center of the conflict. 20 loyal Cretans had barricaded themselves in the house and held out until they were convinced by machine guns to give up. A small store of rifles,*

26 Flieger Abteilung 30. Hudova

Lfd. Nr. des Fluges	Führer	Beobachter	Zweck	Abflug Ort	Abflug Datum	Abflug Tageszeit	Temperatur, Windrichtung, Stärke	Flug Höchste Höhe über dem Abflugort	Weg
1	2	3	4	5	6	7	8	9	10
29.	Uoffz. Keller	L. Asemissen	Arkl. Einschieß.	Hudova	18.X.17	11¹⁵		4-5000	
30.	" Keller	Lt. v.d. Ropp	Aufkl. u. Licht.	Hudova	23.X.17	11¹⁵		4-5000	
31.	" Keller	" v.d. Ropp	Inf.Fl.	Hudova	1.XI.17	4⁴⁵		3-1400	
32.	" Keller	" v.d. Ropp	Aufkl.	Hudova	9.XI.17	11⁰⁰		4200	
33.	" Keller	Lt. v.d. Ropp	Lichtb.	Hudova	14.XI.17	10³⁵		4400	
34.	" Keller	Lt. v.d. Ropp	Aufkl.	Hudova	19.XI	10²⁰		4900	
35.	" Keller	Lt. v.d. Ropp	Aufkl.	Hudova	24.XI	11³⁰		5000	
36.	" Keller	Lt. Temme	Aufkl.	Hudova	26.XI	11⁰⁵		4000	

revolvers, hand grenades, dynamite sticks and fuses was discovered in the house. In between, some weapons that were still wrapped in the French linen in which they had arrived. Countless similar things were discovered on the property by other conspirators. The fighters were taken to prison. They were followed by an angry crowd who yelled at, spat at, and cursed the conspirators. Those accompanying them had great difficulty protecting them from being lynched."[465]

Journalists from the Allied countries turned these stories into legends of massacres and tortures, to incite the official opinion of their countries.[466] There were alleged to be 35 dead, 922 people locked up, and 503 cases of looting. The damage was allegedly almost 7 million drachmas. The publisher of the Venizelist newspapers *Estia* and *Ethnos* were arrested and the presses destroyed.[467]

For the Allied media, these incidents were a godsend. They blew up the incidents beyond all proportion and mixed them up with lies. A typical example was the case of the Venizelist Mayor of Athens, Imanouil Benakis. He was treated a bit roughly during the riots.[468] A French newspaper turned this into a story that he was so mistreated that his right arm had to be amputated. But a few days later, Benakis hand-wrote a letter to Constantine, in which he thanked him for his help. When Constantine demanded a denial from the French newspaper, it did not answer him. This happened to him all the time.[469]

Allied propaganda tried to absolve Venizelos of all guilt, but there is at least one piece of concrete evidence, a letter, that he had had contact with his supporters in Athens: *"... the search of the houses of the Venizelists has produced solid evidence, in the form of a letter from the leader to one of his followers. This states, among other things, that a definitive agreement has been reached between him and the representatives of the Entente. It assures the swift domination of Athens by the full strength of the Entente. The publication of this document,*

Ort	Landung Datum	Tageszeit	km	Durchschnittsgeschwindigkeit	16
	12	13	14	15	
Hudova	17.X.12	1:05			D.W. C.V. 5279/16 (4)
Hudova	28.X.12	1:15			D.F. C.V. 5279/16. (4)
Hudova	1.XI.12	5:40			D.F.W. C.V 5279/16 (4) 2000 m. mit M.G. beschossen. Feindl. Stelung in [...]
Hudova	9.XI.12	12:30			D.F.W. C.V. 5279/16 (4)
Hudova	14.XI.12	11:50			D.F.W. C.V 5279 (4) /16 Luftkampf mit engl. Einsitzer, beim zweiten Angriff er hielt [...] Motor [...] Treffer durch den Vergaser und Gehäuse. Im Flughafen glatt gelandet.
Hudova	19.XI.12	12:00			D.F.W. C.V. N° 2199/12 (5)
Hudova	24.XI	1:15			D.F.W. C.V. 5279/16.
Hudova	26.XI	12:30			D.F.W. C.V. 5279/16.

with an original photo, confirmed the fears that had long prevailed in public opinion. Neither Venizelos' indignant rejection of the authenticity of the document, nor the emphatic affirmation of the Entente ministers that they had never done anything to support his return after the Cretans left Athens, shook the conviction that the coup on December 1st was planned.[470]

In this letter Venizelos wrote: "One obstacle to our last endeavors was the Allies' military weakness in the Balkans. [...] The essential point [...] must be the absolute conviction that the Entente [...] approves our movement and strengthens it through substantial and active support. To the extent that our ultimate domination over Athens and constitutional Greece from the full weight of the Entente which will break that state, is only a matter of time. After all that, what remains of the glorious King [...]? Not even a shadow of his former self. His authority is reduced to fragments, one concession after another. His war teeth were pulled out one by one. The specter of hunger and suffering is already very prominent in ancient Greece. It will be even more terrible once a new and efficient blockade is put in place. The soul of the people has already reached the limit of human endurance. When I decided to take responsibility for the political part of Greece, I was not so foolish as to believe that our great national and political venture would be crowned with success in one, or even several months. I know very well that the confusion in popular opinion, caused by the courage and surprise of the effort, and the prejudices against me and the German agents [...], so long cultivated by the reservists, as well as [...] the blind and total idolization of the people against the person of the generalissimo, can be a heavy blow - that is threatening - but it will be enough to end the risk. I must stress here that we have already reached a final agreement with the representatives of the

Entente."[471]

Admiral du Fournet briefed Paris on the incidents and asked for permission to bomb Athens in order to teach the Greeks a lesson. He asked for two divisions to be sent to occupy Athens and Piraeus. This telegram caused great excitement in Paris, because shortly before its arrival, Romania had collapsed and now there was fear that the *Armée d'Orient* in Saloniki would be threatened from the south. The 16th French colonial division was immediately ordered to Piraeus.

In several telegrams to his minister and to Sarrail the French naval attaché in Athens proposed immediate and severe reprisals. Sarrail responded by asking Paris for permission to march to Athens via Larissa. On December 4th, Paris ordered Sarrail to prepare an offensive against Athens. Sarrail, du Fournet and the naval attaché continued to exert pressure, but in the end Paris did not give its permission because the British and also the French envoy and the military attaché opposed it. Sarrail moved the Italian 35th Division to Kozani and a British brigade to Katerini at least, in order to secure the rear of his army.

When the Greek army minister, in a daily order, praised his troops for their behavior in the fighting in Athens, which had been provoked by the enemies of the crown and the legal government, the French interpreted this as a provocation. On December 6th the French government decided to depose King Constantine and recognize Venizelos as the legal ruler of Greece. Since the British completely rejected this, Paris was forced to give up that project for the time being. But it did not prevent the government in Paris from transferring two divisions to Salonika and asking England and Italy to do the same. Both refused.

On December 8th, Du Fournet carried out an order of his government and declared a blockade of ancient Greece and Thessaly. At the same time, he prepared to bomb Athens. Citizens of Allied states and most of the embassy staff were transferred to Allied ships in Keratsini Bay, off Salamis. The ambassadors of Russia and Italy refused to evacuate their nationals. In their eyes the whole thing was a "weird comedy."

The Greek government protested against the blockade to the diplomatic representatives of the neutral states, saying it had simply been imposed for no reason: *"Greece, at peace with the Entente, has never ceased to give extraordinary evidence of its firm intention to maintain mutual friendships. We are painfully surprised that the same powers are now resorting to measures against Greece that are manifestly contrary to international law and the principles of international rights and freedoms. [...] The government can only formulate the most vigorous and legitimate protest against the use of such measures against a peaceful, neutral people."*[472]

Constantine told the British and Russian ambassadors that he was ready to accept any military agreement that did not pursue political aims: *"He said categorically that he had no further intention of attacking the Allies or declaring war, an assurance he had already given to Lord Kitchener. None of his actions as commander-in-chief of the Greek armies pursued any other purpose than the legitimate defense of Hellas against an invasion of the Venizelists from the north or by the Allies themselves from the sea. [...] Even if the policy chosen by the Entente, starving the Greek nation into submission, forced him to establish communication with the Central Powers in order to secure the necessary food for the people, he would not attack Sarrail's positions in Salonika."* To prove this, he was ready to move his troops southward from Thessaly. With this declaration he hoped to remove the mistrust of the Allies.[473]

The Allies agreed that the Greek army should be transferred to the Peloponnese. They refused to break off diplomatic relations, as France requested. Military action against Greece was also refused and the French government was asked to make this unequivocally clear to Sarrail, which it did. Finally, Admiral Du Fournet was replaced by Admiral Dominique Gauchet.

Lfd. Nr. des Fluges	Führer	Beobachter	Zweck	Abflug Ort	Datum	Tageszeit	Temperatur, Windrichtung, Stärke	Höchste Höhe über dem Abflugort	Flug Weg
1	2	3	4	5	6	7	8	9	10
40	V.F. Keller	L. Lenz	Feindflg	Drama	20.I.	3²¹	/	4300	Saloniki
41	V.Fldw. Keller	L. Wolff	Feindflg	Drama	23.I.	2⁴⁹	/	3000	
42	" Keller	Lt. Gruber	Feindflg	Drama	25.I.	11³⁵	/	3500	
43	" Keller	Lt. Lenz	Feindflg	Drama	30.I.	10⁰⁰	/	4300	
44	" Keller	Lt. Lenz	Feindflg	Drama	1.II.		/	4500	Kutkus Hadzi
45	" Keller	Lt. Frühner	Feindflg	Drama	3.II.	9⁴⁵	/	4300	Bajrem Drama

On December 14th, the envoys of the Entente states made the following demands to the Greek government: The Greek army must retreat to the Peloponnese with all its equipment. Checkpoints will be set up in Corinth and Patras to monitor all troop movements. The withdrawal would have to begin within 24 hours, otherwise Greece would be considered a hostile country. The blockade of Greece would be maintained until these conditions were met and appropriate guarantees were given for the future. *"The blockade of the Greek coasts will remain in place until the Hellenic government has made full reparations for the last attack that was carried out by the Greek troops on the Allied troops in Athens without prior provocation."*[474]

Admiral Gauchet was also ordered to stop any movement of troops from the Peloponnese to the mainland. The blockade ordered by Du Fournet was continued by his successor.

The Greek government had no other choice but to accept the conditions, and took the appropriate action. Greek officers felt humiliated and angry. The Athenian newspapers were hateful to the Allies. The French military in Athens and Salonika used this as an opportunity to inform the government in Paris of this in an *"appropriate manner"*.

If the British barely appeared in those crucial days, it had to do with the government crisis in London. On December 4th, 1916, Premier Asquith announced his resignation. Lloyd George succeeded him on December 7th. Arthur Balfour replaced Grey as Secretary of State. The *War Committee*, now called the *War Cabinet*, did not meet until December 9th. They endorsed the French blockade decision, which was to be upheld until the Greek government compensated the December 1st victims.[475]

The newspapers in Paris and London agitated, based on fictitious stories from Naval Attaché De Roquefeuil.[476] Alluding to the *"Sicilian Vespers"* of 1282, there was talk of an *"Athens Vespers"* or a *"Bartholomew Night"* of the Venizelists, which of course was completely exaggerated in every respect.

[477] Constantine was blamed to have deliberately lured the Allies into a trap. *"But the shooting around Zappeion destroyed the last vestiges of respect for Constantine, both in London and in Paris. Nobody believed him, not even his royal cousin [...] All of Constantine's apologies were immediately rejected."*

The British and the French decided to recognize the Venizelos government. They pulled their envoys from Athens. They allowed the King to remain on his throne for the moment, but the British and French newspapers described him as an even more evil figure than his royal brother in law in Berlin. It was the irony of fate that Du Fournet, a French aristocrat, damaged the Greek throne even more than the revolutionary from Crete in Saloniki.[478]

When the *London Times* correspondent wrote a more objective account, in which he determined that Venizelos was unlikely to be able to lead the Greek people, the article was accompanied by text that gave the impression that the correspondent had been put under pressure. He was fired because his article was inconsistent with UK government policy.[479]

The King's pliability, which angered his followers, was condemned by his enemies in the Allied camp as another example of his duplicity: *"They confirmed [...] that this great deceiver, together with the Emperor, was making secret preparations for war against the Allies. He intended to lull them into a false sense of security by the appearance of submission. Extreme measures are therefore needed, not only to punish him for his past acts, but also to prevent Greece from becoming the base of hostile activity in the near future."*[480] A glance at the actual military situation shows the entire absurdity of these accusations.

When Constantine demanded the establishment of a neutral commission to investigate the events of December 1st and thereafter, the Allied censors made sure that this demand was not made public. The Allied public should only hear what was intended for it, namely that Constantine was a liar, a traitor and a murderer.[481] The French high command feared - for no reason - that the Greek army might launch an attack on the Allied troops in Saloniki. It therefore demanded that the entire royal family and the political and military leadership be arrested and banished to an island where they should be strictly monitored.[482]

In a telegram to the Allies on December 4th, Briand demanded redress: *"The French government is of the opinion that the question of a declaration of war against Greece does not arise, since parts of the Greek nation are fighting on the side of the Allies. But the Allies should demand compensation for the crimes of which they have been victims as well as guarantees for the future. Reparation could be achieved through the abdication of King Constantine [...] who is completely under the influence of his pro-German companions, and the recognition of Venizelos.* A day later the Russian government replied: *"We believe that the French proposals for regime change in Greece, carried out by foreign bayonets, are inappropriate. They are unfeasible in view of the army's and nation's clearly expressed sympathies for the King."* The Italian and English governments agreed. Faced with this opposition, Briand dropped the proposal on December 7th.[483]

At the inter-allied conference in London on December 26th, 1916, in which no military advisers took part, French and English ministers discussed the question of whether to withdraw from Monastir, but did not reach a conclusion. The next day, military and political issues were discussed. We shall come to the former in connection with the end of the autumn offensive. The latter included on the one hand the future Greek form of government, and on the other hand individual demands. In contrast to the French, the British were more in favor of maintaining the monarchy in Greece and believed that coercive measures would only turn the Greeks against the Entente. Final decisions should be made at the Rome conference, scheduled for early January.[484]

Included among the political demands were the following: Meetings of the Reservist League should be banned, all Allied controls restored and all political prisoners released - meaning the Venizelist and the commander of the 1st Corps, General Kallaris, should be retired. Greece must officially apologize to the Allies and honor the fallen Allies in a flag ceremony. On December 31st, the Allied envoys demanded compliance with these demands. In addition, the railway line from Saloniki to the Brallos Pass south of Lamia and the road from Brallos to Itea on the Gulf of Corinth should be available to them. Troops of the Venizelist *Enthniki Amyna* should not be allowed to cross the buffer zone.[485]

The Greek government protested violently. These measures undermined the possibility of self-defense, allowed the revolutionary movement to expand southward, and transferred the rights of the King and government to the Venizelists. The Venizelists and the Allies had more prisoners than the government. The demand for the prisoners' release violated neutrality and the Greek constitution, because they had been legally sentenced. Finally, the government demanded that the blockade be lifted. The French military felt attacked by the clear language and demanded tougher action.[486] These themes were discussed further at the Conference in Rome, on January 5th.

34 Flieger Abt. 20 Drama

Lfd. Nr. des Fluges	Führer	Beobachter	Zweck	Abflug			Temperatur, Windrichtung, Stärke	Flug		
				Ort	Datum	Tageszeit		Höchste Höhe über dem Abflugort	Weg
1	2	3	4	5	6	7	8	9	10
52 v.F.	Keller	Fröhner	Feindfl.	Drama	26.II			3200	Buljak
53 "	Keller	Fröhner	Fernfl.	Drama	27.II			4900	Kavalla
54 "	Keller	Fröhner	Fernfl.	Drama	11.III	6³⁰		4800	Salonik
55 "	Keller	Fröhner	Feindfl.	Drama	12.III	9²²		3500	Buljak
56 "	Keller	Fröhner	Feindfl.	Drama	14.III	10⁵⁴		2500	Drama
57 "	Keller	Fröhner	Feindfl.	Drama	17.III	12¹²		2800	Tachi-See
58 "	Keller	Obltn.Wehmeyer	Bombfl.	Drama	18.III	6⁵²		2200	Buljak

36 Flieger Abt. 20 Drama

Lfd. Nr. des Fluges	Führer	Beobachter	Zweck	Abflug			Temperatur, Windrichtung, Stärke	Flug		
				Ort	Datum	Tageszeit		Höchste Höhe über dem Abflugort	Weg
1	2	3	4	5	6	7	8	9	10
59 v.F.	Keller	L. Fröhner	Fernaufkl.	Drama	24.III	12⁰⁰		4500	Kavalla
60 "	Keller	L. Fröhner	Feindfl.	"	26.III	8⁰⁵		2300	Front
61 "	Keller	L. Daneberg	Feindfl.	Drama	1.IV	8		3500	Stavros
62 "	Keller	L. Fröhner	Feindfl.	Drama	2.IV			4100	Badis
63	Keller	L. Daneberg	Feindfl.	Drama	3.IV	9³⁰		3900	Prasta
64 "	Keller	L. Fröhner	Feindfl.	Drama	11.IV	9³²		4200	Salonik
65 "	Keller	L. Scholl	Feindfl.	Drama	12.IV	11³⁵		3900	Prasta
66 "	Keller	L. Fröhner	Feindfl.	Drama	14.IV			2900	Stavros
67 "	Keller	L. Scholl	Feindfl.	Drama	15.IV	9⁵³		4500	Buljak

35

	Landung		km	Durch-schnitts-geschwin-digkeit	1. Auftrag (kann beigeheftet werden). 2. Ergebnis des Fluges (kann auf Meldekarte, Skizze oder Kartenausschnitt beigeheftet werden). 3. Besondere Vorkommnisse (besondere Bemerkungen und Erfahrungen am Flugzeug und Motor, bei Start und Landung und während des Fluges, Unglücksfälle, Instand-setzungen, Erfahrungen mit Abwurfvorrichtung, Waffen).
Ort	Datum	Tages-zeit			
11	12	13	14	15	16
Drama	26.II				D.F.W. C.V. 3992
Drama	3.III	12:05			Im Luftkampf mit feindl. Eins. Steuerung durchgeschossen. 28 Treffer im Flugzeug. Sind an Start abgegeben.
Drama	11.III	9:00			Von feindl. Eins. von Bad... bis zur Front verfolgt.
Drama	12.III	11:05			D.F.W. C.V.
Drama	14.III	11:30			D.F.W. C.V. Auftrag konnte wegen Wolken nicht...
Drama	17.III	1:15			...
Drama	18.III	8:05			Bombenabw. in eine Baracke.

37

	Landung		km	Durch-schnitts-geschwin-digkeit	1. Auftrag (kann beigeheftet werden). 2. Ergebnis des Fluges (kann auf Meldekarte, Skizze oder Kartenausschnitt beigeheftet werden). 3. Besondere Vorkommnisse (besondere Bemerkungen und Erfahrungen am Flugzeug und Motor, bei Start und Landung und während des Fluges, Unglücksfälle, Instand-setzungen, Erfahrungen mit Abwurfvorrichtung, Waffen).
Ort	Datum	Tages-zeit			
11	12	13	14	15	16
Drama	24.III	2:32			D.F.W. C.V. Fernaufkl. Salonicki
Drama	26.III	9:00			Auftrag wegen Wolken nicht ausgeführt. nicht ausgeführt.
Drama	2.IV				Von Jeindl. Eins. verfolgt.
Drama	4.IV	5:32			Im Luftkampf mit 3 feindl. Eins. 1 Doppeldecker geschossen mis in Flamm...
Drama	11.IV	11:32			Fernaufkl. Von 3 feindl. Einsitzern ange-griffen durch Kurzschluß mit ...
Drama	12.IV	12:42			D.F.W. C.V.
Drama	14.IV				D.F.W. C.V. 9103 Bomben 6 Stück.
Drama	15.IV	11:52			D.F.W. C.V. 9103 Flak beschädigt.

38 Flugz. Abt. 20. Drama

Lfd. Nr. des Fluges	Führer	Beobachter	Zweck	Abflug Ort	Datum	Tageszeit	Temperatur, Windrichtung, Stärke	Höchste Höhe über dem Abflugort	Flug Weg
1	2	3	4	5	6	7	8	9	10
68	Vzf. Keller	Lt. Schiele	Ferndfl.	Drama	13.IV			3000	Struma
69	" Keller	Lt. Fröhner	Ferndfl.	Drama	18.IV	9³⁴		3500	Struma
70	" Keller	Lt. Fröhner	Ferndfl.	Drama	18.IV	2¹²		3200	Struma, Golf von
71	" Keller	Lt. Fröhner	Bombfl.	Drama	19.IV	8⁰⁷		2800	Orfano
72	" Keller	Lt. Fröhner	Fernfl.	Drama	21.IV	9⁰⁷		5000	Wilhelm Babinal
73	" Keller	Lt. Janeberg	"	"	28.IV	3¹⁴			
74									
75									

39

Landung			km	Durchschnittsgeschwindigkeit	1. Auftrag (kann beigeheftet werden). 2. Ergebnis des Fluges (kann auf Meldekarte, Skizze oder Kartenausschnitt beigeheftet werden). 3. Besondere Vorkommnisse (besondere Bemerkungen und Erfahrungen am Flugzeug und Motor, bei Start und Landung und während des Fluges, Unglücksfälle, Instandsetzungen, Erfahrungen mit Abwurfvorrichtung, Waffen).
Ort	Datum	Tageszeit			
11	12	13	14	15	16
Drama	13.IV				6 Bomben auf Mün= D.F.W.C.V 9103 bare Strumamündung
Drama	18.IV	10⁵⁰			Auftrag konnte wegen 5 feindl. Flieg. über Flugplatz Babimal nicht ausgeführt
Drama	18.IV	3¹²			über Weg. tbs 3 feindl. Flugz. Verschwanden sofort beim aufziehen in den Wolken.
Drama	19.IV	9¹⁸			6 Bomben auf 3000 t Dampfer im Golf von Orfano. D.F.W.C V
Drama	21.IV	11¹⁵			D.F.W.C V 9103 Von feindl. Flieg. verfolgt.
		3⁴³			Strumatz Struma u. feindl. Ein=

The Provisional Government

As mentioned earlier, the triumvirate of Venizelos, Koundouriotis and Danglis had arrived in Saloinki on October 9th, took over the leadership of the *Ethniki Amyna* and established a provisional government. Since there were no financial means to support it, only two ministers were named: Immanouil Zymvrakakis, as army minister, and Nikolaos Politis as foreign minister. The remaining ministerial posts were filled by government commissioners. The decrees of the Provisional Government bore the title Kingdom of Greece and a crown as an emblem. On October 14th, Venizelos called on the Allied governments to recognize the Provisional Government.[487]

Venizelos was initially very cautious when he announced his political goals. Knowing that the monarchs of England and Russia supported Constantine, he refrained from attacking the King and the dynasty. He was also aware that the majority of the people of ancient Greece supported the king. The Allies initially found it difficult to recognize Venizelos' Provisional Government. The Russians flatly refused *de jure* recognition. But they were aware that the Allies would not get around *de facto* recognition. So the Russian representative suggested that Venizelos' government should be viewed as a civil authority within the Allied occupied zone, with an advisory role. The Allies found this proposal to be a good one. Grey and Briand thought that Venizelos should be satisfied with such *de facto* recognition. *De jure* recognition was only possible if the Provisional Government had been recognized by some representative body. Italy completely refused to recognize the government. But the non-recognition did not prevent the Allies from equipping the Venizelos troops, and from granting him a loan of 10 million francs for civil purposes. The Boulogne Conference on October 20th confirmed this. Official recognition was again ruled out, even though diplomatic representatives were sent to Venizelos in Saloniki.[488]

"With Greece divided between rival forces of the King and Venizelos, Sarrail had every opportunity to exercise his talent for meddling in political affairs. By the autumn of 1916 his British allies were fully convinced that his attachment to Republican principles was allowing him to narrow his judgment and divert him from military affairs. [The] suspicion that Sarrail was aiming to become governor of a Greek republic, backed by French capital, owes much to the political maneuvers in the final months of 1916. Not just between the Allies, but also with the Greeks themselves. Of course, it is no more the whole truth about Sarrail than the black and white caricatures of the time about the King or the greatest Greek statesman. Constantine was neither a hypocritical traitor who wanted to sell Greece to his German brother-in-law, nor was Venizelos a far-sighted idealist who wanted to roll out the banner of free democracy."[489]

While Briand wanted to postpone *de jure* recognition of Venizelos and de facto recognize both governments, the British were in favor of full recognition of Venizelos' government. That would mean there would be two governments, something that had never happened before, but why not - what had been normal before in this war, anyway? Based on a decision of the War Committee, Grey wrote to Paris on November 7th: *"If Mr. Venizelos were recognized by the King as Prime Minister, this would undoubtedly secure the unity of Greece and the confidence of the Entente Powers. Failure to do so, His Majesty's Government should note that the correct response to recent events, such as the attacks on the Venizelists and the failure of the Greek government to protect them, is an official acknowledgment of the Venizelos government wherever it establishes itself. This situation would undoubtedly be unusual, but certainly no more unusual than the general situation in Greece."*[490]

Given the agreements reached between Bénazet and Constantine, Briand rejected the British proposal. At the same time, London learned that the Greek King was ready to surrender war material in exchange for guarantees. But the British stuck with their recognition course, and Briand with his. The Katerini incident and Admiral Du Fournet's demands regarding the Greek fleet angered the Greek public. Guillemin therefore advised Venizelos not to rush into anything; time was working in his favor. Venizelos, however, did not trust the King and continued to assume that he was pro-German and also expressed this to the French.[491]

The Greek problem was also discussed at the inter-allied conference in Paris on November 15th. Briand reported to the conference the progress made by Bénazet and stressed the value of the war material that the King had promised the Allies. They should therefore stick to the *status quo* of only *de facto* recognition of Venizelos' government. They should provide him with war material, but strive for a reconciliation between Venizelos and Constantine.

Briand did not mention the French promise that Venizelos would not be allowed to advance further south. Asquith agreed that reconciliation should be sought, but England supported Venizelos. The *"strange"* situation should be ended by recognizing Venizelos' government. But Briand prevailed: for the time being, the Venizelos government would not be recognized. Regarding the military situation on the Saloniki bridgehead, it was decided that it should be reinforced, but no offensive was to be launched for the time being. The eastern front had priority.[492]

On November 23rd, the Provisional Government declared war on Bulgaria and a day later on Germany.[493] These declarations of war were meaningless for the situation at the time, because firstly there was nowhere that these Greek troops could attack those of the Central Powers because Allied troops were positioned between them and the Bulgarians. So they could only become active in association with them. Second, the Provisional Government had no army with which to attack. The only purpose of the declarations of war was to induce the Allies to recognize the Provisional Government *de jure*. On the other hand, it was actually an example of irresponsible politics. While Constantine had declared that Greece would enter the war as soon as it was ready for war, Venizelos disregarded such *"trivialities"* and entered the war with his part of Greece - without the slightest preparation. If the Bulgarians had launched an offensive with German support, this could have had terrible consequences for Venizelos' Greece, because it would have become hostile territory. The King's Greece, on the other hand, would have remained neutral. But the Allies' recognition, which had been hoped for with this gesture, did not come.

Recognition came after the *Noemvriana*: On December 2nd 1916 England and France recognized the Venizelos government as the only legal government of Greece. With this, Greece was officially split. On December 7th the Venizelos government once again declared war on the Central Powers.[494] On the same day, Venizelos published a declaration on the riots in Athens, which *"have brought an unbridgeable chasm between the King and the nation. The King did not hesitate to sacrifice the highest interests of Hellenism in order to enforce his arbitrary rule ... and to support Germany to win the war ... from this point on, King Constantine is deposed from the throne."* The King responded by issuing an arrest warrant.

The Athenian Archbishop Theoklitos, who was outraged by Venizelos' persecution of loyal bishops and priests, cursed Venizelos on Christmas Day 1916 in a way that had been customary in the Middle Ages: The archbishop exclaimed: *"Cursed be Elephterios Venizelos, who locked up priests and himself has conspired against his King and his country!"* He threw a stone at a marked spot on the parade ground and called the curse (*anathema*) on Venizelos and his followers, thus banning him from the church. Eight bishops from all the dioceses of ancient Greece repeated his sentence and the ceremonial stone throw. The government had forbidden the meetings, but the people of Athens and Attica came anyway. A total of 60,000 citizens are said to have thrown their stones and shouted *anathema* during the day. In the days that followed, this ritual was repeated in the towns and villages of ancient Greece.[495]

When Venizelos took power in Saloniki, the troops of the *Ethniki Amyna* numbered fewer than 2,000 men. Commander in chief was Colonel Zymvrakakis, who wanted to set up a national army. At first the population was enthusiastic, but when they realized that they themselves, i.e. their male portion, would be the ones to take up arms, the enthusiasm quickly cooled. When recruitment was coerced, wealthy citizens tried to evade military service by taking small fishing boats to an island. But the fishermen received a bonus for each of those they handed over to the authorities. Shiploads of volunteers also arrived, many of them from ancient Greece, which was suffering from the Allied blockade, and wanted to escape hunger in this way. Refugees from East Macedonia volunteered to join the National Army: They wanted to recapture the territory occupied by Bulgaria. By mid-November the National Army numbered 23,000 men. Most of the soldiers from the islands were volunteers and took their service seriously. Those forced into military service from the mainland were unruly. There were even riots on the Chalkidiki, so that the British had to ensure peace and order. The most reliable troops were the Cretans.[496]

The recruiting methods were sometimes brutal. A report which Muslim citizens of Greece presented to the American envoy on November 24th, 1916 contains hair-raising examples. The Muslims were forced to do hard labor building roads. The peasants' herds were driven away. In Verroia, the Muslims had to pay a large sum to buy themselves fre from

Facing Page: A drawing by the Belgian artist Oskar Liedel from the year 1944. It depicts the aerial supremacy of the allies over German aviation. Even the greatest commitment, courage, and bravery of the German aviators could only delay the political and material superiority, until the total defeat of the empire!

LA MAÎTRISE DE L'AIR

serving in the army. Reservists tried to evade service in the Venizelist army by fleeing. A typical example shows what happened next: a lieutenant of the revolutionary army appeared in a village to summon the reservists. *"All the reservists from the village had sought refuge in the area, determined to resist any attempt by the revolutionaries to arrest them. [...] When he saw that his appeal was unsuccessful, [the lieutenant] decided not to give up. He called on the 70-year-old priest [of the village] to hand over the rebels to the revolutionaries without delay. The priest fearlessly responded to the order with this simple sentence: "The reservists will not report until the King calls them." The very angry lieutenant ordered his men to hang the priest, set his house on fire and burn the family alive."*[497] The report contains further examples of unspeakable horror by Venizelos' recruiting officers as well as the French officers who supported them.

The French provided a general to build up the army. But their equipment was poor. The Army of the Provisional Government had just 65 cannons of various calibers with 9,800 rounds and uniforms for a regiment. Since the Provisional Government was unable to take out loans, it had to ask the Allies for financial support. The goal was to raise an army of about 80,000 men. The Allied envoys in Athens pushed through the possibility for volunteers to go to Saloniki. Army personnel were not allowed to do this. First, four divisions were set up, partly from volunteers, partly from conscripts who lived in the areas controlled by the Provisional Government. On October 13th, 1916, the Macedonia Army Corps was set up under the command of Major General Leonidas Paraskevopoulos. On December 16th, two corps of two divisions each were formed from it.[498]

The recruiting methods used to draw soldiers into the service of the Provisional Government Army were also peculiar when it came to pay. A simple soldier in the Greek army received one US cent per day. A soldier in the Venizelist army received five times that amount, and was also reimbursed for travel expenses to Saloniki. The daily wage of a sergeant in the regular army was seven cents (50 cents), that of a lieutenant $6 ($15.50), that of a senior lieutenant $7.50 ($17.50) and that of a captain $8 ($22.50). Senior officers received between $20 and $100. In view of these *"salary discrepancies"* it is astonishing that of the 3,500 officers of the regular army only 200 went to the other side, and of 250,000 men only 12,000 and 100 police officers defected. Sarrail, or France, provided the funds necessary to pay these salaries in the form of a 10 million Drachma loan.[499]

A further *"salary discrepancy"* is to be noted. The Prime Minister of the government in Athens received $160 per month. Venizelos, Koundouriotis and Danglis paid themselves $2,400 per month.[500]

Although the Allies initially only recognized the Provisional Government *de facto*, they supported the establishment of the army. During a conversation between Venizelos and the commander of the British troops in Macedonia, General George Milne, Venizelos declared that he would set up four divisions and drive the Bulgarians out of eastern Macedonia. Milne, who defended the front on the Struma with his troops, thought the Greeks should be tactically and administratively subordinate to him. The British would then also take over the supply of these troops. He refused a dual command; the Greeks would have to be subordinate to either him or the French. London was of the opinion that the Greek troops still to be raised should later be placed under the French command. The three Greek battalions available so far, from which a regiment was formed on December 10th, came under Milne's command.[501]

Milne and Chief of Staff Robertson, who often communicated privately, believed that an attack across the mountains against the Ni -Sofia-Constantinople railway line was logistically hopeless. But on the other hand they either had to attack to get the troops out of the malaria-infested Struma area or withdraw. He thought the first option made more sense because he believed he could beat the Bulgarians.[502] At the conference in Rome, Milne was placed under Sarrail's command, but with the right to turn to his own government in case of doubt.

When inaccurate and exaggerated news reached Saloniki about the events in Athens on December 1st and the fall of Bucharest on December 3rd, Sarrail became nervous. He feared a strong German-Bulgarian attack at Monastiri (Bitola) and a parallel attack by the Athenian army from the south, to the rear of the Allied troops, from Thessaly. He went to Milne and informed him that he had to stop the offensive against Bitola. He wanted to drive the royalist troops out of Thessaly. For this purpose, he wanted to use the 60th British division, the Italian division and some French units that were on their way. Milne actually refused an attack to the south, but since he had learned that there was a larger unit in Thessaly with an estimated 20,000 men and about 100 cannons, he gave his consent to send parts of the 60th Division to Katerini. Winter storms delayed their arrival. But in mid-December an Allied bridge was formed at Katerini, which prevented the Venizelists from marching to Thessaly. The royal troops at Larissa remained calm. For three months the situation was so relaxed that English officers went hunting for woodcock: *"The troops [...] certainly had a peaceful and pleasant interlude. Only a few of their comrades were granted this in the course of the war."*[503] In January, Earl Granville George Leveson-Gower was sent to Saloniki as the diplomatic representative of England.

The whole time, the Venizelist press was free to continue to spread lies about Constantine: *"The Venizelist press office in Saloniki, with its censorship-free channel – while those in Athens were closed except for good news – covered the world with falsehoods, absurd claims and ridiculous allegations. It was claimed that King Constantine received part of his income from the French, British and Russian governments. For this he was accused of great ingratitude because he did not plunge his country into war in return. He was accused of operating a secret radio station in Queen Olga's summer villa in Tatoy, with which he was allegedly in constant contact with Berlin; Statements that he would not even have made in a dream were put into his mouth and circulated publicly. It was claimed that he benevolently obeyed the orders of his brother-in-law, the German ruler, in order to keep his throne. The story of the "treasonable ambush" appeared again and again in the English and French press on December 1st. The old story that he had a secret agreement with Germany to the detriment of the Allies was revived and printed with further embellishments."* [504] According to the old principle of *semper aliquid haeret*, it is hardly surprising that the falsifications of this propaganda shape the image of Constantine to this day.

The Italian Occupation of Northern Epirus

At the end of April 1914, Albania had agreed to negotiations on the future status of Northern Epirus. In May, representatives of Albania and the Greek Northern Piraeus as well as the great powers in Corfu signed a protocol called the *Protocol of Corfu*. The two provinces of Korcë and Gjirokastër formed the autonomous region of Northern Epirus. Since the inhabitants of Northern Epirus insisted on their independence, they were given the greatest possible autonomy: Northern Epirus was allowed to maintain its own armed forces, the official language was Greek, there were postage stamps, and the privileges of the Orthodox Church were recognized. School education would be in Greek. The Albanian central government had the right to appoint the provincial governors and higher officials, but these had to be of Greek descent. An international control commission would monitor compliance with the agreement. The agreement was ratified by the powers and Albania in June 1914.[505]

With the beginning of World War I, the situation in Albania became unstable. Albania broke up into its regions and conflicts arose between them. In October 1914 the Greek army occupied Northern Epirus, with the consent of the Allies. The autonomous government dissolved and Northern Epirus was administered by Greece. Italy and the Allies agreed that the future of Northern Epirus should finally be settled after the end of the war. Prime Minister Venizelos was convinced that Northern Epirus would fall to Greece after the conflict. The royalist successor government annexed Northern Epirus as part of Greece in March 1916. The Northern Epirians were permitted to take part in the Greek parliamentary elections at the beginning of 1916, and sent 16 members to the Greek parliament. However, under pressure from the Allies, the annexation was revoked.[506]

When the Serbian troops left for Albania after their defeat, to be brought to Corfu, the Allies asked Italy to strengthen its bridgehead in Valona (Vlorë). Italy responded immediately and sent two infantry brigades and artillery. In December, it expanded its zone of occupation to Durazzo (Durrës). But in January 1916 the Austrian army occupied Skutari (Shkodër) in northern Albania and advanced on Durazzo, from where the Serbs were embarked to Corfu. When the Austrians attacked Durazzo on February 23rd, the Italians withdrew to Valona. Since the Austrians made slow progress due to the lack of transport, the Italians were able to turn the area around Durazzo into an *"entrenched camp"*, similar to the Allies in Saloniki, located in the north at the mouth of the Vojusa (Vijose), which began at the end of the bay of Valona and ended in the south behind Cheimara (Himarë), so it was about 100 km long.[507]

Since the Italians were advancing inland from Northern Epirus, there were repeated minor clashes between Italian and the Greek troops of the 5th Corps stationed in Ioannina in the area of Cheimara. The demobilization ordered by the Allies in June 1916 also weakened these corps. In July 1916, the Italian general Oreste Bandini took command of the troops in Albania. He was commissioned to occupy Northern Epirus and incorporate it into Albania. Italy justified this with the story of an alleged planned Greek-Bulgarian advance into Albania and promptly received the Allies' approval. From late August to mid-October, the Italians occupied Northern Epirus up to a line from Ag. Saranda (Sarandë), Delvino (Delvinë), Kakavia (Kakavijë), Kleisoura (Këlcyrë), Premeti (Përmet) and Leskovik (Leskoviku). On October 22nd, a French cavalry unit occupied Korytsa (Korçë) and established an autonomous republic there. With the occupation of Korytsa there was a continuous Allied front from Ag. Saranda on the Adriatic Sea to the Struma Estuary on the Aegean Sea. The road connecting these places became an important supply route.[508]

Northern Epirus remained under Italian control until the end of the war. As part of the Paris peace negotiations, Venizelos and Tittoni concluded an agreement according to which Northern Epirus should go to Greece. But then the newly formed Yugoslavia laid claims to part of Albania. Finally, Northern Epirus became Albanian in November 1921. In 1939 Mussolini annexed Albania.[509]

The Allied Autumn Offensive – September 12th–December 16th 1916

At the end of May 1916, four French and one British division stood on the front line outside the *entrenched camp*. Their flanks were secured by cavalry detachments. In the camp itself there were four British divisions and on the Chalkidiki peninsula the 120,000 Serbs were in the process of completing their reorganization. On July 15, Commander-in-Chief Joffre called on the *Armée d'Orient* to work out an offensive plan to cover the Romanian mobilization. Sarrail drafted a plan to attack the village of Huma, about 14 km west of Gevgeli. The attack would be carried out by the Serbs, who were to be supported by a French brigade and a Russian brigade that would join them later. Diversion operations should take place on the other parts of the front. The actual attack was postponed to August 4th because the Serbs did not arrive on time.[510] On the Bulgarian side, the troops were deployed as follows: In the area of Monastiri (Bitola) there was the West Macedonian group with two divisions, in the valleys and on the heights of Axios and Struma there were six divisions and in the area of Xanthi-Alexandroupolis there was one division. Two divisions were in reserve in the Skopje area. The three possible attack routes at Bitola (Bitola Gap), along the Axios and Sturma valleys, had in the meantime been paved. In total there were 200,000 men in Macedonia from 172 battalions with 100 heavy and 800 light guns. The Allied troops numbered 320,000 men in 201 battalions and had 293 heavy and 700 light guns. Due to various difficulties, however, the start of the Allied offensive was postponed to August 20 th.[511]

On August 17th, 1916, the Bulgarians launched a surprise attack from Bitola and advanced on Florina, overrunning the Serbian troops there. Sarrail countered by attacking with the 17th French Colonial Division in the area of Lake Dojran. On August 23rd, he succeeded in occupying the local train station and the surrounding heights. The Bulgarian attack had been announced to the Greek government by the ambassadors of Germany and Bulgaria. In addition, they were promised that the cities of Serres, Drama[512] and Kavala would not be occupied. Since the Greek army was demobilized, an armed resistance was impossible. In addition, Greece was neutral, Prime Minister Zaïmis wanted to negotiate. The attackers occupied Florina on August

Above: Truck driver Johannes Keller is photographed at the end of 1914 in Brussels.

18th, and took the Greek crew as prisoners. Further to the west, they overran the town of Vevi and in the north they occupied the Kajmakcalan Mountains. Serb counterattacks failed.

In the east, Bulgarian troops occupied Sidirokastro and Chrysoupolis. Sarrail strengthend the troops on his left flank, in order to advance on Bitola later.[513]

On September 5th, Sarrail ordered an attack on the left flank from the Edessa area in the direction of Bitola. The aim of the offensive was to push back the opposing troops behind the Cerna Arch (Crna Reka – Erigon). The Serbian troops were to lead

Above: The military passport of Johann Keller.

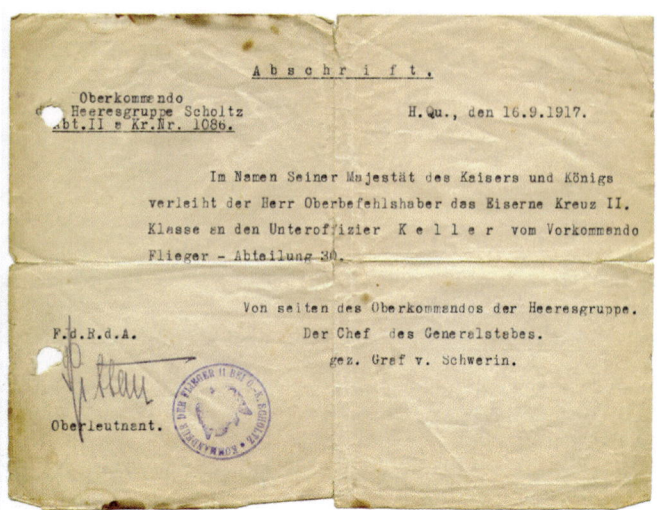

Above: The award certificate for the EK II [Iron Cross 2nd Class] dated September 16th, 1917 for Unteroffizier Johann Keller

the main attack from the villages of Arnissa and Kella north of Lake Vegoritis. The French and the Russians, who had arrived in the meantime, were to carry out an encircling attack along the large Prespa Lake towards Bitola further to the west, beginning south of Florina.[514] The attack began on September 12th. The British diversion attack on the Struma section quickly got stuck.[515] The initial offensive on the left flank was successful. The Serbs forced the first Bulgarian army back to a line from the Prespa lake, via Florina to the Kajmacalan mountains. On September 26th, Lieutenant General Arnold von Winckler took command of this army, which was now the 11th German Army. The situation of this army was very difficult, because the entire supplies had to be carried from the Vardar valley 100km by field railway, motor and horse-drawn carts and ox-carts as well as a cable car, to Prilep. When everything was in place, 750 tonnes could be delivered to the troops every day. Sarrail's troops were supplied directly via the railway line from Saloniki via Florina to Bitola and therefore had no supply problems.[516]

Bitola itself, which at the time had about 60,000 inhabitants, is located in a 20-30 km wide plateau (600m) between the Peristeri massif in the west (2,600m high) and the mountains north of the Voras massif, of which the Kajmakcalan with its 2,500m, is the highest elevation. This plain, known in military history as the *Bitola Gap*, played a huge role in both World War I and World War II. It is the only wide passage through the mountains in northern Greece. The 11th Army's entire front section was about 100km wide and was defended by 65 battalions with 52 batteries. The Bulgarian battalions had suffered heavy casualties in previous battles that had not yet been replenished. The expansion of the defensive positions was unsatisfactory and there were almost no reserves.[517]

When von Winckler took command, the Bulgarian troops defended themselves desperately against the attacking Serbian units on the Kajmakcalan. The mountain was of great strategic importance because the whole Bitola plain could be seen from there. However, on September 30th, the Serbs conquered the summit of the massif. Winckler ordered the immediate re-conquest, but the Bulgarian commanders declared that this was not possible in view of the condition of their troops. Since it was not possible to rearrange the troops in battle with the enemy, Winckler took the right wing back to the Greek border and the left to a line behind the upper Cerna to the east. On October 6th, the Serbs attacked again and two days later advanced on the north bank of the Cerna. The attacks continued until October 15th, but the German artillery and a few machine gun divisions pushed them back,

Above: An EK II.

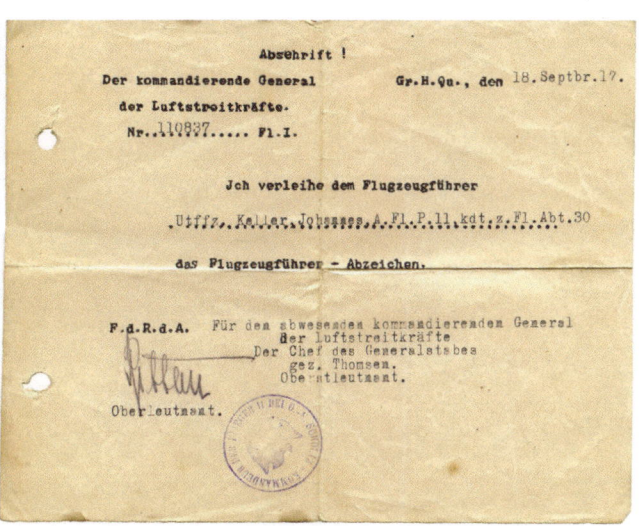

Above: On September 18th, 1917 Unteroffizier Johann Keller received his aviator badge.

Above: German pilot insignia.

albeit with difficulty. On October 18th and 19th, the Serbs managed to break through the Bulgarian lines at Cerna - the few German troops could not prevent it. But shortly afterwards the first German reinforcements arrived and temporarily rescued the situation.[518]

On October 27th, Sarrail attacked again. After hours of artillery fire, the German-Bulgarian troops bloodily repulsed the attack by the Serbs. On November 10th and 11th, further large-scale attacks followed. The German troops defending the left half of the Cerna Arch were able to resist, but on the right wing the Bulgarians backed away from the Serbian attack. A huge hole opened in the front. The German troops were able to withstand the attacks of the next few days, but it was clear that they were threatened with encirclement. Therefore, on November 14th, the front was withdrawn to the heights further north.[519]

On September 12th, the French attacked on the left wing in the direction of Florina. Sarrail had hoped to occupy the city on the first day, but the Bulgarians put up stiff resistance. The local commander, General Victor Cordonnier, could only report that his troops were moving towards Florina. Sarrail was furious, although he himself had

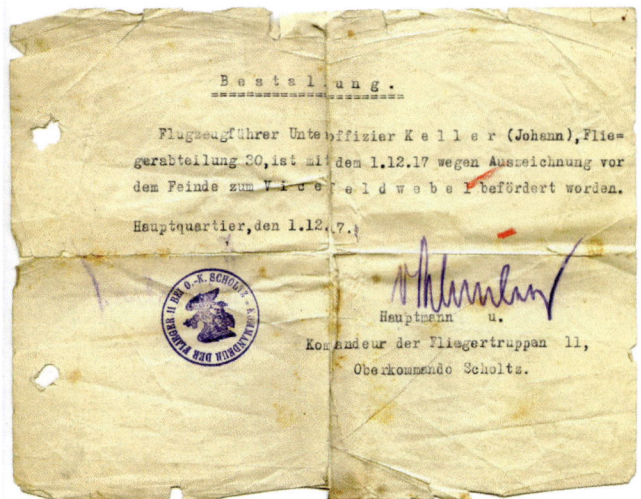

Above: "Appointment: Pilot Unteroffizier Keller (Johann), Fliegerabteilung 30, was promoted to Vice Sergeant on December 1st, 1917 due to exemplary action in front of the enemy."

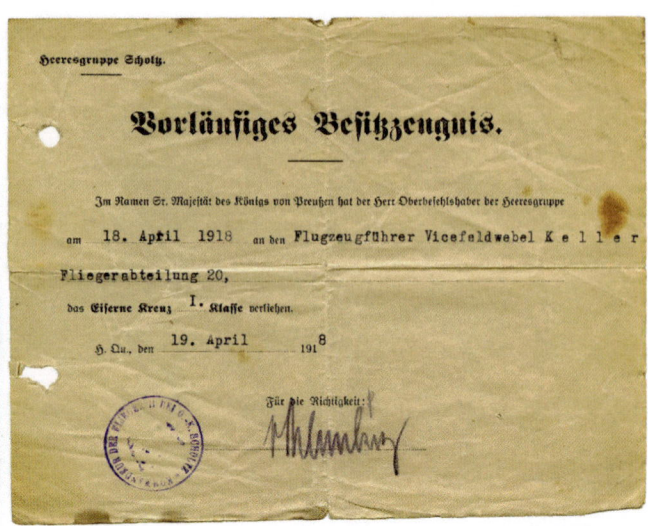

Above: Preliminary certificate of possession – In the name of His Majesty the King of Prussia, the Commander-in-Chief of Army Group [Scholtz] awarded the First Class Iron Cross to the pilot Vice Sergeant Keller, Fliegerabteilung 20, on April 18th, 1918.

contributed by immediately forcing the overtaxed troops to attack. But Sarrail was incapable of self-criticism. He ordered Cordonnier to attack head-on with all his forces. He obeyed and on the evening of September 16th, Cordonnier's troops reached the town of Boresnica, a few kilometers east of Florina, along the railway line. Apparently he believed this was Florina and reported it accordingly. When Sarrail saw the mistake, Cordonnier lost even more of his respect.[520]

Sarrail ordered him to advance further towards Monastiri / Bitola. Cordonnier knew from a visit to the Kajmakcalan Summit that a deep Bulgarian trench system lay ahead of him and that the reinforcements requested - a Russian unit and a French colonial division - had not yet arrived. He informed Sarrail accordingly. The latter nevertheless ordered him to attack on October 3rd. If he did not attack, he would seek Cordonnier's dismissal and replacement in the ministry of the army.[521] Cordonnier replied, *"We are ready to die on the day you choose, but we will not defeat the enemy until the Russians do their part, still without the Colonial Division."* Sarrail appeared on site and there was a violent clash, but the attack was postponed until a monastery above Florina was captured, which happened on October 2nd.[522]

Three days later, Cordonnier flew over the battlefield as an observer in a French reconnaissance plane. That was new for a senior officer in 1916. From the air he could clearly see the trench trap that he was expected to have his men attack. He could just as clearly see that it was possible to find a way around the defensive lines. This led through the foothills in the west and the foothills of the Kajmakcalan, where the Serbs had already established themselves in strength. But he had no chance to put his strategy to the test because when he returned to headquarters he found that an irrefutable order had come in at 3:00 p.m. Sarrail demanded a frontal attack on the Kenali lines on the following day to capitalize on the enemy's dejection after they had taken the monastery. Sarrail was so sure of success that he added instructions to disrupt the Bulgarians' retreat to Prilep after they had left Monastir."[523]

The next day the French and Russians attacked the Bulgarian trenches head-on. Less than 100m was gained and the Allied troops were pushed back. The losses were high. But Sarrail insisted on another attack. Reinforcements were sent and on October 14th the Allies attacked again. *"It was a terrible battlefield. When they encountered continuous (barbed) wire, the French and Russians stalled and when they tried to make their way through, Bulgarian machine guns opened fire on them. In one afternoon, the French lost nearly 1,500 men and the Russians 600. The Bulgarian lines remained untouched. On the extreme right side of the valley, the burning village of Brod testified that the Serbs had fulfilled their mission under all circumstances."*[524]

Cordonnier had had enough of this carnage, and wanted to push through his original plan to encircle the enemy. In Sarrail's eyes, this was

Above: A further honor for Vizefeldwebel Keller on September 11th, 1918: He was awarded the "Warrior Badge in Iron" by Duke Ernst Ludwig von Hessen. This medal is also known as "Bloody Ludwig" because in the statutes of the award, a prerequisite was that the recipient must have possessed the Iron Cross second and Iron Cross first class, as well as having been wounded at least once.

Above: On November 3rd, 1931, the Hungarian „Chancellery of the Medal of Remembrance", the Hungarian memorial medal, is given to retired police officer Keller for his participation on the war front from 1914 to 1918.

insubordination. On October 6th he telegraphed Joffre asking for permission to remove Cordonnier from his post. Four days later, Cordonnier was ordered to vacate his post and return to France. He had been on the Salonika front for exactly 69 days.[525] In his memoirs, Sarrail claimed that Cordonnier behaved in this way because of his cancer surgery: *"I would like to believe that he must have been under that morbid influence when he used the troops under his command and sabotaged the orders he received and did not want to understand."*[526] Obviously he could not imagine that Cordonnier was acting for human reasons and trying to spare his troops unnecessary losses.

One might have expected that this affair would cost Sarrail his post. It showed his bad character. The French headquarters was outraged and said that Sarrail should be executed by firing squad.[527] But his political protectors in the Senate and Chamber stood behind him, and Briand's coalition government relied on them. So Sarrail could not be fired and he could continue his political and military mischief.[528]

Since the forces of the Bulgarian defenders were weakened – there were no reserves and they were threatened by encirclements – the order was issued on November 18th to retreat to a line from the Prespa north of Bitola to about the middle of the Cerna Arch. The attacks continued afterwards, but now there were enough German reserves. The bad weather also hindered supplies for the Allies. Allied attacks stopped on December 20th. The position that had been taken on November 19th could be held. There were also a few minor attacks later, but winter forced both sides to stand still from December 16th to spring 1917.

On the Struma Front, the British, supported by the 1st Battalion of the *Ethniki Amyna*, succeeded in pushing the front to the river itself and forming a small bridgehead between the village of Erakleia and Lake Achinos. On December 7th, the Greek battalion was withdrawn to Saloniki to help set up the 2nd Serres Regiment.[529]

On the Allied side, the occupation of Monastir was celebrated as a great success, especially as the situation in Romania was grim. The British had done little against the Bulgarians in the Struma Valley. Sarrail announced in a daily order that the capture of Monastir was the first French victory since the Battle of the Marne. The Serbs were enthusiastic, after all, Monastir was the first Serbian city to be liberated. The Serbian monks on Mount Athos celebrated a *Te Deum*. The Bulgarian front had been breached, but it had also been closed again. The aim of the operation to support the Romanians had failed. When Bucharest was captured on December 6th, 1916, the Romanian resistance collapsed.[530] With that, the Saloniki operation had basically finally become pointless.

The French government had actually intended to deal a decisive blow to the Bulgarians, but the Allied army was unable to carry it out due to malaria. Malaria had caused more damage than the enemy: it incapacitated around 30–40 percent of the troops.[531] The actual losses of the Allies in the Monastir offensive amounted to 50,000 men, 27,000 for the Serbs and 13,786 for the French. The British lost around 5,000 men. The German losses are said to have amounted to around 8,000 men; those of the Bulgarians are unknown, but are likely to have been significantly higher.[532] So reinforcements had to be sent. At the inter-allied conference in London on December 27th, the French demanded that the armed forces in Salonika should be reinforced and

Above: Bulgarian war memorial medal 1914–1918.

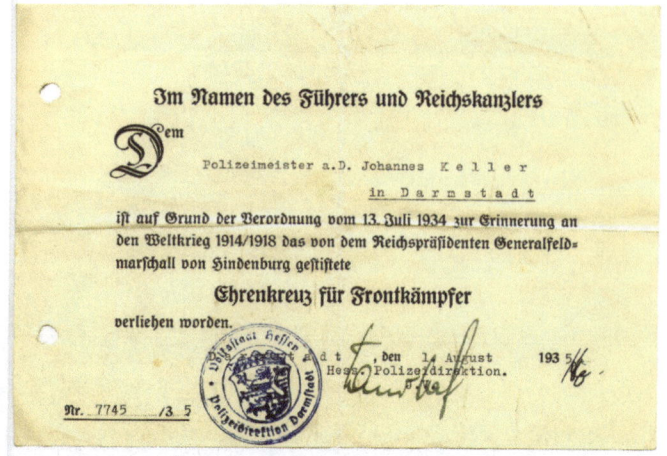

Above & Facing Page, Above: Deutschland during the dictatorship: "In the name of the Fuehrer and Reichs Chancellor [Adolf Hitler]. To the retired police officer Johannes Keller, due to the Order of July 13th 1934 to remember the world war of 1914/1918, the honorary cross for front-line soldiers, donated by the Reichs President General Field Marshall von Hindenburg, is awarded. Darmstadt, August 14th, 1935.

that more decisive action should be taken against the Greeks. Prime Minister Lloyd George declined this request, as an offensive against the Ottomans in Palestine was being planned at the time, and he had neither forces nor material left. Chief of Staff Robertson went even further. He openly stated that he did not believe in a victory in Macedonia and that it would be best to withdraw the troops. Eventually the British agreed to send 36,000 men to Saloniki to compensate for the losses, under the condition that no violent action against Greece would be undertaken without their prior consent.[533]

Indeed, the outcome of the Monastir offensive was poor. In two months of hard fighting, the Allies had gained around 40km of terrain and occupied the city of Monastir / Bitola. The fighting against Rumania had not changed anything, and Germany's supplies to the Ottoman Empire continued unhindered. The Bulgarians had proven to be brave soldiers who, with little German support, were able to hold back the Allied troops.

The future Prime Minister Lloyd George's assessment of the fighting on the Salonika Front is interesting. In a memorandum for the Paris conference in mid-November 1916 published in his memoirs, he sharply condemned the campaigns of 1915 and 1916: *"In retrospect, I consider the case of Serbia to be the most unforgivable and, I fear, the most irreparable of all Allied failures. We now know how important it was to block the German advance to the east. […] In the end we realized how important it was to grab the bridge to the east. But it was too late by then. The Balkans, which could have been a problem, are now a heavy burden. […] In the case of Romania, in 1916 we repeated the grave mistake of 1915 Serbia. […] It was an absolutely inexplicable mistake. […] The Saloniki campaign is a different view of the two most fatal shortcomings that haunt the Entente - indolence and lack of cooperation. A timely implementation of the Saloniki campaign would have secured Serbia and given us the Balkans. The best that can be said so far is that 250,000 Bulgarians and at least twice that number of Turks are tied up. […] The condition of the Salonika Army gives the impression that, for reasons of strategy, the commanding generals have been deprived of any initiative to use the armies under their command too effectively."*[534]

Above & Above Right: Stamps from German aviation departments from WWI.

Above & Above Right: Stamps from German aviation departments from WWI.

Above & Above Right: Stamps from German aviation departments from WWI.

Above: Stamps from German aviation departments from WWI.

Above: Flag decoration of an imperial regiment up to 1918.

This was a pretty realistic assessment of the military situation. The idea that the Balkans as a whole could have been drawn to the side of the Entente is, however, not very realistic. They had been so divided since the Balkan Wars that joint action was out of the question. Lloyd George's assessment of the political situation in Greece, on the other hand, is strongly influenced by the propaganda of the time: *"The history of our negotiations with Greece is a bleak picture of paralyzing indecision. The Greek people are on our side, they have shown their sympathy again and again through their elections. But the King is and always was a friend of the emperor and an enemy of the Entente. He never missed an opportunity to serve the emperor and betray the Entente. He gave valuable information to the enemy, such as our troop strength, our positions, our intentions and our movements."*[535] At another point, Lloyd George alleges Constantine had planned to attack Allied troops from Thessaly.[536]

How Lloyd George came to this ambivalent assessment of the military and political situation is unclear. A seasoned professional politician like Lloyd George could have been expected to be resistant to Allied propaganda. But perhaps that was just wishful thinking, as in the case of France, where it was believed that the very presence of the Saloniki army had prevented the Bulgarians from withdrawing troops from Macedonia and using them against Romania. And this was already a welcome result.[537] Or, contrary to his better judgment, did Lloyd George stick to the official wording that was still in effect at the time? The whole thing is pretty incomprehensible.

The Deposition of Constantine and the New Regime (1917)

The Conference of Rome – January 1917

While the offensive against Monastir / Bitola stalled in Macedonia, politics began to move in England and France. From the beginning of the war until May 1915, the Liberals ruled alone under Premier Asquith. Their most prominent members were Secretary of the Navy Winston Churchill and Secretary of War Herbert Kitchener. The shortage of ammunition and the failed Gallipoli offensive led to the formation of a coalition government with the conservatives in May 1915. In autumn 1915 there was another cabinet crisis in connection with the introduction of general conscription and the failed offensives in Sinai, Gallipoli, Mesopotamia and Saloniki. Asquith sent Kitchener on *a fact-finding tour* of the eastern Mediterranean. Upon his return, he was ousted when General Robertson became Chief of Staff instead of him. Robertson thought victory had to be won in the West.

Kitchener was blamed for the final failure of the Dardanelles operation in late 1915 and the Easter Rising in Ireland in April 1916, as well as the defeat and high losses in the Battle of the Somme. In June 1916, Lloyd George became Secretary of War to succeed Kitchener, who had died on a trip to Russia. Lloyd George was not satisfied with the minister's power, which had been reduced since Kitchener's death, and called for the *status quo ante* to be restored. Asquith initially agreed, but when he realized that Lloyd George was neglecting him and the military in strategic decisions, a conflict broke out and Lloyd George resigned. When Asquith recognized that the Conservatives no longer supported him, he too resigned. With none of the Conservatives willing to take office, David Lloyd George became the new Prime Minister on December 7th, 1916. Unlike Asquith, he was seen as assertive and the public believed that victory was now imminent.

In Paris, eight days later, Briand survived a government crisis only by completely rebuilding his cabinet. At the end of the year Joffre had been appointed Marshal of France and resigned as Commander-in-Chief. General Robert Nivelle took over command in the west. The Grand Quartier-Général in Chantilly was dissolved and Nivelle resided first in Beauvais and later in Compiègne. From then on, the actual management of the war lay with the War Ministry. Sarrail was very pleased with Joffre's exit, as he was now directly subordinate to the War Department. Believing that he now had greater freedom of choice, he applied for permission to attack the Greek army at Larissa, because he still feared their attack on his left flank. Since he believed that his Armée d'Orient was not strong enough to occupy all of Greece, he requested reinforcements. He claimed that if he did not receive them, it was possible that he would have to retreat to a new line of defense and perhaps even evacuate Monastir. Briand couldn't risk angering Sarrail because he had recently had quite a row with the radicals who were known to support Sarrail. If his new cabinet was to survive, he would have to avoid further arguments with them. He therefore tried to evade a decision and, as is well known, sent a delegation to London on December 26th to investigate the British position on Sarrail's proposals and to request the dispatch of two British divisions to Saloniki. It was agreed that the problems at hand should be discussed with the Italian government and the commanders of the Saloniki Front in Rome. Above all, it should be clarified where the next Allied offensive should start, in Italy and / or in Saloniki.[538]

Prime Minister David Lloyd George, Chief of Staff William Robertson and Cabinet Secretary Maurice Hankey left England in the first few days of January. From Paris they traveled to Rome on the same train with Prime Minister and Foreign Minister Aristide Briand and War Minister General Hubert Lyautey. On January 5th, 1917, a three-day conference began in Rome, which also included Sarrail, Milne and the Allied envoys. The conference in Rome was the first real allied conference, because not only the Italians attended, but also the Russians, represented by the Russian envoy and the military attaché. But neither the Allied Serbs nor the Romanians, let alone Venizelos, had been invited. The conference was poorly prepared and the time allotted for it was far too short.

There were different, even contradicting interests. There were divergent opinions amongst the French and English leaders. The *"Westerners"* bet on victory on the Western Front. Robertson was a pure *"Westerner"*. The *"Easterners"* wanted to freeze the front in the west and look for a *"way around it"* in the Balkans. Lloyd George belonged to the latter

Above & Right: 2 Badges.

group, but imagined that the attack would have to come from Italy. He wanted to attack through the Julian Alps with the support of English and French troops and heavy artillery. Briand also bet on the Italians. The Italian chief of staff, General Luigi Cadorna, rejected Lloyd George's proposals; the enemy would notice that the artillery was being brought in, and the tactical surprise would be lost. The Italians did not completely rule out a major attack from Italy; it could be discussed later. They were *"Westerners"* and supported the large-scale attacks planned by Nivelles in the west.[539]

The real reasons for the Italian reaction lay elsewhere: their division in Macedonia was only used to monitor French and British activities, particularly Sarrail's activities in Korytsa. For them, the troops in Albania were more important because they had expansive goals there. If there were a successful major offensive in Macedonia, the Serbs and the Venizelists would gain in importance and could thwart Italian territorial ambitions. So it was no wonder that the Italians remained silent when Lloyd George asked them to send reinforcements to Macedonia.[540]

The conference did not make any clear decisions on the Macedonian front either. The British and Italian military were in favor of abandoning Monastir and drawing back the front in order to shorten it a great deal. The French and the Russian military attaché refused to withdraw *"in the highest interests"* of the Allied coalition, and demanded that a British and two Italian divisions should be sent to Saloniki immediately. If Monastir was surrendered - according to the French - the Serbs could conclude a separate peace with the Austrians and the Bulgarians, a possibility that the British also did not entirely rule out. But the attacks by German submarines made the British more concerned, as 256 ships totaling 662,131 BRT had been sunk in the past six months, 32 percent of which had been British.[541]

Finally, it was agreed to hold Monastir as long as the existence of the *Armée d'Orient* was not threatened, in order to prepare a rear defensive position in the meantime. The British persisted in sending no reinforcements, just a few heavy artillery pieces. Italy also refused to send reinforcements, but was willing to build a road from the port of *Ag. Saranda* on the Albanian coast to the front at Monastir. The French stuck to their course and a little later sent two more divisions.[542]

The conference in Rome, therefore, was unable to create a uniform strategy for the conduct of the war in 1917, nor to determine the further operational

Above: Badge with 2 watchful eagles: "God bless our weapons 1914".

Above: Hat badge of the "Kaiserlicher Aero-Club" (Imperial aero-club) until 1918.

Above & Above Right: A medal made of black painted iron as a patriotic reminder for the surrender of precious metals or jewelry to official purchasing points from 1916: Front: "In the Iron Age 1916". back: "I gave gold as a defense, I took iron in honor."

Above: An EK1 with the engraving "Flying – winning".

Above: The pilots insignia.

Above: The "Iron Half moon" (Turkey).

This Page to Page 311: Patriotic postcards and advertisements in newspapers, in which various German companies showed off their products in the service of the war effort.

Facing Page: Four different advertising stamps, to seal letters.

This Page & Facing Page: Patriotic postcards, stamps or pictures of the era.

procedure on the Saloniki front. Palmer rightly stated: *"In terms of establishing a unified strategy, it was nothing less than a fiasco."*[543]

The only beneficiary of the conference was Sarrail. Sarrail's reputation had suffered under the Cordonnier situation, as well as that of the division commander Charles de Lardemelle, who was also relieved of his command without a good reason.[544] When there was no answer to the question of how to proceed in Saloniki, Sarrail went to see Lloyd George on January 7th. He wrote about his first impression of Sarrail: *"He was one of the rare personalities about whom one cannot form a balanced opinion. For his partisans he was a brilliant general, for his critics a great charlatan. Joffre said he did nothing during the fighting in France to justify the view that he was a capable general. [...] One thing I am sure of - the official military leadership here and in France despised him. To them he was a political general. Lloyd George was impressed by Sarrail's charm!"*[545]

Sarrail feared a German attack from the north and a simultaneous attack by the Greek royalists from the south and therefore repeatedly demanded the military occupation of Thessaly. In his conversation with Lloyd George he called for a free hand throughout Greece, *"All he asked of the British and French governments was that they close their eyes or look away for a fortnight while he tended to the Greek royalists. If the Greeks wage a guerrilla war against him, he would shoot all the prisoners."*[546] Lloyd George refused, because once the Greek army was in the Peloponnese, it was harmless. It would be crazy to open yet another front instead of waiting for the retreat. In addition, the blockade would ensure an accelerated withdrawal. Sarrail was persuaded and promised Lloyd George that he would not do anything without informing him beforehand.[547]

On the one hand, the majority of the conference saw the Saloniki regime as the better government,

but on the other hand they did not want to sever relations with Athens. They wanted to prevent the rebels from marching south from Saloniki, but demanded the accelerated withdrawal of the regular troops to the Peloponnese under close control of the Allies. Allegedly there was fear of an attack by the Greek army in the rear. In order to keep the Greek government cooperative, they were ready to guarantee that there would be no march south and that the blockade, which had now lasted over a month, would be lifted in the foreseeable future. On January 6th, 1917, the Allied envoys presented these to the Greek government, demanding that they be accepted within 48 hours.[548] If the Greek government did not accept these conditions, diplomatic relations would be severed, and Sarrail would take all military measures he deemed sensible.[549]

The demands repeated those of December 14th and 31st, according to which the Greek army should be transported to the Peloponnese as quickly as possible, with the Allies retaining the right to control this. If it was not accepted within 48 hours, the Allies would take other measures to protect their troops. *"For their part, the Allies would be willing to respect the Greek government's wish not to enter the war. We would not allow the Venizelists to invade or take control of areas that are still under the control of the royal government. We would also lift the blockade as soon as our demands are satisfactorily met."*[550]

The government in Athens was in a desperate position. The blockade made itself felt and food became scarce. It had no way of defending itself against the pressure of the Allies. So it had no choice but to accept the conditions. It even had to release the convicted political prisoners and could only hope that the other side would do the same. The Greece ruled by Athens was no longer a sovereign state; it was an Allied-controlled province, an occupied state. On January 10th, after an overnight meeting, the Privy Council decided to surrender unconditionally. According to the British war correspondent and author Frederick George Abbott, Greece *"would empty the cup of humiliation to its bitter residue."*[551]

This Page & Facing Page, Bottom Right: Patriotic postcards, stamps or pictures of the era.

The troops were transferred to the Peloponnese under the strictest control of the Allies. The reservist league was dissolved. The jailed Venizelists were released and a commission was set up to determine their compensation. *"In short, according to the unanimous testimony of Entente diplomats and publicists, Greece reliably accommodated and seriously fulfilled each of the demands. Despite repeated requests from the government that the blockade should be lifted in accordance with the promises made, it was not only extended, but also tightened over the course of the following months."*[552]

On January 25th, the Greek government issued a formal apology to the four Entente envoys for the unfortunate events of December 1st. Three days later, a Greek army unit marched with lowered flags to the Zappeion and saluted the Allied flags raised there as well as the naval and army units, and the envoys. The Greek unit was led by the King's brother. The commander of the 1st Corps and the cabinet ministers were also present. It was a deeply humiliating scene.[553]

Despite this display, Sarrail claimed that the Greek government had disregarded the January 6th ultimatum. The transport of the troops to the Peloponnese was moving slower and slower. To the north of the Isthmus of Corinth there were large numbers of troops, and the gendarmerie in Thessaly was reinforced by army units. Partisan units were allegedly set up in the neutral zone. Secret weapon stashes were everywhere. The goal was to stab the Allies in the back. The French officer in charge of monitoring troop movements into the Peloponnese agreed with Sarrail's view and recommended reinforcing the blockade. On the basis of these reports, the French General Headquarters came to the conclusion that an attack on the Allied troops was being planned in Greece.[554]

The situation with the partisan units was as follows: Army command officers who were concerned about the situation in the neutral zone

Above: The Goddess of Victory lures an aviator to fame and glory, but behind it, Death is already there.

Above: King Constantine, King of Greece.

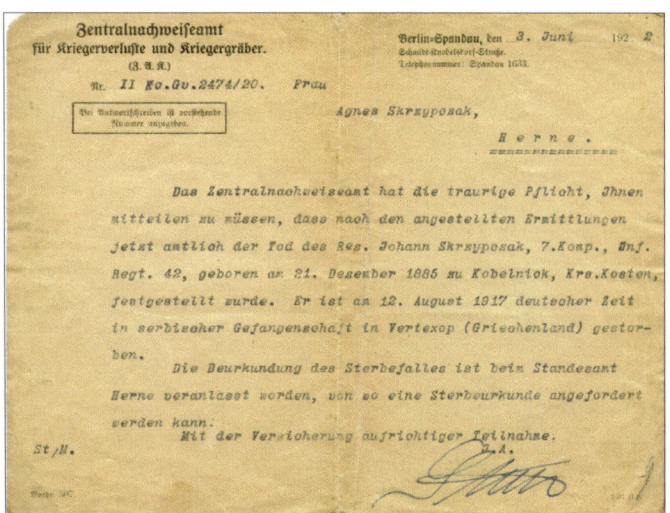

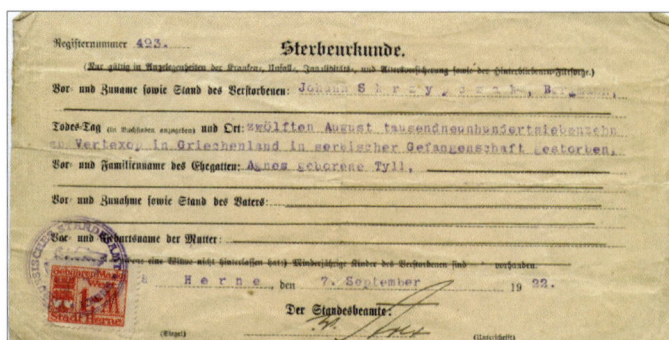

This Page: Death notice for a German soldier who died in Serbian captivity.

This Page: Patriotic postcards, stamps or pictures of the era.

had set up partisan groups with the support of the local residents and the Reservist League. Their main aim was to prevent the infiltration of armed forces of the Provisional Government. The hard core of these groups were the partisan groups from Northern Epirus, which had been reestablished there after the Italian invasion in autumn 1916. The activities of these groups were directed against the attacks by the Albanians who tried to *de-Hellenicize* the country. The Athens government supported the partisan groups in the neutral zone. It was also their job to create propaganda against the Venizelist government. They had strict orders to avoid clashes with the Allied troops. Since the Senegalese French colonial

Above: Newspaper advertisement, war bond poster.

troops behaved criminally, looted, pillaged and raped, armed clashes resulted. These were reinterpreted by Sarrail as attacks on the Allied forces.[555]

This Page & Facing Page: Newspaper advertisements, war bond posters.

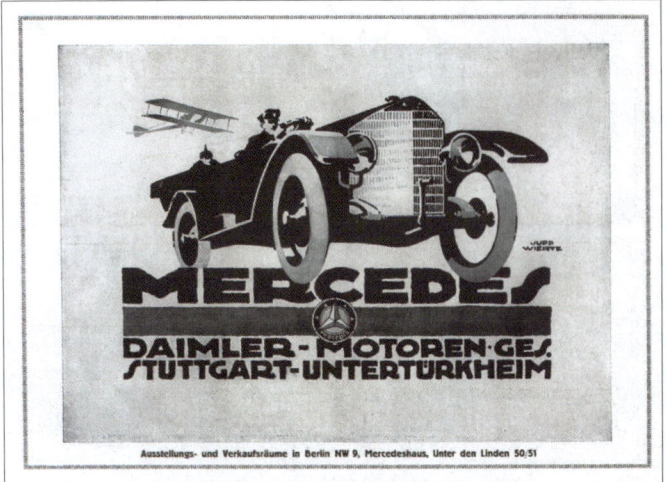

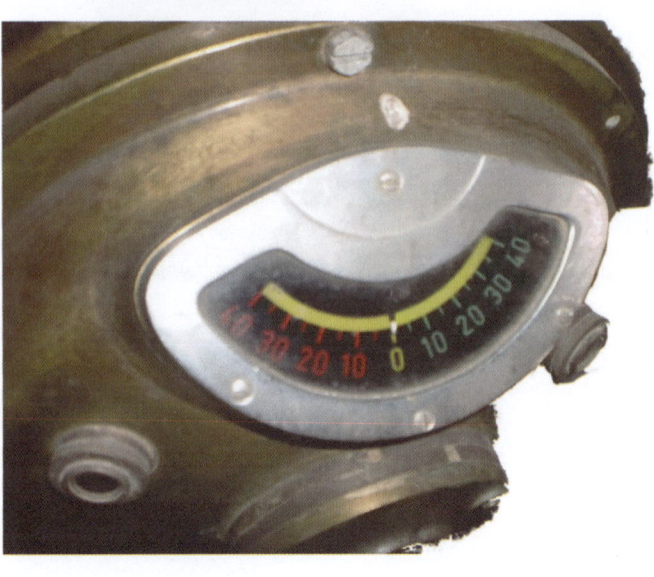

Above: Death notice of a German soldier, who fell on October 29th, 1918 in Macedonia.

Above & Facing Page, Middle & Bottom Right: An airship steering position from its cockpit.

Below: Postcard of the era from Germany.

Above, Below, & Facing Page: Postcards of the era from Germany, England and France.

SALONIQUE - Aéroplane Français

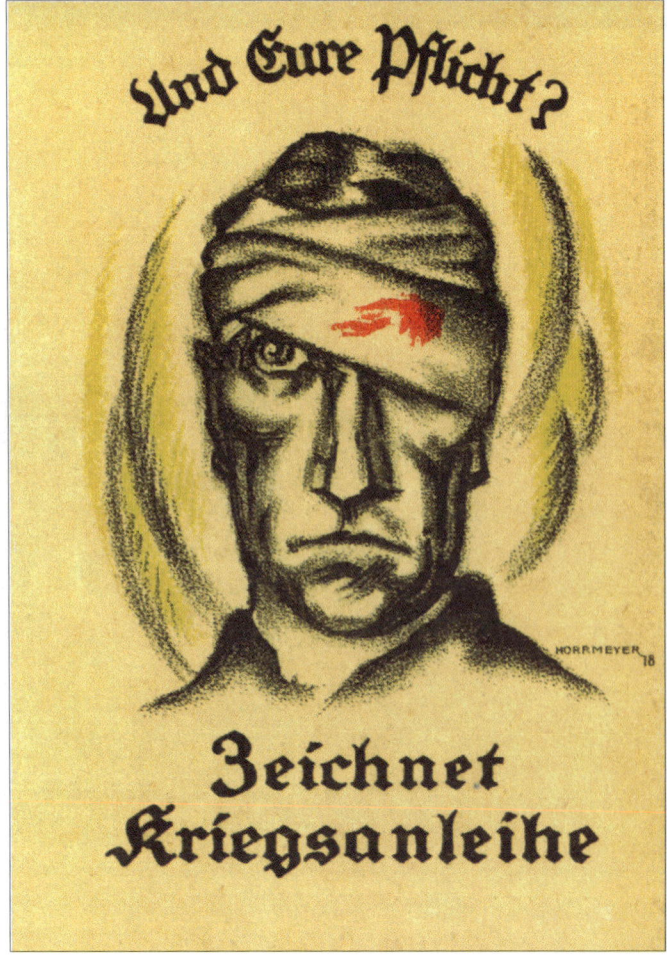

315

Die besten
Glückwünsche
zum
Geburtstage
Sender Familien
Heinr. Behle

Above, Below, & Facing Page: Postcards of the era from Germany, England and France.

Facing Page, Below Right: The rudder of a German DFW CV, an aircraft built at the LVG (Luftverkehrs Gesellschaft) in license, that was shot down on May 30th, 1918 by Sergant Marcel Guillet. The occupants Richter und Tonstein lost their lives during this reconnaissance flight. Photo source: Slg. Daniel Porret, authorization Renaud MANSUY (Aiolfi & Partners).

Blockade and Further Repression

The Allies had promised that the blockade would soon be lifted and that the Venizelists would not be allowed to advance south, but they did not keep their promises. From the formation of the modern Greek state in 1830 until the 1950s, Greece was never able to supply itself with grain. Blockades inevitably led to famine. This was true in World War I and worse in World War II.[556] When the blockade was imposed on December 8th, 1916, knowledgeable journalists from the Allied countries had predicted that hunger would follow in less than a week, especially since the Allies had completely cut off the Greeks' sea traffic. In fact, famine began to be felt as early as the end of December.

Hunger hit the poor first. The authorities and private aid organizations tried to provide the hungry with food and to keep prices low, but the extent of the misery was so great that this aid had little effect. But the cessation of shipping also affected small industries, which could no longer produce due to a lack of raw materials, and had to lay off their workers. *"As the factories emptied of workers, the streets were filled with beggars. The necessary distortion of the flour caused epidemics of dysentery and poisoning, especially among children and the elderly, resulting in many deaths in these groups. Doctors attributed many infant deaths to a lack of breast milk in the women's breasts. Bread, the staple food of the Greeks, was disappearing. All classes of society now use carob and herbs. On February 23rd, a lady from the highest levels of Athenian society wrote to a friend in London: If we were in England, we would all be punished for cruelty to animals. Since there is no flour, we make our little bread rations from oats and rather rotten oats that were reserved for the carriage horses. The poor things have nothing to eat, they have become a collection of apocalyptic beasts. We walk as much as we can because they can't carry us anymore."*[557]

Fish was another Greek staple, *"French cruisers stopped the fishing boats and asked the crew if they had joined the Venizelos rebellion. If the answer was negative, they sank the ship and confiscated the catch. Often accompanied by theft of property and violence against the owner, as well as abuse against their ruler. To the protests of the pitiful fishermen, the French commanders replied: If you want to be left alone, all you have to do is get rid of your King."*[558] It was no wonder that the Greeks had the impression that the Allies wanted to damage the King's Greece with the blockade. This was supported by the following episode: When a few ships with flour arrived, they were confiscated by the Allies and forwarded to Saloniki. There the bread baked from it was offered to the King's followers when they switched sides, which many loyal royal followers refused and replied that they would rather die. As the King shared the distress of his subjects, he gained greater esteem among his followers than ever before. Venizelos became an object of hate for many, after all, in their opinion, he alone was to blame for their misery. In many villages the inhabitants cursed Venizelos as a traitor. The hatred continued to grow. As the war correspondent Abbott reports, there are said to have been children's graves where the gravestones read: *"Here lies my child, starved to death by Venizelos."* The censors managed to suppress these facts so that the British public did not know about them.[559]

The Allies had also promised to prevent the Venizelist revolution from spreading by sea or land. When Venizelos had the island of Kythira off the southern Peloponnese occupied by his people on January 10th, the Greek government protested and the Allies sort of kept their promises by declaring the island to be autonomous. But when shortly afterwards, between February and May 1917, the Venizelists took over the islands of Kefalonia, Zakynthos and Kerkyra as well as Lefkas, the Allies had already forgotten their promises. The methods used were horrible: First, a French ship with food appeared at the affected island. The islanders could only buy it if they promised to opt for Venizelos. In most cases, the residents initially rejected this blackmail, but hunger eventually forced them to submit. To keep them under control, the Venizelists received some of the confiscated Greek warships.[560]

Actually, the Venizelists had no business in the neutral zone and in Thessaly, because the Allies had promised to keep them out of that region. But this did not prevent Sarrail's patrols from becoming active there as well. *"General Sarrail's patrols plundered the villages, tormented the rural people and did not even spare the honor of their wives. Anyone who knows the passionate views of Greek peasants towards female chastity can imagine the indignation these outrages aroused. In this case the horror was compounded by the fact that some of the perpetrators were half-wild Africans. In one case (Feb. 22), the colored lechers paid for their lust with their lives: one of their patrols was surprised and massacred."*[561]

Sarrail had no qualms in cracking down on royalist partisans, whom he called *Komitatschis*. Bulgarian irregulars were usually known by this name. *"Mass executions were one of the methods of military tyranny that General Sarrail took delight in, without scruples and with a certain pride."* Again and again he had people shot and houses destroyed or grain confiscated. In a justifying statement, he stated: *"To sum up, the Greek government organizes and supports gangs. The security of our army in the Orient demands that they be suppressed. I gave orders to kill all of these non-combatants. These commands were followed and their execution will continue."*[562] This was precisely the justification used by the Germans in Belgium to justify repressive measures that the Allies classified as war crimes. But in Greece the alleged *partisan phenomenon* did not exist at all. There is no evidence that stands up to scrutiny that the royal government formed partisan gangs to attack the Allies and to cooperate with the German troops. So they were propaganda claims. But there is an indication that these partisan groups were initially equipped by the French secret service to secure the borders to the north.[563]

On the other hand, the forced recruitment measures of the Venizelists in order to attract *"volunteers"* are well-documented. Venizelist gendarmes surrounded villages on the mainland and the islands and recruited the young men. Anyone who tried to flee was shot. Those who hid were found and forcibly recruited. Many fled on the first night. Despite these methods of violence, Venizelos managed to recruit only about 10,000 men in the first six months.[564]

The growing army in Saloniki and the constant German submarine attacks meant that the supply of food to the Allied troops in the course of the spring caused problems. It was therefore not surprising that Sarrail developed the plan to confiscate the grain harvest in Thessaly. This would, on the one hand, make it easier to supply the Allied troops and on the other hand put the royalists under pressure because they would suffer even more hunger. In April, Sarrail received permission from his government, that had been requested since January, to occupy Thessaly and confiscate the harvest. The offensive against Bulgaria forced him to postpone the occupation until May.[565]

In February 1917 the tension between the two parts of Greece rose to a boiling point. On February 5th, Elliot recommended to London that the blockade should be lifted. In his opinion, at least grain and coal should be allowed into the country. But London didn't respond. On March 7th, London asked him to protest to the Greek government against the terrorization of the Venizelists in Epirus. On April 3rd, he confirmed that the Venizelists were being attacked in Athens.[566]

Constantine was concerned about the developments and therefore dismissed Prime Minister Lambros and on February 5th, 1917 reappointed Zaïmis as Prime Minister in his place. He hoped that this would invalidate the propaganda claim that the Greek government was *Germanophile*. Zaïmis began negotiations with the Allies on behalf of the King and tried to bring about a reconciliation with Venizelos. But Venizelos did not want a *rapprochement*[567] and even increased the propaganda machine against Constantine: *"At a signal from the conductor [Venizelos], all the instruments in the orchestra joined the familiar melody. The entire press of France and England played the song of slander and fairy tales. They came out at regular intervals and with constant monotony: conspiracies and secret scripts, bizarre stories about German intrigues, constant repetition of compromised or compromising names; designed to suppress any attempt at accommodation that did not include the return and rule of the great Cretan [Venizelos]."*[568]

Perhaps Zaïmis could have done something earlier, but in mid-March 1917 the Briand government fell and Alexandre Ribot succeeded him on March 20th, 1917. Minister of War was Paul Painlevé. The new French prime minister had even less affection for the Greek monarchy and was influenced by public opinion in France, which was in the meantime extremely hostile to King Constantine. In the eyes of the public and of Ribot and Sarrail, Constantine was in close contact with Kaiser Wilhelm.[569] Minister of War Painlevé did not entirely agree with Nivelle's plans for a major offensive in the West, but initially let him have his way.

On April 19th, a Franco-English-Italian conference was held in Saint Jean de Maurienne. The main topic was the future division of the Ottoman Empire. London and Paris agreed that Italy should receive a large part of southwestern Anatolia, including Smyrna. Venizelos' *Megali Idea* was beginning to become extremely questionable. Prime Minister Ribot accused the Greeks of failing to keep their promise to move the army and war material to the Peloponnese. The Allies were to give France a free hand, which included the deposition of Constantine, the occupation of Thessaly, and the confiscation of the crops there. Lloyd George and his Italian colleague Sidney Sonnino agreed in principle, but only in case there was another massacre of Venizelists in Athens. The monarchy, they said, should be retained. The question of the harvest was to be sorted out later.[570]

The change in French leadership naturally had repercussions on Greek politics. Venizelos began to steer an anti-dynastic course; some of his followers even propagated the proclamation of a republic. In April, Venizelos twice told the diplomatic representative of England in Salonika that the King must abdicate if he wanted to save the dynasty. If he abdicated now, his eldest son could succeed him; later that would be impossible.[571]

In Athens, Elliot saw the situation more calmly: the King had kept all promises and promises, albeit with some delay. None of this was any reason to withdraw the guarantees given to the Lambros government. But the impeachment of King Constantine would not be bad, provided it could be done without violence. An Allied and Venizelist invasion would certainly spark armed resistance.[572]

But now the problem of the grain harvest came back to the fore. At that time, Thessaly was the only area of Greece that produced a considerable grain harvest. The harvest would be enough to replenish the government's grain stores that had been emptied during the blockade and thus ensure the nutrition of ancient Greece for the next six or seven months.[573]

The argument put forward by Cyril Falls in his official history, that if the grain had been in the possession of the royalists, German submarine commanders would have had a free hand to torpedo any grain ship, since it would have been destined for Saloniki, is somewhat inaccurate: How should a submarine commander know what a ship had loaded? He was only interested in whether it was an enemy ship or not. This is where Falls' propaganda legends seem to have been set up.[574] In truth, Venizelos also spread rumors, when he claimed to the British diplomatic representative that an officers' committee had been established, that was to bring in the harvest with the help of the reservists. He did not say how this was supposed to happen. If this proves impossible, the harvest should be torched. The people of Thessaly stood behind him and if the Allies gave him a free hand he would bring in the harvest. On May 1st, the Foreign Office learned that Sarrail was planning to confiscate the harvest for the *Armée d'Orient*, and aptly commented: *"Politically, it is more desirable than if the royalists should have it. But it is morally unacceptable."*[575]

Next, a large demonstration of 30,000 people took place in Saloniki, calling for the King to be deposed. Venizelos said to the British diplomatic representative, unbelievably, that he had had nothing to do with it. The demonstrators had allegedly acted independently to force him to act. He still wanted a reunification of Greece with the heir to the throne as monarch. That all this was pure hypocrisy was evidenced by the statements of a Venizelist newspaper in London that Venizelos, as the savior of Greece, would heal the country from the *cancer of the dynasty*; the healing process would have to be achieved with *"Prussian methods"*.[576]

These maneuvers continued through May: *"Throughout May this concert of subtlety and slander continued: sometimes very quietly, then swelling to a great noise, like a huge, dirty river, through all the Allied capitals. This adds to the people's unspoken desire for new sensations, sharpens their nerves and fills their hatred and fear of King Constantine. At the end of the month the curtain rose and Mr. Venizelos stepped out to make the statement his propaganda has prepared our heads for: "I flatly reject all thoughts of reconciliation, steadfastly and conclusively." His followers joined this choir.*[577]

War correspondent Abbott commented: *"With that, the great Cretan and his companions have finally given up pretending that their plot was not directed against the King. Or that they intended to postpone closing their demands on him until after the war. Their relief must have matched their tension: it is not hypocrisy but the need for agreement that worries the hypocrite. But their burst of openness was mainly a gauge of the attitudes of the powers from which they derived their importance."*[578]

At the inter-allied conference in Paris on May 4th, the French suggested that Sarrail send troops to Thessaly to secure the harvest. The British agreed and were ready to send 500 men. They agreed to offer the Greeks to lift the blockade if they agreed to buy up the harvest and then to distribute it equally to the royalists and Venizelists. The whole thing would be strictly controlled. Venizelist troops should not participate; but should there be an uprising in Thessaly in favor of the Provisional Government, they would be allowed to send troops to Thessaly. These were supposed to ensure peace and order after the establishment of a Venizelist regime in Thessaly. Sarrail was to carry out this mission. Lloyd George reminded though, that military action could only take place with the consent of the Allied governments. But going forward, the Allies in Greece would only speak with one voice - that of France. The French government agreed to withdraw its envoy Guillemin and replace him with a more capable man. The British promised to recall the envoy Elliot and leave the official business of the embassy to the *Chargé d'Affaires*.[579]

Constantine's Last Interview as King

At the beginning of June, Constantine gave an interview to the correspondent for the New York Times, Adamantios Polyzoidis, under the condition that it was only to be published after he gave his approval or if he was dead or deposed. The Entente deposed the King, thus allowing publication on June 14, 1917. The King's statements provide deep insights into his thoughts and feelings and should therefore be quoted in more detail.

"I never feared for my life; [...] I also never tried very hard for my throne, and if I wanted to keep both, then I do it for the fate of Greece and the fate of the Greek people. The only ones I care for and who are close to my heart. [...] Yes, the Greek people love their King and if I ever lose the throne, it is not because the Greek people take it from me. [...] I know it is the Entente who want nothing more from me, not the Greek people. The attempt to push me out is as old as my first objection to the Dardanelles expedition. [...] Certainly I could be the most popular of all kings as far as the Entente Allies were concerned, if I had joined their struggle and led my people into ruin and annihilation. Of course, I would not lose anything, no matter how great the sacrifices and suffering of my people, because that is the lot of kings. [...] I would be accommodated comfortably wherever the Greek capital was relocated after Greece was reduced to nothing in a crushing defeat."

The reporter wanted to know if defeat had been certain. Constantine replied that after two weeks the Greek army would have collapsed when the German, Austrian, Bulgarian and Ottoman troops attacked. In addition, the Turks would have taken action against the Greeks living in Asia Minor.

"Forcing Greece into war would have been the easiest way to my personal fame and advantage [...] but I, the absolutist, the autocrat, who believes in the divine rights of the king, as my opponents tend to call me, I strictly adhered to a pacifist policy. Simply because all the people of Greece who will have to fight when the war comes are against this war. They are against sacrificing themselves in a futile attempt that is of no use to anyone. They call this dispute a struggle for the rights of the weak and the oppressed, and they want us to believe that Greece is neither weak nor oppressed. In fact, we are hardly better off than Belgium. Does it take place to uphold our constitutional freedoms? Nonsense! The present war takes little interest in such minor matters. Your freedom and your constitution only count if they have any material use for the Entente. If your parliament stands for war, it's good. If it votes for peace, it is at most a gang of crooks paid for by Germany. These grandiose names for lofty ideals and popular basic rights are only of value if they are used to shake up a people and march them into the slaughterhouse, which in our time is called a "front". If people want to sit still for the same ideals and mind their own business, then they are nothing."

When the journalist asked the King about his relations with the Emperor and Germany, Constantine replied: "You are the newspaper man [...] and you know how easy it is to bring a lie to life when you have all the means at your disposal necessary for dissemination. The party most affected by the lies is silenced and denied freedom of expression and the benefit of being heard. [...] The general feeling [in Greece] was never for the Germans, just as little was the feeling for the Entente, even if they were preferred. And no one was ever ready to commit suicide for the Allied cause. We were originally victims of the Allied envoys in Athens. [...] The ambassador of France [Mr. Guillemin] and the ambassador of Great Britain [Sir Francis Elliot] behaved more like Venizelist provincial princes than like representatives of the best intentions of their own countries. They just want to take Mr. Venizelos to where I am now. Is that what their own governments want? I can't know, but I doubt it. The ambassadors from Russia [Prince Demidoff] and Italy [Count Bosdari] assure me that they are benevolent and naturally the case of Greece they cannot be. Some of the neutral ambassadors are absolutely non-aligned, but the rest are Venizelists. I am sorry to add that the American envoy [Garret Dropper] falls into the last category." When asked about the possible outcome of the war, Constantine expected a remis. When the reporter wanted to know what the King thought of the Roupel case, Constantine replied: "The liberation of Greece is much more valuable than all the Roupels [Fort Roupel] in the world. Indeed, the liberation of Greece is far more valuable than the Greek throne and the life of King Constantine."[580]

These were clear words that realistically assessed the situation. They proved once again how right Constantine's decision to keep Greece neutral had been. In a sense, his statements formed his creed. Unfortunately, this interview was not noticed by the public or political actors at the time, or later by research.

The Deposition of King Constantine

Due to the failed spring offensive, the British announced at the Paris conference in early May that they intended to significantly reduce the number of their troops. With that, the British General Staff had finally prevailed. One division and two brigades would be withdrawn. And with that decision, the British *de facto* left the decisions about Greece more and more to the French.[581] On May 28th, the British Secretary of War presented the Cabinet with the latest French proposals, about which his colleague Painlevé had informed him. The French wanted to march into Thessaly, depose the King and set up a Venizelian government. The French felt no hostility to Constantine or any particular inclination towards Venizelos, but the former was hostile to them and the latter was the only politician they could rely on. The advisers to the Navy and the Army flatly rejected this plan. A lively and controversial discussion ensued, and they decided to invite the French leadership and their military advisers to a conference in London.[582]

But in the meantime things had happened that greatly changed the previous balance of power. The Russian tsar abdicated in mid-March 1917. He was a direct cousin and friend of Constantine and in the past had repeatedly supported the King and slowed down the French. The Kerensky government promised to continue the Tsar's policy on Greece, but the fact was that Constantine had lost one of his most important supporters. The change of government in France to Pivot tightened the anti-Constantine course. When the US entered the war on April 6th, the Greek government lost its most important neutral ally. Finally, on April 19th, 1917, at the St. Jean-de-Maurienne Conference, Italy received binding pledges for territorial gains in southwest Asia Minor that reached as far as Smyrna. This made Venizelos' ambitions in the same region obsolete, so Rome was now ready to support him.

The conference began on May 28th and lasted two days. On the French side, Prime Minister Ribot and Minister of War Painlevé took part. The French had long wanted to overthrow King Constantine. At first they had hoped that this would happen with a revolution, through Venizelos, but then they realized that this wouldn't happen because the forces supporting the King were too strong. But since the French politicians were aware that a deposition of the Greek monarch, who was, after all, the brother-in-law of Kaiser Wilhelm II, would be well-received by the French public, they continued to push it. In addition, toppling the King would distract the public from the fact that the government's policy towards Bulgaria had failed.

When the French presented their ideas to the conference, *"our allies on this side of the English Channel found scruples, they repositioned themselves to show that these scruples had little weight against their intentions. Even after a fundamental agreement was reached, there was still uncertainty about the dethronement of the king without bloodshed. But the French did not share this uncertainty and after three days of tough negotiations they conveyed their views to the English ambassadors. It was agreed that this operation should be carried out without war."* The only military action the British agreed to was the establishment of harvest control posts and the occupation of the Isthmus of Corinth to prevent the King from attempting to draw troops from the Peloponnese. *"After we had soothed their consciences, the British envoys refrained from further investigation."*[583]

Finally, a solution to the question of Constantine's removal from the throne was found. Lloyd George invited Secretary of War Painlevé and a few others to his country house: *"We sat under a tree in the garden for several hours on a beautiful summer night to discuss the whole problem. We came to the unanimous decision that the French, who had troops nearby, should land them in Piraeus. Then they should occupy Athens, dethrone the King and install either the heir to the throne – Prince George – or his second son – Prince Alexander."*[584]

With that, Constantine's fate was sealed. Controlling the harvest in Thessaly was not a major problem, but how could one remove King Constantine, who had fulfilled all the demands of the Allies, albeit slowly. A nonviolent way had to be found in order to avoid war. Senator Charles Jonnart, who had been Governor General of Algeria for nine years and had been a member of the Senate and chairman of the Foreign Affairs Committee since 1914, was appointed High Commissioner of the three protecting powers of Greece – France, Great Britain and Russia. As an agent of the three powers, he would be superior even to Sarrail, and he was ready to use any diplomatic and military means to accomplish his mission.

Jonnart, who was a lawyer by training, was looking for a legal path. He claimed that the King, clearly of his own free will, broke the constitution guaranteed by France, Britain, and Russia. In doing so, he lost the confidence of the protecting powers, and they therefore considered themselves freed from their obligations towards him, which had arisen from their protective rights. Jonnart does not seem to have been aware of all the facts of Greek history: In Greek history there were three treaties which discussed protecting powers: The earliest is that of 1830, from which the Greek state emerged. It said that the three powers guarantee the existence of the monarchical and independent state. The second contract was from 1832 and the corresponding passage was almost identical. The third treaty, of 1863, adopted the wording from 1832. The guarantees therefore all related to the political existence of Greece and not to the state or form of government. The first article of the 1830 Protocol states: *"Greece should form an independent state. It should enjoy all the rights – political, administrative and economic that are part of complete independence."* Nowhere is there any mention of the protecting powers having the right to interfere in the internal affairs of Greece, because Greece is defined as an independent state.[585]

The imminent intervention of the Allies had just as little a legal basis as any previous one in previous years, against which the Greek government had protested again and again. Abbott correctly writes: *"Legitimacy was named to obscure the imperatives of politics: the impeachment of King Constantine was ordered not because it was legitimate, but because France demanded and England, for good reasons, could not allow France to do it alone. What Russia thought of this action soon became astonishingly clear. Petrograd not only withdrew its troops from the show, but made quick work of the arguments about "guarantees" and "protection". It firmly stated that "the choice of the form of government in Greece and its administrative organization belong exclusively to the Greek people."*[586]

On May 1st, the French government ordered Sarrail to speed up preparations for the occupation of Thessaly and the isthmus of Corinth. On the same day, Russia and Serbia announced that they rejected any violent action.[587] Jonnart embarked on a French destroyer in Brindisi on June 4th, 1917. In Corfu he had a conversation with Admiral Dominique Gauchet, the commander in chief of the Allied fleets in the Mediterranean and successor to du Fournet, in order to create the necessary maritime conditions for his operation. Then he sailed through the Corinth Canal to Salamis. On board the cruiser, he spoke to the French and English ambassadors. During these discussions, Jonnart must have found out that Zaïmis was considering resigning. Had it come to that, a military cabinet would certainly have been established, which would certainly have resisted his actions. In order to prevent this, he managed to get information from Zaïmis through unknown channels that he would wait until Jonnart returned from Saloniki before he might resign.[588]

Elliot also informed Jonnart that the British Foreign Secretary believed that King Constantine should be told that his impeachment was temporary and only for the duration of the war. In his conversation with Jonnart, the Russian envoy expressed himself completely negatively. Jonnart learned from the French military attaché that the reservists, led by General Dousmanis, intended to occupy Thessaly.[589]

Then Jonnart went to Sarrail in Saloniki to talk to him about the necessary preparations for the army: the simultaneous occupation of Thessaly and the Isthmus of Corinth and a landing of troops near Athens. Jonnart then conferred with Venizelos, who thought Jonnart's previous measures were good, and suggested that the public should be given some time to calm down after the deposition of Constantine. There was talk of several months. Only then would he, Venizelos, return to Athens. Venizelos was also of the opinion that Prime Minister Zaïmis should remain in office in the meantime. If the latter resigns sooner, a military government would likely be established and armed clashes could ensue.[590]

On June 6th, Lloyd George asked his colleague Ribot to recall Sarrail immediately. He had been responsible for the failure of the offensives in Macedonia and no longer enjoyed the trust of the British government. The French government demanded that the release be postponed until the end of the forthcoming operation in Thessaly, Athens and Corinth.[591]

On the morning of June 9th, Jonnart left Saloniki, accompanied by General Charles Regnault, who was to command the troops at Corinth and Athens. In the evening he arrived at the roadstead near Salamis and boarded the ironclad ship *Justice*. That night, a telegram arrived from *Quai d'Orsay* informing Jonnart that the British were protesting against allowing troops to disembark in Piraeus - contrary to what had been agreed to at the London Conference - before the ultimatum was given to King Constantine. He should take the British reactions into account for every step. Regnault, on the other hand, thought that the resolutions of Saloniki should be adhered to, and that the delivery of the ultimatum and the landing of the troops should be carried out in a synchronous fashion.[592]

In Paris, on that day, a protest note had been received from the British government. In it the

British refused a violent deposition of the King. According to the English view, it would contradict the agreements of the London Conference. It would be better to convince the King to leave the country until the end of the war, with his son acting as viceroy. Russia, Italy and Serbia supported this position. Paris informed Jonnart of this and let him know that the British would probably do nothing against a *fait accompli*.[593]

On June 10th, Jonnart and Zaïmis met on board the *Bruix* in Piraeus. Jonnart now had to get Zaïmis to stay in office until Venizelos could take power in Athens. He knew, however, that Zaïmis would not do this if he knew what role he was actually supposed to play. Therefore, at the first meeting, Jonnart only spoke about the purchase and control of the Thessalian harvest and the occupation of the Isthmus of Corinth. Zaïmis had heard rumors of further demands and was therefore pleasantly surprised by Jonnart's lesser demands, and agreed. In order to gain his trust, Jonnart spoke of the great future that the protecting powers had in store for Greece. Under the pretext that he was expecting new instructions from Paris, Jonnart ended the call and made another appointment for the next morning.[594]

When Zaïmis appeared for the second meeting the next day, Jonnart handed him the ultimatum, in which it was stated that the protective powers of Greece had decided to restore the country's unity. *"His Majesty King Constantine obviously broke the constitution of his own accord, for which France, Great Britain and Russia are the guarantors. I have the honor to declare to Your Excellency that the King has lost the confidence of the Protecting Powers. They see themselves freed from all obligations that arise from the property rights, as far as they concern the King."* The King must abdicate within 24 hours, appoint his second son Alexander as his successor and leave the country.[595] The Allies and Venizelos rejected the legitimate heir to the throne, Prince Georg, because he had completed his military training in Germany. On the one hand, Constantine was asked to breach the constitution, as this provided for a *primogenitur succession*, on the other hand, the argumentation and call for resignation were based on false arguments: the protective powers guaranteed the existence of Greece as a state, but not its constitution.

Zaïmis was horrified and overwhelmed with emotion. *"Mr. Jonnart spoke eloquently and urgently. The powers only sought the unity and freedom of Greece, the greatness of Greece now divided, partially torn and in a state of anarchy on the eve of a civil war. The High Commissioner would do everything in his power to ensure that the transfer of power could be carried out as peacefully as possible. He made an urgent appeal to the prime minister's patriotism."* When the war was over, the powers would not mind if it were the desire of the Greek people for Constantine to return to the throne. Zaïmis would be permitted to announce this publicly. The powers had no intention of *"bringing Venizelos back: once the unity of Greece was achieved, the Saloniki government would disappear. Later, Mr. Venizelos could legally return to office after a new election. On the other hand, if the ultimatum were not kept, they threatened to bring the entire dynasty down, the forcible establishment of a republic and the immediate return of Venizelos. The conversation ended with the dark explanation, if the decree were not followed exactly to the letter, he would do the same thing with Athens as the Germans would have done with his hometown Arras, namely reduce it to a heap of rubble."*[596]

Zaïmis realized that he had no chance against Jonnart and agreed to deliver the ultimatum to the King and to remain in office for the time being. After the King had received the message, he convened the so-called Privy Council, a meeting of all former prime ministers. Dimitrios Rallis, Stefanos Dragoumis, Stefanos Skouloudis, Dimitrios Gounaris, Spyridon Lambros, Nikolaos Kalogeropoulos and Nikolaos Stratos showed up. The King declared that he had decided to leave the country with the crown prince in order to spare him the disaster that the Allies threatened. But only Stratos and Zaïmis were in favor of submitting to the Allies. The others admitted that Greece had no chance of defending itself, but it would be better for the King to be driven out by force than for him to submit. He had been humiliated enough in the past. But the King stuck to his decision and made everyone promise that they would do their best to ensure that there was no disturbance that might worsen the situation. At the end of the session, everyone left the room with tears in their eyes.[597]

At the Cabinet meeting that afternoon the King maintained his position despite the opposition of the ministers. Zaïmis was tasked with formulating the official answer to Jonnart. Jonnart was informed of the content and was satisfied. The next morning it was presented to Jonnart. It said very briefly that the King, who always had the best interests of Greece in mind, had decided to leave the country with the crown prince and that he had designated Prince Alexander as his successor.[598] Constantine did not mention the abdication with a single word.

So far, the impeachment had gone smoothly. But now protest was stirring in Athens. For the previous two years the Athenians had suffered repeated humiliations and had become sensitive. When Jonnart appeared in Athens on June 9th, the rumor mill began to turn. On June 10th, it was reported that the government had been given an extreme ultimatum. Groups formed and had heated discussions. Zaïmis published a statement, the text of which was intended to reassure the Athenians. This worked for a few hours, but when Zaïmis returned from his second meeting with Jonnart, it was whispered that the King's abdication had been requested. The short-term convocation of the Privy Council intensified the rumors. When the worried faces were seen at the end of the session, tensions increased. When information about the occupation of the Isthmus of Corinth and reinforcement of the fleet at Salamis arrived in the city, and people learned that the King was about to leave, the shops closed, the church bells rang as at a funeral, and crowds of people streamed from all sides to the royal palace to prevent the King's departure. The people around the palace asked for their monarch and shouted: Don't go! Citizens' deputies and the Athenian garrison went to see the King and asked him not to go. But Constantine stuck to his decision. He was born and raised in Athens and is a Greek *"to his bones"*. But in this situation the security of the country had priority. He must do his duty and leave. Zaïmis tried in vain to calm the masses with Allied promises that the King would return after the war.[599]

"Nothing could alleviate the widespread pain. As darkness fell, Athens was a strange sight: silent figures marched, one after the other, to the place where King Constantine spent the last night in his capital. They made their hopeless pilgrimage without the slightest noise, and as they walked they met groups who returned just as noiselessly. "It was," said an eyewitness, "as if the people of Athens were visiting a grave or a prisoner." The crowd kept vigil all night. At around 4:30 a.m., a vehicle was seen driving to a side entrance of the palace. The crowd recognized the King's chauffeur and suspected he had come to pick up the King and his royal family, who were at the door. The guards threw themselves on the ground to show that the vehicle would have to run over their bodies. The King and the royal family withdrew and the car drove away empty. Two more attempts to leave the palace were also unsuccessful. The crowd would not allow a door to be opened. Unified and quiet, those present set up a guard."[600]

The crowd didn't leave the next day either. A printed appeal from the King did nothing, much less an appeal from the new King. The crowd wasn't impressed. The majority of Athenians loved their King, the Venizelists, of course, did not. The paradox is that the royalists loved a King who was not at all Greek by blood, but he felt himself to be one. His father was a Dane and his mother a Russian. Hardly any other Greek politician or military man was so loved by his supporters and so hated by his opponents as Constantine.

In the afternoon Jonnart informed Zaïmis that the King's departure could not be delayed any longer. If the Greek police were unable to disperse the crowd, he would call in some machine gun companies from Piraeus. At around 5 p.m., the royal family managed to leave the palace through the main entrance after a deception. They drove to the summer residence in Tatoi in a car. On September 14th, Constantine and his family embarked in the small port of Oropos and set course for Italy. Their final destination was Switzerland.

But the Allied propaganda continued. *"Press reports from London and Paris reported that King Constantine and Queen Sophie went to Germany when they left Greece. Crown Prince George allegedly had volunteered for service in the German army. A constant stream of propaganda turned out to be just as baseless as the allegations against King Constantine. This was partly to excuse France's behavior towards him, that Great Britain tolerated."*[601]

The Russian envoy reported to Petersburg on June 16th: *"I have to regret it internally that the Saloniki expedition planned for the liberation of Serbia has degenerated into a campaign against disarmed, peaceful Greece [...] I have the courage to note that the course taken by France with regard to Greece in no way corresponds to my concepts of political morality and, in my opinion, hardly arises from general reasons of expediency. But now forcing Venizelos upon ancient Greece in connection with maintaining the unheard-of six-month blockade is not only not in accordance with the solemn promises of the powers [...], but in my opinion completely contradicts the highest principles of freedom proclaimed by the [Entente] association of small nations and the right to self-determination."*[602]

In Athens all remained calm after the King's departure, and Jonnart's announcement that the blockade had been lifted may have contributed. In a proclamation, he announced that from now on, constitutional conditions would be met in Greece. The protecting powers would ensure that civil liberties were respected. Mobilization was not planned.[603] Although the blockade was formally lifted, the supply situation remained poor. There were no transport ships. Even in November 1917, the *Daily Mail* had to admit that the famine became more widespread every day. Even when Venizelos presented himself to the Allied capitals, little changed. The Allies continued to use the confiscated Greek ships for their own purposes.[604]

In Thessaly, on the other hand, there was bloodshed. A freshly assembled French division advanced in three columns from Grevena, Servia and Katerini through the Tempi Valley towards Larissa. The commander of the 1st Greek Division had orders to defend himself against irregulars. Should Allied troops penetrate into Thessaly, any clash should be avoided, even if they were accompanied by Venizelist units. The advance began on June 10th. At first everything remained very peaceful. But when French cavalry entered Larissa, tensions rose. When the Greek general met the French general on the outskirts, the latter declared him and his staff under arrest. If a shot was fired, he would order Larissa to be shot at in return. The French troops then entered the city and disarmed the Greek units. When the Evzones regiment was disarmed, and the French troops asked for the officers' swords, these felt that their honor had been attacked. Under the leadership of the regimental commander, around 100 soldiers tried to move south. They were pursued by Moroccan spahis and a skirmish broke out in which both sides suffered losses. The French quickly surrounded the Evzones, and 49 officers and 269 men were captured. On June 13th, Velestino, Volos and Trikala were occupied. The arrested officers were taken to Katerini and imprisoned. The soldiers were sent to Litochoro to build roads. When the regiment stationed in Trikala heard of this, it left and marched through the mountains to Lamia. On June 26th, Allied cavalry occupied Lamia and Itea. This turned Thessaly into occupied territory. The official history of the Greek General Staff states: *"The behavior of the French volunteers, especially that of the Africans, was appalling. The residents suffered terribly from the looting, the destruction of the farms, the rape of their wives and other acts during the occupation."*[605] In the official English version, however, the incident was downplayed and alleged that the people of Thessaly had shown themselves to be Venizelos-friendly.[606]

The Greeks had trouble not only with the French, but also with the Italians. As already described, the Italians occupied northern Epirus in September 1916. Since January 1917, they had been advancing further south and invaded the Greek city of Epirus. At the beginning of April, they reached Delvinaki, Konitsa and Filiates and drove out the Greek administration. According to rumors, they then wanted to advance on Igoumenitsa, Paramythia and Preveza. On June 8th, an Italian cavalry regiment appeared near the village of Lykostomo, a few kilometers north of Ioannina. The local Greek commander opposed them and reported the incident to the 9th Division under General Georgios Mavrogiannis. He consulted with Athens and sent one of his two regiments to reinforce Lykostomo. He himself went forward to make it clear to the Italian commander that he could not occupy Ioannina. If he tried anyway, he would face resistance. The Italian was not interested, but gave him until the afternoon to consult the Greek government. At 11.30 a.m. the Athenian answer came: he shouldn't resist the occupation of Ioannina. Mavrogiannis informed the Italian, who told him that he and his troops, as well as the Greek authorities, had to leave Ioannina by 6 p.m. At 6:30 p.m. Italian cavalry occupied the city.[607]

On June 9th, the 9th Division reached Arta. The Preveza garrison was ordered to withdraw to Amfilochia. Many soldiers from the Epirus region deserted and returned to their villages with their weapons in order to protect their families from attacks by Italian soldiers. On the same day, the Italians occupied the Metsovon pass and Paramythia. The Italian advance angered the French and English, because Epirus was actually supposed to come under French control. In order to enforce these claims, French troops tried to land in Preveza on June 12th but the Italians prevented them from doing so. The landing was only successful on June 16th. The French commander asked the Greek unit there either to join the forces of Venizelos or to withdraw to the south. The garrison withdrew to Arta. On June 10th, the French occupied Filipias on the road near Arta. Finally, on June 16th, 1917, the Italians published a proclamation decreeing the unity and independence of Albania. The Albanians were called upon to organize themselves accordingly. The proclamation was signed by the Italian king and the military commander of the Italian troops in Albania. This sparked a conflict that lasted until World War II.[608]

The occupation of the Isthmus of Corinth went, to the amazement of the French, without the slightest difficulty. The Greek garrison greeted the newcomers in a friendly manner, and the French realized how ill-informed they were. In response to the British protest, French troops landed not in Piraeus, but at Eleusis, which did not change the actual situation. On June 12th, the troops went ashore. In order to break any resistance, the battleship *Verité* had entered Piraeus with General Regnault on board. The crew was on high alert and the guns were aimed at the city. Immediately after landing in Eleusis, the French troops marched on Athens and Piraeus and occupied important strategic positions. The Greek troops withdrew to their barracks and the population looked on indifferently. On the evening of June 12th, Jonnart received a telegram of praise from Prime Minister Ribot: *"The results you have achieved will crush any protest from the British government. If you read my telegrams you will understand that no matter what the outcome, I would have protected them. I could not allow the British Government to assume that at the time of their departure from Paris we had already agreed to ignore our commitments to London. They made their decision after understanding the situation. Success justifies your actions, and the British Government owes you thanks which we hope will not be denied to you."*[609] In reality, as he had predicted, the British accepted the *fait accompli*.

The American war correspondent Hibben saw things a little differently in May 1920: *"... the constitution of a small but brave and noble people was ruined by the joint efforts of two of our allies. The neutrality of a small country has been violated, the will of the people destroyed, its law broken, its citizens persecuted and the press silenced. A government was imposed on a free people through violence, and through violence a government was and is kept in unrestricted power to this day."*[610] In his opinion, Greece had become a French colony.

Jonnarts promises quickly proved to be empty. He had two lists of people against whom he now moved. Venizelos had drawn both proscription lists up.[611] The first list contained 30 names, including ex-prime minister Gounaris, former mayor of Athens Spyros Merkouris and his son Georgios, MP Georgios Pesmazoglou, ex-chief of staff Dousmanis and Colonel Metaxas, and writer Ion Dragoumis, who were handed over to the French military. The people on that list were then deported to Corsica. Ion Dragoumis had written an article criticizing the Entente and Venizelos.[612] Ex-Foreign Minister Streit had suspected that he was on this list and had gone into exile with the King. On the second list were 130 names, including the two ex-prime ministers Skouloudis and Lambros, six ex-ministers, a general, an admiral, other high-ranking officers, lawyers and journalists. They were all monitored by the police from that moment on, and were under house arrest. The three brothers of the King were exiled to Switzerland with their families. Members of the lower classes who, according to the French secret service, had taken part in the *Noemvriana* events were arrested and brought to justice. In order not to compromise himself, Jonnart forced Zaïmis to carry out this purge. Zaïmis had resisted at first, but Jonnart had imposed such pressure on him that he submitted and obeyed.[613]

If any one of these "suspects" tried to escape, the police tracked him down and his property was confiscated. Many royalists were arrested and taken to a camp on the island of Lesbos. A total of 600 royalists from all over Greece were interned there. Jonnart had promised that there would be no reprisals but nobody cared, and neither did Jonnart himself.[614]

In Jonnart's opinion, these reprisals were necessary because it had meanwhile been decided to bring Venizelos to Athens earlier than originally planned. In Jonnart's opinion, the Athenians were friendly towards himself and France. When he showed himself to the Athenians, they shouted *"Vive la France"*. Abbott's comment is correct: *"At all times, beginning with the time of the Roman consul Flaminius, that Greeks could be found who loved their liberators more than freedom."* Venizelos knew his countrymen better and was not sure whether he would really be welcome in Athens. The French could certainly install him with their presence of power, but he did not want to be imposed and therefore preferred to delay his return a little longer. But Sarrail wanted him out of Saloniki because he'd had enough of him by now. So Venizelos, accompanied by Andreas Michalokopoulos, arrived in Piraeus on June 21st on board a French warship.

There they were lodged on the same ship as Jonnart. On June 22nd, Jonnart held a meeting with Prime Ministers Zaïmis and Venizelos. It quickly became clear that a reunification of Greece could only take place if Venizelos took over the government. Legally, this could only be done through new elections. But neither Jonnart nor Venizelos wanted that, since neither of them was sure that they would win the desired majority. Therefore, they used a constitutionally dubious trick: the parliament, which had been elected on June 13th, 1915 and in which Venizelos held the majority of the deputies, would be reconvened on the grounds that it had been dissolved illegally. Jonnart correctly conveyed this proposal to Zaïmis, who agreed and did what was expected of him - he declared that he was stepping down. Jonnart thanked him and the new King Alexander entrusted Venizelos with the formation of the new government. He asked for two days' time to allow his cabinet colleagues come from Saloniki. Zaïmis then postponed the declaration of his resignation to June 24th, 1917.[615]

When it became known in Athens that Zaïmis had resigned and Venizelos had been entrusted with forming a government, around 3,000 citizens gathered in Syntagma Square and marched to

Zappeion. There the crowd protested loudly against Venizelos and the Allies. Venizelos felt that the purge of Athens had not been thorough enough and asked General Regnault to order the immediate occupation of Athens by French troops. Anyone who opposed the Allies should be shot. The occupation took place on June 25th. The surroundings of the Lykavittos, the Acropolis, the Pnyx, the Theseion and the stadium were each occupied by a battalion. Two battalions each were held as reserves on Filopappos Hill and in the Makrigiannis district. Three batteries stood between Syngrou Boulevard and Filopappos Hill. A little later, Venizelos' "*Life Guards*", the 9th Cretan Regiment, and 400 Cretan gendarmes from Saloniki arrived in Athens. Venizelos was sworn in as the new Prime Minister on June 27th, 1917. The French troops, with the exception of those on the Pnyx, the Lykavittos, and the Acropolis, were withdrawn.[616]

Venizelos' return had to be accompanied with a solemn ceremony. Venizelos hesitated, knowing he would always be blamed for being brought back to power by the Allies, but he could not avoid the show. He was to officially go ashore on the morning of June 27th, 1917: *"At the agreed hour, the French troops entered Athens with their machine guns and occupied the most important places on the route that Venizelos was to take. The area around the palace, where he was supposed to swear his oath of office, and the palace itself were guarded by 400 Cretan gendarmes. His loyal bodyguards came from Saloniki. Notwithstanding all these precautions, Mr. Venizelos and his ministers objected to facing the enthusiasm of their fellow citizens. So they drove straight to the palace at top speed and avoided the central route. Then they drove to the Hotel Grande Bretagne, in whose corridors his Cretan supporters also stood guard. Thanks to this vigilance, as General Regnault observed, no assassins were seen on any street corners, nor in the passageways of the palace and the hotel, as the prime minister and his friends had feared. But Mr. Venizelos took no chances as far as his own life was concerned: a Cretan regiment landed in the afternoon to replace the foreign battalions."*[617]

On the evening of June 27th, an organized crowd gathered in front of the *Hotel Grande Bretagne*. Venizelos appeared on the balcony and made one of his eloquent speeches. He condemned the previous system and promised improvement. *"He would make everything right again and he would bring the system back to vigorous activity. Any impurity would be purged, and pure, fresh blood would circulate throughout the body of politics, giving health to every fiber of the state. With regard to external affairs, he felt it needless to say that Greece's place was on the side of the powers that were fighting for democracy."*[618] A day later Venizelos recalled diplomatic representatives of Greece from Berlin and Sofia, saying that the Revolutionary Government of Saloniki had declared war on Bulgaria and Germany on November 23rd and 24th, 1916. Greece's neutrality was over. With this step Venizelos hoped to get closer to his dreams of the *Megali Idea* and to be rewarded with territory at the peace conference, but he did not have any written contractual commitments from the Allies. He had done what Constantine had always consciously avoided; he had led Greece into the war at his own risk and without Allied backing and security.

When Jonnart left Athens on July 7th, 1917, he had every reason to be satisfied with his work. He had successfully installed the protégé of France as Prime Minister and chased the alleged *German friend* Constantine from the throne. He had ruthlessly pushed through the interests of France in every respect and confronted the British with a *fait accompli*. Greece's neutrality was eliminated; the country would now fight on the Allied side. Nobody cared that he had done all of this with brutal force and that Greece was practically an occupied country with a collaborative government. Greece now had only one government, but that did not end the division of the country.

The New Regime

Venizelos actually had wanted to get rid of the monarchy, but he was unable to do that. However, King Alexander was inexperienced and had a weak character, and he signed everything that was presented to him. The revitalized parliament, in which Venizelos had the majority, was nothing more than an institution that validated his decisions.[619] In the throne speech on the occasion of the reopening of parliament, written by Venizelos, the King announced that Greece had offered the powers leading the war its own weak forces to those who fought for the rights of nationalities and the freedoms of peoples.[620] Alexander, therefore simply repeated the war declaration that Venizelos had already given.

Venizelos knew that domestically he would be in a much more difficult position. Only if he succeeded in eliminating the old regime could he implement his plans. To do this, it was necessary for him to rid himself of its supporters, and for that he needed a suitable instrument, namely martial law. He had parliament agree to this. When the opposition fiercely opposed the restriction of freedoms, *"[he] emphasized the fact that Parliament was determined to crush any attempt at overthrow with an iron fist."*[621] The Parliament was a willing tool in this endeavor.

In order to break the resistance from the judiciary, Venizelos announced that he would clean it up and shield it from political interference. All courts were cleared of royalists, and the dismissed judges were replaced by members of the Venizelist party. This reversed the 1909 reform carried out by Venizelos himself: *"In this way, the harmful link between legal and political forces that had been abolished in 1909, perhaps the most beneficial achievement of the reconstruction period, was restored. Venizelism became the indispensable condition for going to court with an opportunity to experience justice."*[622]

The clergy had a similar experience. Those who participated in the curse were exiled to a monastery unless they submitted and revoked their teachings. The civil service was also cleaned up. *"All the prefects [nomarchs] and many minor functionaries were fired. Schoolmasters were let go by the hundreds. The National University, National Library, National Museum, and National Bank all underwent a careful cleanup. In each department, the worst traditions of the penal system that existed before 1909 were refreshed and revived. [...] About two thousand army and naval officers, from generals and admirals down, were sent into retirement or placed under arrest. And an almost hysterical need to threaten every civilian with violence who, in his opinion, was behaving reprehensibly and whose position was dangerous for the new order: a tactic full of brutality in the treatment of Mr. Venizelos' main opponents."*[623]

Others were jailed for years, including a former minister who was incarcerated until 1920. With the exception of Admiral Koundouriotis, all members of Skouloudi's cabinet were charged with treason for leaving Fort Roupel to the enemy. The 82-year-old Skouloudis and the 77-year-old Dragoumis remained under the surveillance of gendarmes for the rest of their lives. The members of the Lambro cabinet fared not much better. The officers of the Kavala garrison who *"treacherously"* surrendered were brought to justice. Their trial dragged on for years without judgment. Obviously, opponents of the regime were to be morally disqualified. Anyone who had collided politically with Venizelos in the past experienced his revenge. *"He acted like he wanted to enjoy their humiliation and he held them down to take advantage of their helplessness."*[624]

Even the exiles were punished. *"Arch traitor"* Constantine's pension was cancelled. Nobody from the royal family was allowed to travel to any land of the Entente. Visiting them was forbidden. But not only royalists were persecuted. Even peasants who had a picture of their King in their house were punished. The persecution of oppositionists even took on grotesque features: *"A woman was dragged into a police station because her parrot was heard whistling the Constantine March."* Anyone who baptized their son *"Konstantinos"* had to expect trouble.[625]

What Abbott calls the *Spoil's system* is nothing other than Greek clientelism. The new system gained followers through the distribution of money and the creation of public service jobs - a practice that continues today but was first used on a large scale in 1917. *"In addition to the steadily growing swarm of native parasites, profiteers, casual workers and adventurers who thrive on the profit of the people, a less numerous, but no less voracious crowd of foreign donors, licensees and contract hunters marched. The interests of the state were frivolously exchanged for the support of the party from the Entente capitals."* All of this led to a steadily growing national debt - a well-known phenomenon again today.[626]

Abbott's judgment is correct, if somewhat sharply phrased: *"Such a terrorization of opponents and preferential treatment of the supporters, such a promotion of the oppression and tolerance of corruption, such a prostitution of the judiciary, such a cynical indifference to moral principles - unprecedented even in the History of Greece - made the Cretan regency hated and despicable. What made the government even more hateful in the eyes of the people was the fact that it had been forced upon them by foreign powers and that it functioned under false pretenses. As free men, they resented the violence done to their freedom. But as intelligent men, they would hold less against open violence than the desecration of the concept of freedom, which adds mockery to anger and means a daily insult to their intelligence."*[627]

Of course, Venizelos was not responsible for everything that happened on his behalf. Much happened that he would have protested against. But he had started it all. He had drawn up Jonnarts Proscription lists; Jonnart would not have had the knowledge to create such lists. Venizelos began inflating the civil service by creating two more ministerial posts in his cabinet than had been the case previously. But once he started, he couldn't stop it, he became a prisoner of his own system. At the same time, however, he programmed the next change in power, which would be just as radical as his own. None of this had much to do with democracy. It led to a permanent conflict between the two camps, which continued until the Second World War and, after a shift to the left, lasted until after the civil war.

On December 18th, 1934, Metaxas wrote in daily newspaper in Athens about the humiliations that Greece had to endure until it was finally forced into war: *"Mr. Venizelos could ask us: what does it all matter if he succeeds in emerging from the war with a Greece double in size and strength? That is the "big lie" and we should prove it to him when we discuss the Treaty of Sevres [...] But even if Greece had come out of the war with additional territory, which was not the case, due to Mr. Venizelos' policy of using foreign troops to subjugate his fellow citizens. Greece was morally broken, it has lost its national pride and self-respect. The habit of being mercenaries penetrated his blood - even towards its inhabitants - and the total, exclusive devotion to the fatherland had diminished. Such moral decline cannot even help preserve territorial gains."* A little later he added: *"Perhaps they will tell us that the successes from the victory will cover up all these grievances. Do you think so? Do you think there will be a success great enough to conceal the ruins that such desecrations and misfortunes did to Greece?"*[628]

The Allied Spring Offensive 1917

In February 1917, Sarrail presented his new offensive plan, which was based on orders from the French High Command in Chantilly and was intended to bring the breakthrough on the Macedonian front. The basic idea looked like this: there would be a primary and a secondary offensive, as well as some diversion operations. The main attack would be led by Serbian and Allied troops east of Monastir through the Veternik Mountains. His goal was to reach the Axios (Vardar) north of the Greek-Serbian border and thus stab the enemy in the back. The side offensive would start from the Monastir area. Their destination was the city of Prilep. A flanking attack on the left bank of the Great Prespa Lake was intended to aid this operation. Diversion attacks were to take place in several places along the front lines. Since the weather made it impossible to start the operation before mid-March, the operation could be logistically and tactically well prepared.[629]

In mid-February the front line north of Korytsa was shifted further to the north, so that it now ran between Lake Ochrid and the Great Prespa Lake, which allowed the Allies to fully control the Strait of Ag. Saranda on the Adriatic to Monastir. In mid-March the attack began on the *Height 1248* west of Monastir, which was an important observation point. The fighting was tough and the casualties were heavy, but by mid-April the heights were in Allied hands. The actual offensive had to be postponed again and again due to bad weather conditions. Finally, Sarrail decided on the launch: The British army was to attack on April 24th between the Axios (Vardar) and Lake Dojran. The Franco-Italian army on the left flank was to begin its advance on Prilep on May 5th, as was the French 122nd division west of the Axios (Vardar). Finally, on May 8th, the Serbian army was to attack.[630]

The strength of the defenders on the front from Lake Prespa to Lake Dojran amounted to 163 battalions, of which 143 were Bulgarian and 20 German. From Lake Dojran to Orfanou Bay there were 46 Bulgarian and 9 Turkish battalions. Altogether there were a total of 218 battalions. On the Allied side, the forces were roughly equally large, assuming that the approximately 22 deployed divisions together had a little over 200 battalions.[631]

Despite the difficult conditions with heavy losses,[632] nothing was achieved during the fighting other than a few position shifts. 450 officers and 12,500 men paid for the conquests of a few hundred meters of land and the occupation of some trenches with their lives. The opponent's losses, however, were considerably lower. Almost no prisoners were taken. At the end of the fighting, the units were exhausted and further offensive operations were temporarily ruled out. The original plan had been good, but it was badly executed. The attacks were unsynchronized one after the other. Nowhere did the attackers achieve a local superiority. The Bulgarians again turned out to be good soldiers. If things got sticky at any point, a German intervention reserve was quickly on the spot. With the Allies, cracks between the nationalities became apparent. The Russians had been infected by the revolutionary movement at home. The Italians were not very interested in these operations. The Venizelian units were brave, but not yet fully trained. The British wanted to leave and the French no longer trusted their commander. Sarrail's loss of confidence weighed heavily. *"The tragedy of the army was that there was no hope of any decisive success under its commander-in-chief, although he was experienced, resourceful, and in some ways remarkably gifted."*[633]

On May 23rd, 1917, Sarrail stopped the attacks. He ordered the troops to pretend towards the enemy that the attacks would soon resume. But then, British units were relocated to Egypt. French troops were detached from the front, and went to occupy the neutral zone at Grevena and Katerini. Others had to bring Thessaly under their control, and still others had to secure Athens and the Isthmus of Corinth. The troops also had to take part in the deposition of King Constantine. Politics suddenly took precedence. Any offensive operations were ruled out under these circumstances. The argument that has been heard over and over again that the very existence of the Allied army in Macedonia tied up a large number of opposing troops is only partially true, because Bulgarian troops were for the most part on this front; the few German troops were negligible. It is a long-shot argument that was supposed to make sense of the futility of the failed Allied Macedonia operation.

One might have expected that the Germans would take advantage of this situation after Romania was done. But now the war suddenly changed fundamentally. On January 31st, 1917, Berlin had announced the unrestricted submarine war. The USA then broke off diplomatic relations with the German Empire and declared war on Germany on March 2nd. On March 8th, there were riots in

Petersburg. On March 12th the Russian army joined the revolutionaries. On March 15th, the czar was deposed and a provisional government took power. The workers' and soldiers' councils in Petersburg declared the war over. The Germans sent Lenin to Russia on April 16th, 1917, and within a short time Russia was no longer a factor in the war.

Further Military Developments

At the beginning of 1917 the Venizelist army numbered 1,497 officers and 53,271 men. Of these, 612 officers and 24,091 men were on duty. There were three divisions: one from Serres, one from Crete and one from the islands. The gendarmerie, which also had military training and thus combatant status, consisted only of Cretans and had 247 officers and 5,361 men. A large proportion of the soldiers volunteered, but there were also some conscripted classes. The Provisional Government actually wanted to call in entire conscription classes in the areas it controlled, but there was also a dearth of means of transport and equipment, especially artillery and machine guns. The Serres division took part in Sarrail's spring offensive.[634]

After Venizelos took over the government, the Serres division was transferred to Athens on July 25th. Venizelos then demanded that Allied controls be abolished and the Salamis naval base returned to the Greek government, which they did; the light fleet units were returned and the British were charged with their reorganization. After the arrival of the Serres division, French troops were relocated from Regnault to Macedonia. Allied troops evacuated the neutral zone. All of Greece, with the exception of the actual combat zone, was again placed under the authority of the Greek government. Epirus remained occupied by the Italians, although the Allies had decided to release it. It was only when the new Greek Foreign Minister Nikolaos Politis protested violently at the Paris conference in July 1917 that the Italians finally agreed, after some back and forth, to evacuate Epirus. But they stayed in Korytsa.[635]

The Italians had found another twist on how to back up their claims on Epirus, namely with the involvement of the *Koutsovlach* or *Aromanians*. Shepherds who spoke the Aromanian language lived all over the northern Pindus area. Aromanian is an East Latin language, like Romanian. To this day, it has not yet been finally determined to which ethnic group the Aromanians belong. For the Romanians, they are compatriots who have migrated south. But Aromanian is as different from Romanian as French is from Italian. Certain syntactic peculiarities of Aromanian led to the assumption that the Koutsovlachs originally spoke Greek and then adopted Latin as their lingua franca in the past, so that they are actually Greeks.

Be that as it may, since Aromanian is related to Italian and Romanian, the Italians in World War I and the Romanians in World War II derived claims to power. The Aromanians were to join the great Latin fatherland of Italy in 1917. To this end, the Italians appointed consuls in all Koutsovlach villages to carry out pro-Italian propaganda, which they did very intensively. The Greek government lodged a massive protest and succeeded in closing all consulates, after the Italians had dragged their feet for some time, with the exception of Ioannina.[636]

Venizelos repeated the declaration of war on Germany after he came to power, but with the few troops at his disposal he could not do anything, and the army had to be rebuilt. In order to do this, resistance in the Greek population – which did not want to go to war - had to be overcome. The rebuilding also had to be financed. In addition, Venizelos was not ready to integrate the remnants of the old army that still existed. He feared a royalist coup and therefore dismissed 2,000 royalist officers. This was a loss of competence and knowledge that no army could easily cope with. It was therefore no wonder that rebuilding the new army took so long.

But not only was the army cleared of its royalist officers, but also the Navy. The British Legation reported to the Foreign Office: *"The new Secretary of the Navy [Koundouriotis] is busy switching out all of the Navy personnel. [...] After the departure of the Venizelist officers [...] many of the remaining officers loyal to the King were promoted, now that the Venizelist officers are returning, the number is too great. In addition, there is difficulty with the arbitrary retirement of a number of officers loyal to the King who are being retired solely for political reasons."* When the British envoy complained about the mass dismissal of royalist officers, Venizelos replied that the approach had been very moderate: of 134 officers fired, only 15 had ended up in prison. In his opinion, others were to blame.[637]

Venizelos opted for a successive approach: initially he pulled in only age groups that lived in areas that were Venizelistically oriented. The financial problem was solved when France granted Greece a 50 million gold franc loan in August 1917. The Greek envoy spoke to Marshal Ferdinand Foch about the reorganization of the Greek army. It was agreed that England should supply the heavy artillery. But England refused because it had other obligations. Venizelos then decided to negotiate with the Allies directly on site in order to obtain the necessary equipment. At a conference in Paris under the new Prime Minister Georges Clemenceau,

Venizelos pledged that Greece would deploy 300,000 men on the Macedonian front. In return, Greece was to receive a loan of 750 million gold francs.

In June the existing French military mission was greatly expanded. It was commissioned to reorganize the army. The mission, enlarged by sixty officers, began its work on June 30th, 1917 under the leadership of the French military attaché Paul Braquet. The aim was to enlarge the Greek army, but Venizelos initially refused to introduce a general mobilization. He wanted emotions to calm down for a while. Six or seven divisions were to be established by the end of 1917. Later the number would be increased to ten. The French staff of the mission estimated that in order to make the divisions operational, an additional 160 artillery and 100 mountain pieces were required. The divisions were to have three regiments of three battalions each. Each battalion would consist of three infantry companies and one machine gun company. Each division was to have an artillery battalion with two batteries each, a grenade launcher battalion and two engineer companies.[638]

At the end of 1917, General Braquet was transferred to the Western Front. The new head of the military mission was General P. E. Bordeaux, who had already served in General Joseph-Paul Eydoux's French military mission prior to the Balkan Wars. Bordeaux initially wanted to build up officer units, which in turn would create the conditions for the formation of larger units. Six divisions would be set up by the end of the year. When Venizelos received the French 750 million gold franc loan, mobilization accelerated. In addition, the officer corps was cleared of political opponents. Command posts were filled with Venizelos' supporters. Three corps were to be set up, but the ramp-up was slow. By April 1918, only four divisions could be established. Clemenceau pushed for their transfer to the Macedonian front.

One of the reasons for the delay was the lack of trust between the Greek soldiers and the French. The 38th *Evzones Regiment*, for example, clashed with the Moroccan Spahis during the occupation of Thessaly in May 1917, and losses had occurred. All officers, including the division commander, had been arrested. Rebuilding mutual trust took longer than rebuilding the units. Only in early 1918 were seven divisions mobilized. It was hoped to be able to mobilize three more divisions by June 1918.[639] But these troops were not combat-ready. Training as well as experienced officers were missing.

Again and again there were protests against the deployment of Greek troops on the Macedonian front, in which soldiers and royalist-minded officers also took part. The latter were immediately dismissed, which made the shortage of officers ever greater. When attempts were made to expand the mobilization in February 1918, mutinies and uprisings broke out, which were brutally and violently suppressed.[640]

The situation of the Allied armies in Macedonia in the summer of 1917 was neither satisfactory in terms of strength nor in terms of morale. The French units had too few officers and professional soldiers. Because of the mutinies in the west, every French soldier was given a week's leave every four months. With the troops deployed overseas, the implementation of this arrangement was, of course, impossible. It was decided that that soldiers who had been overseas for more than 18 months would be allowed to visit their families. But then new difficulties arose on the Saloniki front. German submarines had greatly reduced the available transport space. Of the 195,000 men in Saloniki, 20,000 had the right to visit their homeland. When Sarrail refused the visits, rioting ensued, which, unlike riots in the West, ended bloodlessly. Sarrail understood that something had to be done and therefore had rail transport expanded across Greece in order to shorten the route and avoid the threat of submarines from Constantinople.[641]

The Serbian army had borne the brunt of the attacks in the spring and was worn out. They called for the front to be shortened so that some units could recover. The British downsized their army when the 60th Infantry and Cavalry Divisions were transferred to Egypt. The Italian high command wanted to detach the 35th division from the front at the Cerna, and move it west to the other Italian divisions in northern Epirus. Due to the revolutionary propaganda of the Bolsheviks, morale amongst Russian troops was badly damaged, and the Greek army could not be fully operational until 1918.[642] The mood among the troops therefore, was very bad.

So far there had been an outlet for the bad mood: the city of Saloniki. The city offered pleasures that every soldier's heart longed for. But on Saturday, August 18th, 1917, a fire broke out in the old town, which was mainly of wood construction. By Tuesday, August 21st, the entire old town had burned down. Although there were hardly any deaths, there was much material loss, for example the entire quinine supply of the army was incinerated. Much worse, however, was the loss of the entertainment district: *"Saloniki has never been an elegant city, but in its kitschy way it was a happy place. And the men from the trenches, who only came through the port when they arrived, had always hoped that they might get* *a few days off to go down to Saloniki to celebrate and see if the rumors they had heard about the Odeon and the skating rink were true. Now all those cabarets and music halls were gone. Just like the hotels and Flocca`s, the "elegant café" where young officers took the nurses. [...] Montmartre - or was it Babylon! - it went up in smoke. For the remainder of the campaign, Saloniki remained a desolate place."*[643]

At the inter-allied conference in Paris at the end of July 1917, Lloyd George announced that he would withdraw more English divisions as soon as the Greeks reached the Saloniki front, but that they would not completely withdraw. The French and the Serbs were angry. The Italians, Greeks, Russians and Romanians remained silent. It was decided to discuss the matter again at the next conference in London. Chief of Staff Robertson was still opposed to the Saloniki enterprise: *"The Saloniki forces will not contribute significantly to the benefit of the war, [...] the Saloniki expedition was strategically bad from the beginning. We cannot use the large army that we have there for an offensive..."* At the conference in London that took place a little later, Lloyd George came under such pressure that he was again prepared to compromise: they only wanted ore to pull out more troops in agreement with the other Allies.[644]

In August and September there were small advances now and again, which were not very successful. The most successful attack was that against the town of Pogradec, south of Lake Ochrid in Albania, but which was ultimately insignificant from an operational perspective. The Macedonia front as a whole became meaningless. The weight of the war shifted again to the north and east. In April, the *Nivelles offensive* ended in massive defeat and huge losses. In May and June there was a large mutiny in the French army, involving over a million men and 54 divisions. The French army was breaking apart. Only the presence of the British prevented the front from collapsing. The submarine war, however, lost its importance due to the convoy tactics. The Austrians inflicted a crushing defeat on the Italians at the Battle of Caporetto in Slovenia, taking 180,000 Italians prisoner and capturing 1,500 artillery pieces. The Italians had to withdraw as far back as the Piave River. The Eastern Front collapsed completely and the Germans threatened Petersburg. On October 26th the communist revolution took place and on November 14th Lenin formed a new government. This made it clear that Russia would soon conclude a separate peace.

The losses and mutiny led to the fall of the Ribot government on September 7th, as the socialists refused to continue to accept him. The previous Minister of War, the socialist Painlevé, became the new Prime Minister. He fired Nivelles and made Philippe Pétain his successor. Although Painlevé managed to set up a joint Allied High Command (*Supreme Allied Council*), on November 13th, he lost the parliament's confidence. Clemenceau became the new prime minister. With that, Sarrail lost his political backing and it was only a matter of time before he would be replaced.

In the fall of 1917, Sarrail reorganized his troops. The *Armée d'Orient* stood on the left flank from Lake Ochrid to the eastern arc of Cerna. It was followed by the Serbian army north of the Voras Mountains. The Axios (Vardar) area was secured by mixed French-Greek units. The British defended the right wing as far as the Gulf of Orfanos. These troops' priority was to hold their positions. As far as the available means allowed, positions for trench warfare were expanded. There was a first line of defense and a second line about 15 km behind it. The *Entrenched Camp* of Saloniki was the last position. The supply route from south to north worked, more or less, through roads in the valleys and the two railroad lines. A major handicap was the lack of an east-west road connection parallel to the front. The supplies from France came via Saloniki and Ag. Saranda, the latter road connection being heavily dependent on the weather. There were ammunition supplies, but food supplies were scarce.[645]

Worse, though, were the relationships between the commanders of the individual armies and Commander-in-Chief Sarrail, which were more than tense. In addition, the Allied governments constantly demanded Serrail be replaced. As long as the socialists were in the government, they held a protective hand over him. But when Clemenceau became prime minister and formed a government without the socialists, his end neared. Clemenceau also strived for a good relationship with the Allies and was therefore ready to replace Sarrail. He also thought that Sarrail's leadership of the troops in Saloniki was lax, and he instructed Pétain to relieve him. But Pétain was not in the mood for a confrontation with the radical socialists and asked Clemenceau to do it himself. He actually wanted to appoint General Franchet d'Espérey - with the consent of President Poincaré - but the latter refused the transfer.[646]

Therefore, on December 10th Sarrail was replaced by General Adolphe Guillaumat. Sarrail left Saloniki on December 22nd, 1917. A terrible period in Greek history had ended. Sarrail returned to his home town of Montauban, and began writing his memoirs.[647]

German-Greek Relations 1916–1917

After Kaiser Wilhelm's unsuccessful attempt on August 4th, 1914 to draw Constantine to join the war on the German side, the Germans had initially come to terms with Constantine's policy of neutrality. They also did not interpret the Allied landing in Saloniki as Greek partisanship for the Allies. Only when the Greek side had to make more and more concessions under Allied pressure did the German leadership begin to have doubts. It was believed that Constantine wanted to maintain neutrality, but would he remain *"master of the situation"*?[648]

Shortly before the end of the Central Powers' campaign in Serbia, Constantine had let the Germans know that they were not to enter his country. On November 5th, 1915, Falkenhayn replied via the German military attaché in Athens, Major Ernst von Falkenhausen: *"If Greece declares that it wants to disarm Serbs who cross the border and force Entente troops to leave immediately, the pursuit would not be continued over the border."* On November 24th, Constantine declared himself incapable of doing this and demanded that only German troops be used in the pursuit on Greek soil. Participation by Bulgarian troops is *"excluded."* Falkenhayn replied on December 1st: *"Germany, Austria and Bulgaria are allies. In fulfillment of an alliance purpose, they were attacked by the Entente through Greek territory. So it goes without saying that they will defend themselves against the attacker together and with full parity with the Allied troops."* Falkenhayn reserved full freedom of action in the event that the operation was to continue on Greek soil. The Bulgarians would be invited to respect the Greek border.[649]

Falkenhayn suspected that the retreating Allied troops would fight after crossing the Greek border near Saloniki. The question was whether to follow them and attack them. After the evacuation of Serbia, there was actually no longer any *"imperative political necessity"*. However, if the Allies were now to be driven out of the Balkans entirely, this would be of great moral importance. Expulsion was now easier than later. It could also happen that Greece would be pushed onto the enemy side in the event of an attack. Although Austria showed no interest in entering Macedonia, at the end of December, Falkenhayn was of the opinion that the German and Bulgarian forces available were sufficient to occupy Saloniki. Falkenhayn favored the attack for two reasons: he was preparing to attack Verdun, so he could use a diversionary attack in Macedonia. News from Athens said that a rapid advance by German and Bulgarian forces into Saloniki would not trigger a Greek declaration of war. Greece's entry into the war was ruled out due to the supply situation, as State Secretary Gottlieb von Jagow from the Foreign Office informed the Supreme Army Command after consulting the Greek envoy. So Falkenhayn decided to extend the pursuit to Greek territory as well. But he abandoned the plan at the end of December due to transport problems.

In November 1915, the military attaché Falkenhausen had warned that a Bulgarian invasion could lead to the overthrow of the monarch. In January he reported that the King and General Staff evidently had no objection to an early attack on Saloniki. Apparently the King hoped to end the increasing Allied interference in the internal affairs of Greece and to be able to return to real neutrality. But the unsolved problems of the transport forced another postponement. Falkenhayn believed he could get the Greek government to participate in the operation against Saloniki. On February 7th, 1916, he asked Military Attaché Falkenhausen whether he thought it possible for Greece to participate. The latter answered in the negative simply because of the lack of food supply; the Entente allowed very little grain into the country, so that after a few days, a famine would break out. Two days later, Constantine said the same to the German ambassador: *"In view of the economic leverage available to the Entente, it is impossible for Greece to take the side of the Central Powers anyway."*[650]

The German military leadership now wanted to take precautions that there would be no incidents between the Bulgarian and Greek armed forces in the event of an invasion. Negotiations were held during the second half of February. Germany and Greece would each issue a declaration on the Bulgarian-German invasion of Greek territory. But now the situation on the western front meant that Germany had to forego an attack on Saloniki. Instead, some key locations in Greek territory should be occupied, such as the *Roupel-Pass*. Obviously Berlin underestimated the effect of the project because the German military attaché warned: *"I would like to point out again that the occupation of the Roupel Pass can result in a complete diversion of Greece and even abandonment of Greek neutrality in favor of the Entente."* The envoy also sounded the same warning. Thereupon, on May 14th, Falkenhayn asked the commander-in-chief in the Balkans, Field Marshal August von Mackensen, whether the advantages of occupying the *Roupel-Enge* were as great as the disadvantages of Greece taking the side of the Entente. But the stubborn Mackensen insisted on the occupation, which took place on May 26th. As agreed, Greece protested violently against it. State Secretary Jagow then stated that the invasion was not a breach of neutrality, since this area of Greece was occupied by the Entente.[651]

The Allies were not deceived by the exchange of notes and accused Athens of a secret acquiescence to the Central Powers. At the beginning of June, a blockade was imposed and the demobilization of the Greek army was forced. *"In the eyes of the Greek population, hostile propaganda ensured that responsibility for the difficult supply situation was attributed to the King as the bearer of foreign policy and that the failure of the so often announced German-Bulgarian attack on Saloniki was interpreted as a sign of the weakness of the German-led confederation."*[652] The massive reaction of the Allies caused the German leadership to refrain from further occupation for the time being. When the Allies attacked German-Bulgarian positions on the Vardar in August, these concerns disappeared and the Greek government was informed of the impending counterattack, which could also move onto Greek territory. It was guaranteed that the occupation would be lifted at the end of the war. In Athens this declaration was received with calm satisfaction.[653]

When the war with Romania began, the German ambassador reported from Athens that if Romania attacked Bulgaria too, it was possible that Greece would join the war on the side of the Entente. The increasing reprisals of the Allies against Greece were attributed to the King through their propaganda in order to drive a wedge between King and his people. The expulsion of the Central Powers' diplomats and the resignation of Zaïmi's cabinet were additional means of forcing Greece to side with the Allies. On September 8th, Hindenburg feared that Greece would enter the war against Germany shortly. The German moves against the Army Corps near Kavala, which was sent to Görlitz, caused deep resentment in Athens. Entente propaganda exploited this affair intensely. But even when Venizelos set up his Provisional Government in Saloniki, Constantine stuck to his policy of neutrality. At the beginning of October, he informed Kaiser Wilhelm accordingly.

The events of autumn 1916 up to the *Noemvriana* led Constantine to wish for a German-Bulgarian attack on Saloniki. He hoped that with the expulsion of the Allies, the increasing political pressure that was being exerted on him would also ease. Because then the Allies would no longer be able to exert any military pressure on him either. This development was closely monitored in the German Supreme Army Command. It suspected that Constantine might even be ready to put his army at the disposal of the danger that threatened from Saloniki. Since the German side feared the abandonment of neutrality and the transition of Greece to the Entente camp, they were ready to attack Saloniki. So it was no longer just a question of neutrality, but also of Greece's hoped-for entry into the war on the side of the Central Powers. At the end of December 1916, a request to this effect was sent to Constantine.[654]

On January 2, 1917, a telegram reply to Falkenhausen arrived from Queen Sophie, which said: *"Due to the continuation of the blockade, we only have bread for a few days, other food deliveries are also decreasing. Therefore, a war against the Entente is no longer an option. Negotiations are ongoing. I consider the game lost if an attack doesn't take place immediately. After that it will be too late."*[655]

On January 4th, 1917, Constantine's answer followed: Greece only had bread for 14 days. *"His government asks for the clearest and most specific information possible as to whether and when we could take action against Sarrail, since it would have to set up its policy accordingly."* Hindenburg replied on January 6th that a German attack could only be carried out with a prospect of success *"if Greece advanced with stronger forces in the general direction of Monastir and thereby relieved our front. [,,,] The ultimate success is based on this cooperation alone."* In order to convince Constantine of these proposals, the previous military attaché Falkenhausen was sent on a secret mission to Greece. But on January 8th, a military assessment of the situation by Constantine arrived, which he had already formulated in December.[656]

sAfter an overview of the development of the last two years, Constantine spoke about the current situation: *"In addition, the lack of food, raw materials and coal must be taken into account. It should also be remembered that in the present situation the declaration of war must precede the mobilization and the Entente will probably endeavor to plunge Greece into war in order to destroy it before the German attack begins. The new note calls for complete disarmament through the transfer of all artillery and all mobilization material to the Peloponnese."* On January 10th, 1917, Constantine's rejection came at Hindenburg's suggestion. A fight was impossible under the given circumstances.[657]

In the meantime, Falkenhausen had arrived in Larissa by air on January 9th. Queen Sophie telephoned him to inform him of the Allies' 48-hour ultimatum and that the cabinet wanted to accept because of the blockade. The King was at a meeting of the Privy Council. Falkenhausen asked Sophie to put the King on the phone, which she then did. Falkenhausen tried to change his mind, but Constantine refused. The Privy Council decided to bow to the ultimatum. Falkenhausen had another opportunity to telephone the King on January 10th. Completely depressed, the King said that he had to accept the demands of the Entente. That was the end of Falkenhausen's mission.[658]

The answer of King Constantine was the answer of the Greek government to the German authority responsible for these questions, the Supreme Army Command. This is where the military and political decisions were made. Wilhelm II had lost more and more power since the outbreak of war and at the beginning of 1917, and had little more to say, neither militarily nor politically. Sophie's much-quoted telegram to her brother of January 9th, 1917, which was published in the supplement to the Greek White Paper, must therefore be seen as a private letter in which she expressed her personal opinion towards her brother. Sophie had no influence whatsoever on politics. The letter in the version of the white paper read:

"I thank you very much for your telegram, but in the absence of sufficient food supplies for the duration of such an undertaking, as well as ammunition and other things, we are forced to forego an offensive. You understand my situation! How much I suffer! I thank you from the bottom of my heart for your loving words, for the [illegible] terrible circumstances. May the nefarious pigs receive the punishment they deserve! I hug you lovingly. Your lonely and sad sister who hopes for better times, Sophie."[659]

After talking to Falkenhausen on January 9th, she is said to have sent another telegram to her brother on January 10th. There are two versions of this. The text in the supplement to the Greek White Paper reads: "I am grateful and glad that I at least spoke to Falkenhausen in Larissa by phone today and that I heard from you directly. I am [illegible] that the ultimatum was accepted, but unfortunately we were forced to accept it, even though we wanted to wage war on Germany's side. For two reasons, one because of the political advantages and the other because of the need to get rid of our angry enemies. Likewise, to respond to the sympathy that the Hellenic people have already shown towards the German matter. The shortage of food and ammunition for the duration of the campaign and especially the shortage of heavy artillery around [illegible] the fortified and prepared positions of our enemies in the narrow passes north of Thessaly and [the danger] that at any time threatens our capital and our only means of communication, as well as the English fleet, the reports that they are ready to launch a campaign against Greece from Malta force us, to our great regret, to abandon this plan. I hope you do not lose sight of the fact that in the pursuit of our plans, if connected by rail, Greece, because of its geographical position, could become a useful and valuable aid to our beloved fatherland. Between the peoples we should always continue our work in the interests of Germany against our enemies, you can be absolutely sure of that. I am proud that the indescribable sufferings and fears we have endured and continue to suffer in order to maintain neutrality have enabled us to provide important services. Loving. Sophie."[660]

Mühlmann published a different version:

"I am saddened that the ultimatum has been accepted, unfortunately we have been compelled to accept this, although we wished to fight on the side of Germany, both because of the political advantages and because of the need to get rid of our bitter enemies. and to return the sympathy already shown by the Greek people for the German cause. However, the lack of food and ammunition for the duration of the campaign, and especially the lack of heavy artillery with which to supply the fortified positions, in order to prevent the enemy from proceeding against the gorges north of Thessaly, which was an imminent danger both for the capital and for our only means of communication, through the English armed forces, which are assembled on the island of Malta and destined for the campaign against Greece, forces us, to our great regret, to renounce this plan ... "[661]

Both texts, the English as well as the German, sound like letters, but not like telegrams. If one compares this text with those that Sophie sent earlier, it becomes clear that her earlier texts were written briefly and in the typical telegram style. On the whole, the content of the two texts is the same. However, the German text contains grammatical, lexical, syntactic and content errors which are extremely irritating and which raise doubts that the text was really written by Sophie. A native speaker does not make such errors. The syntax of both texts is also reminiscent of Greek.

The fortified positions north of Thessaly would have been in the neutral zone. Since the Allies controlled the Greek telegraph system, the text must have been transmitted by radio. Shortness would have been the order of the day, as in Sophie's other telegrams, and no complicated sentences. The question also arises as to where Mühlmann got the German version, but this can hardly be answered. Mühlmann's other quotations are verifiable and it is actually astonishing that he did not notice the "*German*" of this telegram. It is also irritating that Sophie comments on military details. It is known that she was interested and engaged in social issues, but not military ones. British troops were in the country and did not have to come from Malta. Malta was a Royal Navy base.662 The above errors lead to the conclusion that this text of the telegram came from the poisonous Venizelist propaganda shop.

The English text is a main argument that there was a conspiratorial relationship between King Constantine and Kaiser Wilhelm.663 However, such a relationship would have been rather pointless, because Wilhelm II no longer had any influence on the conduct of the war; Hindenburg and Ludendorff decided on this and Constantine was in touch with them. Both texts are useless as evidence of Constantine's pro-German stance. In view of these *Gravamina*, the suspicion arises that the author of this text wanted to incriminate the exiled King later. It should also be noted that Willmore's Anti-Constantine pamphlet appeared in the same year as the Greek White Paper. The propaganda against Constantine continued even after the end of the war.

In 1918, the *Daily Mail* special correspondent G. Ward Price published "*The Story of the Salonica Army.*" Price was the second most important man in the *Rothermere* press empire after the owner. His reports from the Dardanelles and the Saloniki Front influenced world opinion. He wrote of Constantine: "*There are no reasons to defend King Constantine, he acted unconstitutionally, fraudulently and treasonously; in addition to being disingenuous with his Serb allies, he did all he could to make sure that our attempts to help the Serbs petered out. He was also wrong in his most credible argument that he was doing everything for the good of his people. But I think it is wrong to imagine, as many seem to do, that his opposition to us was aroused by sheer resistance and German stubbornness. The refusal to recognize King Constantine or to show consideration for the interests of his subordinates, the Greeks, had the disadvantage at the time that it aroused the constant expectation in England that he would suddenly recognize the error of his ways. He would turn and be ready to cooperate with Venizelos at the behest of the Allies. This seemed like a conclusion so meaningful and inevitable that we were always inclined to be patient and moderate and give it another chance. But the King's misguided hostility to the Entente has many causes. (...) First of all, the King was, of course, a stubborn man. As a sign of this, one only had to look at his angular, fleshy and heavy head. He was also steeped in the doctrine of the divine right of Kings. (...) And all these influences, assumptions and prejudices taken together, made King Constantine Germany's vassal. They were strengthened and driven by his wife, this capable and strong-willed woman, the Emperor's sister, Queen Sophia.*"664 The description of Constantine and Sophie are a piece of clumsy propaganda.

With the end of Falkenhausen's visit, direct contacts between Germany and the Greek royal family broke off. Since Germany no longer had a diplomatic mission in Athens, there was no longer any direct information about the situation in Greece. The only direct contact with the Greek government was through the Greek embassy in Berlin, and when Venizelos took power in Athens on June 27th, 1917, this too ended. German-Greek relations were over for the time being.

The Collapse of the Macedonian Front (1918)

The Spring Offensive

General Adolphe Guillaumat, the new commander of the Allied troops in Macedonia, was still relatively young at the age of 54, but he was experienced. Before the First World War he had already participated in twelve campaigns in North Africa and Indochina, as well as China. At the beginning of the First World War he had been in command of the 33rd Infantry Division and in December 1914 he became chief of the 4th Infantry Division. From February 1915 to December 1916 he led the 1st Army Corps and then became commander of the 2nd Army. He had distinguished himself in Verdun. He listened to others and incorporated their views into his decisions, but was then stuck with them. General Milne had the impression that Guillaumat had his command under control in a week.[665]

On December 30th, 1917, he informed Milne of his orders, signed by Foch and Clemenceau. From now on, not only Saloniki was to be defended, but all of Greece, south of the current front line. Should one have to retreat, the *Entrenched Camp* and the area east of the Pindus must be held. Once the organization of the defense was completed, one had to begin thinking about and planning offensive actions. It was also important to keep the ports of Valona and Corfu as supply bases. Before the Greek army was deployed, its task would have to be reconsidered. These orders meant that from now on ancient Greece was more important than Saloniki. The *Armée d'Orient* became a separate entity and was no longer the commander-in-chief's personal force.[666]

On the basis of these orders, Guillaumat reorganized his troops. He moved the 35th Italian Division to the west so that it could join the Italian troops in Albania, as the Italian leadership had long requested. The Russian units had so far secured the western and eastern banks of the Great Prespa Lake. But in the wake of the Russian Revolution a very large number of soldiers had deserted, and the fighting strength of the Russian units had been considerably reduced as a result. So he replaced them with the 156th French Division, and had the Russians brought to Tunis. The deployment of the troops in February 1918 looked like this: The area of operations of the *Armée d'Orient* under General Paul Henrys began on the Albanian river Apsus west of Korytsa, where the 35th Italian division was located. It reached as far as the eastern arm of the Cerna. Including the 43,336 Italians, there were a total of 159,835 men. It was joined by the Serbian army, whose area of operations extended as far as the village of Enotia. The Serbian army had 134,143 men. They were followed eastwards to the then still existing Lake Ardhan east of the Axios (Vardar), by the First Division Group. The group consisted of the 122nd French division and the three Greek divisions under the command of Major General Immanouil Zymprakakis. The group numbered a total of 74,378 men, of which the Greeks alone were 53,740 men. The British Army formed the right wing to the Gulf of Orfano. under General Milne, with 97,463 men.[667]

According to Foch's orders, Guillaumat had two tasks: he was to defend ancient Greece and hold Saloniki. But in the meantime, events had occurred that fundamentally changed the previous situation. On November 28th, the Soviet government made an offer of peace to the Central Powers. Armistice negotiations began on December 3rd, leading to a provisional armistice on December 15th. Official armistice negotiations began on January 8th, 1918. The war in the east had ended. On March 3rd the peace of Brest-Litovsk was signed. Freed from the two-front war, the Supreme Army Command had already decided in November to carry out a decisive major offensive on the western front. Hindenburg and Ludendorff had been preparing this offensive since January 1918, which was given the code name "Michael" and was to begin in March.

On the basis of the armistice between Germany and Soviet Russia, it was also clear to the Allies that Germany was preparing a major offensive in the West; at the same time, however, it was feared that the Macedonian front would also be fortified and that a major attack could be imminent. So far there had been no overall unified defense plan. In view of the new situation, this was now urgently needed. The armies, structured according to the orders of Foch, were given appropriate tasks to defend ancient Greece and Salonika. Guillaumat had a triple-staggered trench system built. At the same time, communication channels were repaired and new ones were built. Narrow gauge railways were installed in the Florina and Polykastro area. In the Axios (Vardar) valley, a new road was built on each side of the river.⁶⁶⁸

In early March, Guillaumat began to think about offensive operations. In a report to Foch he wrote that at the moment large-scale operations, such as the retaking of Serbian territory or the defeat of the Bulgarian army, were unthinkable. An operational breakthrough on the Macedonian front would only be possible if a resounding success had been achieved in the west or in Italy. At that moment the front could only be straightened here and there or the enemy occupied so that his troops would be tied up or he would even have to send reinforcements. One could attack near Monastir and in the Cerna area in the direction of Prilep and on the Dojran-Vardar front with the aim of Rabrovo. An advance through the Struma valley against Serres and Polykastro was also conceivable. The main attack would be carried out along the Struma and the Vardar. He ruled out an attack over the mountains between Cerna and Vardar for offensive operations.⁶⁶⁹

In the meantime, the Germans had increased the number of their divisions on the Western Front from 147 to 191, compared to only 178 Allied divisions. The great spring offensive began on March 21st. At the end of March, the German troops had advanced 60 km to the west across a breadth of 80 km. In view of this situation, the Allies agreed that Foch should be the joint commander in chief. Since he feared that the Germans might bring Bulgarian troops to the western front, he ordered Guillaumat on April 4th to do something to keep the Bulgarian troops on the Macedonian front. He suggested smaller attacks. Guillaumat asked his army commanders to make appropriate proposals.⁶⁷⁰

London did not think much of Foch's ideas. On April 12th, the British Chief of Staff replied that an offensive in Macedonia would make little sense. The Germans withdrew their troops from there and the Bulgarians would not dare to attack without German support. There was no way they would send their troops to the Western Front. The Germans concentrated all their forces in the west. To defeat them, they would have to do the same thing. They had to press ahead with the deployment of the Greek troops so that their own could be withdrawn and relocated to the west. Foch replied that only minor attacks were planned in Macedonia. In the meantime, Guillaumat had ordered a number of minor operations.⁶⁷¹

The division group consisting of the three Greek divisions and a French division was to attack the area west of the Axios (Vardar) near the village of Skra on April 6th, and the British army was to become active in the Dojran sector on April 7th. The *Armée d'Orient* was supposed to advance from Lake Ochrid and Great Prespa towards the Cerna on April 8th. Finally, the Serbian army was supposed to take action on April 11th in the mountains north of the Voras Mountains. At the same time, the commanders were to prepare for a major offensive, which should not be started until the conditions were favorable.⁶⁷²

These local attacks went as planned. A joint attack by the Italians and the French in Albania, conquered the Ostrovitsa Mountains. The French attack from Monastir / Bitola to the northeast was unsuccessful. The Serbian attack against the Dobropolje heights brought just small territorial gains. In the attack on Skra, the Greek troops deployed there proved to be successful and the attacks by the British also went according to plan.⁶⁷³

When the Germans advanced far into France in March, the Allies scraped together all reserves in order to throw them at the front. They even considered withdrawing units from the Macedonia front. But they didn't do that because half of the men had malaria. During the next German offensive, the subject came back onto the table. The British military wanted to withdraw two divisions from Saloniki. But when the French protested, the idea was dropped again. At the meeting of the Supreme War Council on May 2nd, Lloyd George suggested that of the 12 battalions of the divisions had deployed in Macedonia, three could be drawn out in order to send them to France. They would be replaced by Indian battalions. Clemenceau agreed and even wanted to replace French divisions with Greek ones, provided Guillaumat thought it made sense. It was decided to consider the matter on location, and two officers, one French and one English, were sent to Saloniki. The British reported at the end of May that the French had withdrawn 12,000 men and sent them to France without informing the British. Lloyd George took this as an opportunity for his part, as proposed, to replace English battalions with Indian ones and to send the British to France. At this point in time, no one thought of a major offensive.⁶⁷⁴

The two dispatched officers conferred with Guillaumat, Milne and the Italian commander. It was decided to send 12 British battalions to France and replace them with Indian ones. A French division was to be dissolved in coordination with Guillaumat and replaced by a new Greek one. The Italians would make no changes.⁶⁷⁵

All the while, Guillaumat was working on a plan for a major offensive. Before he could present it, however, he was recalled to France. The German advance on Paris in March had led Clemenceau to look for a general who was trusted and who, as governor of Paris, could organize the defense of the capital. The decision favored Guillaumat. In addition, the French government wanted him close by as a substitute for Foch or Pétain in case something should happen to either of them. On the other hand, Clemenceau needed a scapegoat for the setbacks on the Macedonian front; General Louis Franchet d'Espérey was chosen. He was recalled and transferred to the Saloniki Front. On June 9th, Guillaumat hastily left Saloniki. He did not even have time to introduce his successor to his new post. General Henrys took over command for the transition period. Guillaumat's successor, Franchet d'Espérey, arrived in Saloniki on June 17th.[676]

When he arrived in Paris, Guillaumat presented a first draft of his offensive plan for Macedonia: the situation had changed since March 1st. The Germans had allegedly withdrawn troops from Macedonia; the Bulgarian army showed signs of demoralization; the Greek troops had shown themselves courageous and combative in attacking Skra. All of this justified the abandonment of the previous caution and made it possible to think about a major offensive, for which he named three directions of attack. This paper was never officially discussed, but it formed the basis for his successor's operational planning.[677]

Rebuilding a New Greek Army

When Venizelos took power in Athens, he had the National Defense Corps (*ethniki amyna*), which consisted of three divisions. These were the division of Serres, Crete, and the islands. The royal army was mostly demobilized and was largely located in the Peloponnese. The corps consisted largely of Venizelian volunteers. Its morale and physical condition were good. At first Venizelos did not dare to call up reservists, as many of them were royalist-minded. When the 1st Division was set up in Larissa, primarily recruits of certain age groups were called into the ranks. The French military mission was responsible for training.[678]

The mobilization of the *"New Royal Army"*, as it was called, could of course only really be tackled once Great Britain, France and the USA each provided a loan of £10 million in December 1917. In February 1918, an Allied Military Commission and - for obvious reasons - an Allied Finance Commission were set up to oversee the spending of these funds. Virtually everything had to be bought in France or England, from weapons to uniforms to food. Since everything was transported to Greece by ship, this delayed the build-up of the armed forces. In March and April three divisions were set up: the 13th in Chalkis, the 2nd in Athens and the 9th in Ioannina. The 13th and 2nd divisions consisted of reservists and formed the 1st Corps with the 1st Division, which was transferred to Macedonia in June and placed under the British. The 9th Division remained stationed in Epirus due to a lack of transport and was not under Guillaumat's responsibility.[679]

The next three divisions, which would form the II Corps, were to be set up in the Peloponnese. But this was the area where the King's loyalists were strongest. The British were initially against this, especially since they were supposed to take over the equipment. But when Clemenceau insisted, they agreed on April 29th. The three divisions were therefore mobilized, which went off without any problems. From the end of June, the 3rd Division was in Patras, the 4th Division in Nafplion and the 14th Division in Kalamata. In Epirus the V Corps was formed from the 9th and 8th Divisions in Preveza. The divisions each consisted of three regiments, each of which consisted of three battalions. A battalion consisted of three infantry companies and one machine gun company. Each corps had an artillery regiment with four batteries of four guns each. On the army level there was a cavalry brigade made up of two regiments, each with a heavy battery. There was even an air force supporting the corps from the air. Towards the end of the year the Greek army numbered around 270,000 men, of which around 160,000 were on the Macedonian front and 15,000 in Epirus. The combat strength of the Greek troops in Saloniki was 90,000 men. The French military commission had done a good job.[680]

Of the force, Falls said: *"In general, the Greek soldier appears brave and dashing to British observers, but unpredictable, probably better at offense than defense. He is exposed to bouts of depression and political influence if held in reserve for too long or on a very calm front. Few troops improved faster in active combat conditions. The rural population, especially the mountain and island inhabitants, were superior to the urban population. But in general the troops proved that only the Serbs were superior to them as marchers. Despite the dismissal of many of Constantine's sympathizers, the older regimental officers appeared reasonably efficient, the younger ones reckless and careless."*[681]

Despite all these reconstruction successes, it must not be overlooked that this army, even if it was again called the *"Royal Army"*, was actually a Venizelist party army. From now on the Greek army was not a national army that stayed out of politics and stood above it, so to speak, but a political and politicizing army that, depending on which political grouping was in power, belonged to one or the other. This laid the foundation for the numerous military coups in the 1920s.

The Serbian army also grew with the arrival of Bosnians, Croatians and Slovenians – former members of the Austrian army who had suffered Russian imprisonment and there discovered their Slavic roots. A total of 16,000 men were recruited and brought to Saloniki.[682]

The mobilization in the Peloponnese did not go smoothly. On June 10th, a mutiny broke out in Servia at the 3rd Battalion of the 12th Infantry Regiment of the 3rd Division. The mutineers came from the Nomarchies Achaia and Elis. The men rejected Greek participation in the war because of their royalist inclination. The whole regiment soon joined the mutineers. The mutineers went south without their officers, shouted anti-war slogans and shot into the air. However, many of them later returned to their units. The division was informed and mobilized an Evzones regiment and a regiment from Crete, as well as two companies from Ioannina, who managed to arrest the mutineers. The leaders were brought before a military tribunal in Kozani on June 29th. Five officers and eight soldiers were sentenced to death and shot.[683]

One can assume that this was not an isolated incident. At that time there was a latent civil war in Greece. The country was politically torn and the two camps were split by a deep hatred of each other, which reached into the families and divided them. This went down in Greek history as *Etnikos Dichasmos*.

Ahead of the Big Offensive

In order to correctly classify and interpret the events on the Macedonia front, it is first necessary to look at developments on the western front. The German offensives in March and April led to a large territorial gain, but no strategic breakthrough. The offensive in May brought the Germans back to the Marne; Paris was only 90 kilometers away by road, or 62 kilometers as the crow flies. The British Cabinet discussed a possible evacuation of its own troops, but the Marne Line was stabilized with the assistance of American troops. On June 5th, the Supreme Army Command broke off the offensive because of the high casualties, counterattacks by the Allies and supply problems. The next offensive (*Operation Gneisenau*) had to be broken off on June 14th for the same reasons. The second Piave battle from June 15–22 ended in defeat for the Austrians. The real turning point of the war on the Western Front was the second Battle of the Marne. It began on July 15th when the Germans attacked with all the forces they had left. The attack made good progress at first, but on July 18th the French and Americans launched a counterattack through which the Germans lost all of the ground they had conquered in May and June. From then on the initiative rested with the Allies. The Allied tank attack near Amiens on August 8th, 1918, in which the German army suffered a heavy defeat, was described by Ludendorff as the "*dark day of the German Army*".[684] But he was still not prepared to accept the consequences.[685]

Franchet d'Espérey arrived in Saloniki on June 17th, 1918. On June 22nd he was informed by the War Department that instructions would shortly be sent to him by courier. He should work out an offensive plan based on Guillaumat's plan. The directive was expressed on the same day by the ministry in cooperation with Foch and Guillaumat. Clemenceau signed it on June 23rd and it arrived in Salonika on July 2nd.[686]

The directive stated that Germany was doing all it could to get to a decision in France. This made it possible for the Allies on the periphery, including in Macedonia, to take the offensive. Germany could hardly do anything for Bulgaria and Bulgaria itself was in an internal crisis. The aim of the Allies should be to break through the Bulgarian lines of defense so that the Serbs and Greeks could gain access to their lost territories. However, this could only be achieved with the use of all Allied forces. "*However, such a project could neither be carried out immediately nor through a single attempt. The general offensive must be preceded by several partial offensives, extended over a certain period of time and with increasing strength, all of which lead to a final decision phase.*" These small operations would enable the great offensive to begin before the fall.[687]

In the first days of July, the *Supreme War Council* met in Versailles. It asked the military to draw up a report on whether an offensive in the Balkans would make sense or not. The governments or the Council would discuss the results. The British, as usual, were against an offensive. Nevertheless, the military experts met on July 11th in the Trianon Palais in Versailles. This time Guillaumat was also there. He thought an offensive would be successful. After a long discussion it was decided: "*It is appropriate to consider the question of a general offensive in the Balkans with a view to the effect that may arise on the situation in Bulgaria. It is undesirable to carry out this offensive unless it leads to an outcome of more than local importance.*" Clemenceau forwarded this resolution to Franchet d'Espérey on July 18th and asked him to continue with his preparations. On July 19th, Guillaumat stated that if they started immediately, preparations would take two months. The offensive could start on October 1st.[688]

On July 22nd, General Milne, unaware of Franchet d'Espérey's plans, reported to London. He wrote that in view of the fact that the Bulgarians were tired of the war, the Austrians had difficulties in Italy and the Germans were fully engaged in France, a major action could be started in Macedonia which would have far-reaching results. He suggested first conquering Drama and Kavalla and then occupying the Roupel-Enge through a flank operation. He thought Guillaumat's plan to advance in the Vardar Valley had failed. Even if this flank operation did not succeed, it would lead to the relocation of so many Bulgarian troops that the Serbian and French armies could advance further west. The timing of this operation would have to be determined by the *Supreme War Council*.[689]

On July 25th, Milne reported to London that Franchet d'Espérey had asked him to prepare an offensive at the end of September. The *War Office* thought little of these plans. The Chief of the Imperial General Staff (CIGS), General Henry Wilson, flatly rejected an offensive in the Balkans. He wanted to replace the British currently deployed there with Indian troops. He informed Milne that *"an offensive would not be desirable, except at a time when Germany could not reinforce Bulgaria from the western front. This condition will not exist in September."* After a brief back and forth between Wilson and Clemenceau, a decision was reached on August 3rd, 1918, when the military representatives of France, England, Italy and the USA met in Versailles. They made the following resolution: *"It is necessary to press ahead with the preparations for an offensive in Macedonia with great speed on the agreed basis in order to enable the Allied armies in the east to carry out the attack by October 1, 1918 at the latest."*

However, these preparations would have to take place without support from the West. The actual decision to attack should be taken by Franchet d'Espérey, unless things happened that prompted the *Supreme War Council* to intervene. The British representative agreed, but doubted that this offensive would bring much in view of the troops' health.[690]

Little happened until September 4th. A conference was held that day at 10 Downing Street, attended by Lloyd George, the Secretary of War, the Secretary of State in the Foreign Office, Chief of Staff Wilson, the French Ambassador and General Guillaumat. Lloyd George asked the latter about the quality of the troops in Saloniki and the situation with the Bulgarians, and wanted to know how he assessed the situation. Guillaumat answered precisely. He still thought only a limited offensive was possible. But when Lloyd George wanted to know whether he thought an advance to Sofia was possible in six months, he said yes. Lloyd George was persuaded and after a brief consultation with his compatriots he announced: *"The British government gives its consent to the proposal insofar as it affects them or the British troops."*[691]

All that was needed now was the approval of the Italian government. Guillaumat traveled to Rome and managed to convince the Italian government. In the meantime, Franchet d'Espérey reported on September 5th that they could strike in the middle of the month. On September 10th he received a telegram from Clemenceau in which it was stated: *"You may therefore take up the fight if you think it appropriate."*[692]

The German casualties in the Battle of Amiens amounted to 75,000 men, of which 50,000 were prisoners. The allied incursion was considerable at about 20 km, but the moral effect was enormous: large parts of the army had no will to continue fighting. In academic research there is talk of a *"fight strike"*. In fact, from March to June 1918 the number of the army's operational soldiers shrank from 5.1 to 4.3 million. More than 200,000 men were sidelined every month (fallen, wounded and sick). There were around 70,000 convalescents as well as young recruits, born in the year 1900 and newly arrived, who, however, were not ready for action until the autumn.[693]

In Ludendorff's view, the continuation of the war after August 8th assumed the character of an *"irresponsible game of hazards"*. *"The fate of the German people was too high for me to gamble. The war was to be ended."* [694] He did not recognize or did not want to recognize that his spring offensives had already been a game of hazards. He suppressed the fact that, strictly speaking, the war had already been lost on July 18th, when the French stopped the last offensive and went over to the counterattack. On August 13th, the *Supreme Army Command (OHL)* at least came to the conclusion that the initiative in the war had probably passed to the enemy. But it was not yet ready to admit that the war could no longer be won. Peace negotiations should only be offered when the situation improved.[695]

Throughout August, the Allies kept pushing the Germans back. At the beginning of September, they returned to the position from which they had started the offensive in March; this position went down in history as the *Hindenburg line*. On September 2nd, the OHL ordered the retreat to the somewhat shorter *Siegfried position*. The Allies followed and on September 27th the *Siegfried Line* was broken for the first time. The German troops were weakened by losses that could not be replaced, desertion and capture, poor food and, above all, illness. The first wave of the Spanish flu raged among the soldiers. It was clear that the Allies could hardly be stopped any longer.[696]

The Allied attack on the *Siegfried Line*, in particular, which began on September 25th, 1918 and led to the first breakthrough on September 27th, was such a hard blow for Ludendorff that he was unable to inform the government over the impending military catastrophe. The new Foreign Minister Paul von Hintze found out about it in a roundabout way and traveled to the *main headquarters*.[697] In the meantime the news arrived that the alliance of the Central Powers was on the verge of collapse. At the beginning of September, Hindenburg had expressed the opinion that none of the participating states should make peace offers to their opponents on their own. *"It was a mistake to believe that something essential for a single state or for our whole could be improved in this way. The Turkish Grand Vizier, who was in Spa in the first half of September, judged the situation just as we did. Tsar Ferdinand [of Bulgaria] also spoke at the same time that his country's efforts for peace outside the common alliance were out of the question. [...] For the reasons given, I did not feel compelled to consider the Austro-Hungarian attempt to unilaterally suggest a peaceful settlement with the Entente in mid-September to be a positive one."*[698]

In reality, the situation in Austria-Hungary was even more catastrophic than among the Germans. The Army suffered under famine, desertion, suicides and diseases. The Sixtus Affair had shaken domestic politics.[699] On August 21, at the Belluno conference, the deputy chief of staff explained the hopeless situation to the shocked military leaders. On September 14th, Austria sent a first note to the Allies requesting the start of peace negotiations. When this remained unanswered, the government sent a new one on October 4th. At the beginning of October, Austria-Hungary began to break down into its component parts. On October 27th, the Italians and the Americans successfully attacked Vittorio. A day later, the still-ruling government asked for a ceasefire.[700]

There was also more bad military news. On September 19th, the British attacked Palestine and broke through the Turkish lines with almost no resistance. The defeat on the Syrian front was only a matter of time. When Bulgaria collapsed, Turkey was cut off from German supplies and the Allies threatened to attack Istanbul from Greece.[701] The Turks sent the prisoner of war General Charles Townsend to the British headquarters in the village of Moudros on the island of Limnos, in order to let them know that the Ottoman Empire wished to enter into armistice negotiations. The armistice was signed on October 30th, 1918.[702]

The Allied offensive against Bulgaria began on September 14th. Later, we will go into the course of the operation and the reasons why the Bulgarian army completely collapsed after a few days. It must be noted here, however, that Bulgaria sought an unconditional armistice on September 26th. This was completed on September 29th. The news of this request reached the OHL on September 28th and triggered hectic reactions.

The bad news that came in every day naturally weighed on the OHL, especially Hindenburg and Ludendorff. While Hindenburg kept calm, Ludendorff found it difficult to take bad news. Chancellor Bethmann Hollweg had stated earlier that Ludendorff was only great when he was successful. If something went wrong, he would lose his nerve.[703] The news of Bulgaria's collapse was the famous last drop that brought the cup to overflowing.

Keegan writes about what happened next: *"Now,* *however, he completely lost [his nerve] and let his paranoid anger run free 'against the Kaiser, the Reichstag, the navy and the home front'. His staff closed the door to muffle his screaming until he gradually regained his self-control. At 6:00 am he showed up to go down one floor to Hindenburg's room. There he told the old field marshal that there was now no other option than to strive for an armistice. The position in the west had been broken through, the army would not fight, the people had lost their courage and the politicians wanted peace. Hindenburg silently took Ludendorff's hands and they parted like men who had buried their deepest hopes"*[704] Ludendorff had not suffered a nervous breakdown, as has often been recounted, but had a massive tantrum. If he had had a nervous breakdown, he could hardly have recovered so quickly.

On September 29th, Hindenburg and Ludendorff informed the Kaiser and the Reich government of the situation: the army had reached the end of its strength and one would have to immediately make peace and begin armistice negotiations with the Allies. Wilhelm II took the matter up calmly. He knew that the army was at the end of its tether. Hintze was assigned to seek a truce and peace.[705]

On October 3rd, 1918, Hindenburg followed up: The OHL insists on the *"immediate issue of an offer of peace to our enemies. [...] As a result of the collapse of the Macedonian front, the weakening of our western reserves that has become necessary and as a result of the impossibility of replacing the very considerable losses which occurred in the battles of the last few days, there is no longer any prospect of peace to impose upon the enemy."*[706]

In his memoirs, Hindenburg wrote that the Western Front *"wavered but did not fall. At that time, however, a wide void was being torn in our entire war front. Bulgaria had collapsed."*[707]

This statement is one of the two most famous attempts to cover up the defeat. Even better known in Germany is the so-called *stab in the back* legend, according to which the army that was undefeated in the field was deprived of victory by a stab in the back from home. The claim that the collapse of Bulgaria caused the German defeat is absurd. The Turkish defeat was just as important, or rather, just as unimportant for the war situation in the West as the Bulgarian surrender. The decision was had been made in the west. The wars on the periphery had practically no influence. Hindenburg did not mention that the German army was finished and would have collapsed within a short time. This interpretation is also supported by the fact that Ludendorff did not rage against his allies, but against his own people.

The collapse of Bulgaria became something of an alibi for the defeat. Ludendorff compared the collapse of the Macedonian front with the Allied tank attack on August 8th and equated the significance of the collapse in Macedonia with that of the *"dark day of the German Army."*[708] The official history of the First World War published after the war by the Reich archive and later by the Reich Ministry of War or the High Command of the Wehrmacht did not reach into the year 1918, but the preliminary study in the series *"Schlachten des Weltkrieges"*, which was also published by the Reich archive, describes the collapse of the Macedonian front in this way: *"With the collapse of the Macedonian front began the collapse of all the iron walls that Germany had built with human corpses in its struggle for life and kept tenacious for years. There was a breach. The fate of the German fighters in Macedonia soon became Germany's fate."*[709]

Perhaps one should note at this point that at the beginning of the year, 22 German battalions and 72 German batteries still stood at the Macedonian front. With their machine-gun companies, they had, as it were, formed the corset bars of the front. When Ludendorff was planning his major offensives in the west, most of the German troops were withdrawn from Macedonia. At the end of August 1918 there were only 3 German battalions and 32 batteries on the Macedonian front.[710] In other words, much blame for the collapse of the Bulgarian front can be placed on Ludendorff himself. We will get to internal Bulgarian reasons later.

The collapse of the Bulgarian army not only led to legends and myths on the German side, but also amongst the Allies. We will also talk about these developments later, and see how they continued into the Second World War.

Planning the Allied Offensive

At the beginning of July 1918, Franchet d'Espérey began planning the great offensive. He knew that there were only two sensible points of attack on the northern Greek border if one wanted to advance northwards towards Skopje: the Vardar valley and the heights accompanying it on both sides and the plateau at Bitola / Monastir. It was precisely at these two points that the earlier attacks had begun. Since the Bulgarians had well-developed positions there and were awaiting the attacks, any offensive there would have become extremely bloody undertakings. Further to the east, the mountain ranges running parallel to the border prevented any major attack. West of the Vardar there was a mountain range called Moglena, of which the Kajmakçalan was a part, which had been occupied by the Serbs since 1916. Most of the mountains, Dobropolje, Veternik and Koziakas, had been fortified by the Bulgarians, but the trenches were not in a staggered system and could therefore be broken through. The most important thing, however, was that the Bulgarians did not expect an attack there. If the surprise attack succeeded, the road connection from the Vardar valley to Monastir would be interrupted, so that no reinforcements could come from that direction.[711]

The attack would take place in two phases. In the first phase, two Serbian and two French divisions, to which 80 more would be assigned, in addition to the 84 existing heavy artillery of the Serbs, would occupy the three heights. Thereafter, four more Serbian divisions and the cavalry of both armies were to advance further north to the villages of Kafadar and Negotin. The Serb general Živojin Miši would lead the attack. The attackers had 75 battalions, 756 machine guns and 580 cannons and grenade launchers. The Bulgarians numbered 26 battalions and had 245 machine guns, 146 artillery pieces and grenade launchers. So the attackers were three times as strong.[712]

The *Armée d'Orient* would attack at Monastir and the Cerna to tie up the troops there. In the Vardar-Dojran sector, the company's own artillery would first set up a continuous barrage that would force the enemy into cover. Thereafter, an attack would be targeted in order to prevent the enemy from regrouping and building a front against the units that had broken through in the west. In the Struma sector, the Greeks, reinforced with a few French units, were to attack the Roupel-Enge and advance to Livunovo in Bulgaria. After a successful breakthrough, the *Armée d'Orient* would advance on Skopje, with the Serbian units advancing on Veles-Štip, the 1st division group heading to Hudova and occupying the heights of Gradec Planina, the British army penetrating into the Strumica valley and the Greeks along the Struma advancing north. The attack was scheduled for September 15th.[713]

At the beginning of September, the Allied armies numbered a total of 28 divisions, eight French, six Serbian, four British, one Italian and nine Greek; In addition, the 9th Greek Division stationed in Ioannina was on its way to Florina. They were divided into four armies, the British with four British and five Greek divisions, the 1st division group with one French division and two Greek divisions, the Serbian army with six Serbian and two French divisions and the *Armée d'Orient* with five French, an Italian and a Greek division. Since the Greek divisions were divided amongst the various groups, there was no Greek high command.[714]

The Bulgarian front stretched from the Adriatic Sea to Komotini. It consisted of two army groups, each with two armies. On the right wing of the Bulgarians, the 11th German Army was under the command of General Kuno von Steuben, which was made up of the LXII. and the LXI. Corps. Even if this army and the corps were called German, only the commanders and the staffs were German. The troops were Bulgarians. The headquarters of the 11th Army was in Prilep. It was followed to the east by the 1st Bulgarian Army with headquarters in Valandovo, and which consisted of five divisions. The 2nd Bulgarian army had its headquarters in Livunovo and was three divisions strong. The 4th Bulgarian Army had its headquarters in Xanthi and consisted of one division and six battalions. The 11th Army and the 1st Bulgarian Army formed an army group under General Friedrich von Scholtz with headquarters in Skopje. The total strength of the Bulgarian army was 380,000 men, including 24,000 Germans. They were organized in 252 Bulgarian and three German battalions. Its artillery had 1,345 guns, including 297 heavy ones. The Allies numbered 291 battalions and had 1,522 guns. The Bulgarians had more machine guns than the Allies, but these were old and out of date. Only the few German machine gun companies had modern machine guns. In the air, the Allies were vastly superior with 200 aircraft to the enemy's 80 aircraft.[715]

At first glance, the forces appeared to be equal, but the Bulgarians had to use their forces to secure a stretch of more than 250 km as the crow flies.

The Allied forces were able to collect their troops in certain places, so that they were far superior to the Bulgarians. The Allied troops were also better fed and better dressed. However, the Spanish flu hit both armies. The Bulgarian army, but also the population, was tired of war. Together, including the Balkan Wars, they had had five years of war behind them, if you exclude the year 1914. On the German side, too, it was recognized that Bulgaria, like the other allies, was at the end of its rope. Since the beginning of 1918, the Bulgarian government had been informed that the German leadership was also trying to negotiate a peace. It was recognized that the German offensives in the west had failed. The *Sixtus affair* indicated that things were not going well in Austria-Hungary either. Shortly before the start of the Allied offensive in Macedonia on September 13th, Emperor Karl informed Tsar Ferdinand that the Germans could no longer count on sustained resistance on the western front. A day later, the Austrian Foreign Minister proposed that a general peace conference be convened. This initiative showed the Bulgarian government that it might be left alone and that victory for the Central Powers was out of the question. On top of that, relations between Bulgaria and its ally the Ottoman Empire were strained.[716]

In addition, the Bulgarians were angered with German policy on the Dobruja question, as they had not been granted the entire region. When the defeat of the Central Powers became apparent in the summer, Tsar Ferdinand I dismissed the pro-German Prime Minister Radoslawow and on June 21st, 1918 appointed Aleksandar Malinov, who had already been Prime Minister between 1908 and 1911. When his attempt to conclude an armistice with the Allies was unsuccessful, he encouraged army leadership to continue fighting. But when the Macedonian front collapsed, he sent a delegation to Allied headquarters to negotiate a ceasefire, which was also concluded on September 29th. Overall, it can be said that in the summer of 1918 the Bulgarian army was no longer that of the battle for Monastir / Bitola and Lake Dojran.[717]

The Attack

The offensive began at 8 a.m. on September 14th, 1918, with artillery bombarding along the entire front line. The shelling was particularly severe in the Dobro Polje sector, where the main attack was to take place the following day. During the night, French-Serbian patrols reported that the wire barriers had been more or less much destroyed. At 5:30 a.m. on September 15th, the infantry attack began, in two directions. The 2nd Serbian Army and the French 122nd Division attacked northwest towards the mountains of Dobro Polje and Sokol. The 17th French colonial division and a Serbian division advanced northeast against the Kamene and Veternik Mountains. Around 4 p.m. the Dobro Polje ridge was occupied. At about the same time, the Veternik ridge was occupied. That night the Sokol height was also conquered. Around 6 p.m. two more divisions advanced and penetrated the gap in the Bulgarian front. Thus, on the evening of the first day, the opposing positions were occupied across a width of 11 km. Smaller attacks on the other parts of the front prevented Bulgarian troops from being withdrawn and relocated.[718]

The heavy artillery fire on September 14th had resulted in few losses among the Bulgarians, but it had shaken the troops' morale. When the infantry attack began the next day, on the one hand there were mass desertions and on the other, entire units retreated without waiting for orders. Reinforcements came very slowly and there were too few. In view of this situation, the Bulgarian command ordered to withdraw to a line further north. The losses of the front troops that day are said to have been 40-50 percent, most of them deserters and prisoners of war. Reinforcements were sent forward during the night; there were five Bulgarian regiments, two German battalions and five batteries.[719]

On September 16th the offensive continued with undiminished strength. The 2nd Serbian Army occupied the dominant Koziakas Heights. A Bulgarian counterattack was rebuffed. The 1st Serbian Army, which had occupied the Sokol Heights that night, advanced on the village of Gradesnitsa. The resistance there was overcome together with units of the *Armée d'Orient*. Further east, French and Greek regiments conquered the heights near the village of Peukoto. On the evening of that day the breach in the Bulgarian front was 25 km wide and on average 7 km deep. The Bulgarian troops moved to the north in a disorderly manner. It was clear that the front would soon collapse.[720]

The offensive continued on September 17th. The penetration depth was 15 km. Nothing stood in the way of an advance into the Vardar Valley. When they withdrew, the Bulgarian troops left behind a lot of valuable war material. Reinforcements only arrived in bits and pieces, so that it was no longer possible to stop the Allies despite the terrain, that was favorable for defense.[721]

On September 18th, the advance of the Serbian divisions west of the Vardar Valley continued, and the British began their planned advance on Lake Dojran. The Serbs made good progress and the *Armée d'Orient* covered their left flank by advancing on both banks of the eastern Cerna arm. The 1st division group advanced on the right flank. The 2nd and 3rd Bulgarian divisions, against which the main attack was carried out, moved north-west towards Prilep and north-east towards Konopište. Franchet d'Espérey ordered the Serbian troops to continue their advance. The 1st Serbian Army advanced on Belavonitsa and Prilep and the 2nd Serbian Army on Kafadar, a few kilometers from Negotin in the Vardar Valley. The *Armée d'Orient* continued to secure the flank by also pushing forward. The army group did the same, advancing on Huma. Both advances continued on September 19th.[722]

While the attack by the Serbian units went well, the attacks in the British sector had problems. From an aerial survey, they knew that the Bulgarians had strongly fortified the section of the front between Vardar and Lake Dojran. When the attack began on September 18th, the attacking 22nd British and Greek Serres divisions were unsuccessful, and suffered heavy losses. In particular, the 3rd regiment of the Greek division had a high number of dead and wounded and had to be removed from the front. On September 19th, the attackers were hardly more successful. The British and Greek losses were high: The British lost 165 officers and 3,155 men, the Greeks 173 officers and 2,154 men. The Bulgarians lost 83 officers and 2,643 men. The offensive on Lake Dojran had failed. The Greeks were of the opinion that it was due to the inadequate artillery preparation of the British.[723]

The Serbian divisions continued to advance successfully north on September 20th. The commanders reported that the Bulgarian resistance was weakening. Franchet d'Espérey expected that the 2nd Serbian Army would soon reach the Vardar and ordered it to build a bridgehead on the east bank from which the Bulgarians could retreat. On

September 21st, the 2nd Army reached Negotin on the Vardar. The 1st Army pushed further north. The Allies took 7,000 prisoners and more than 100 guns, and large amounts of ammunition fell into their hands as well. Aerial reconnaissance reported that the Bulgarians were withdrawing, leaving scorched earth behind them. The 1st Army advanced on Prilep but on the British and Greek fronts east of the Struma, little moved.[724]

In the days that followed, the Serbian troops pushed forward. The 1st Army reached the road from Prilep to Gradsko. The 2nd Army succeeded in building a bridgehead east of the Vardar. The *Armée d'Orient* advanced in the west towards Skopje. Only the British and the Greeks on Lake Dojran and further east made no progress. But on the rest of the front a collapse of the Bulgarian army became apparent on September 25th. The troops didn't feel like fighting anymore. When officers tried to prevent them from deserting, the soldiers threatened them. The deserters even attacked the Bulgarian headquarters in Küstendil, which led to a breakdown in communications. On the same day, the Bulgarian cabinet met under the chairmanship of Tsar Ferdinand. The Chief of Staff of the Bulgarian Army described the situation as serious but not hopeless. But when news of the deserters' attack on Headquarters became known, it was determined that resistance would be pointless, and the Cabinet decided to enter into negotiations with the enemy for a separate peace.[725]

On September 28th, a Bulgarian delegation appeared at the British lines. It consisted of the finance minister, a general, a senior diplomat and two senior officers. Milne ordered them to receive accompanied safe passage to Saloniki. But this did not stop the Allies from advancing further. A day later, the French occupied Skopje. After the conquest of Skopje, Franchet d'Espérey moved his headquarters to the city. But prior to that he received the Bulgarian delegation. He contacted the French government to discuss the terms of the armistice. The next day these were passed on to the delegation.[726]

According to the terms, hostilities would end on September 30th, 1918. Bulgaria would evacuate all Greek and Serbian territories. It would demobilize its army with the exception of three divisions. Two of them would secure the border to Romania and Turkey. The third division was responsible for internal communication. All weapons and equipment of the demobilized units would be stored in depots under Allied command. Allied troops should have the right to pass through Bulgaria unhindered and to occupy certain strategic points. The Bulgarian soldiers of the 11th German Army who were west of Skopje would lay down their weapons and would be considered prisoners of war. The German and Austrian troops on Bulgarian soil would have to leave the country within four weeks.[727]

The Bulgarian troops east of Skopje were disarmed without major problems. The troops west of them only learned about the armistice on the night of October 2nd and were not disarmed until October 3rd. The German units of the 11th Army moved north to continue fighting. During their offensive and the subsequent pursuit, the leading formations of the Allied troops had covered 130 km. They had taken 80,000 prisoners, including 1,600 officers and five generals. During the fighting they captured 800 pieces of artillery and countless grenade launchers, machine guns and huge amounts of war material. The Allies had lost a total of 15,000 soldiers, of which 3,500 had fallen.[728]

With the signing of the armistice on September 30th, 1918, the war on the Macedonia front, meaning the conflict between Bulgaria and the Allies, came to an end. There were still nine German battalions and a few batteries in the Bulgarian war zone. If the Austrians and the Germans had wanted to, they could easily have built a defensive front in the Balkans with German troops from Russia and Romania. After all, there were 41 German and 18 Austrian divisions there. There were also 2 Turkish divisions on European soil.[729] But on the one hand the necessary will was lacking, because Kaiser Karl had already made a peace offer to the Allies on September 14th, and on the other hand Austria-Hungary was about to disintegrate into its constituent parts. So it was possible for the Serbs to advance to Belgrade in the following few weeks.

The question arises as to why the breakthrough through the Bulgarian front happened in September 1918 and not earlier. There are two main reasons: the first and probably the most important is the withdrawal of German troops from this front. The remaining three battalions could do nothing. As already mentioned, the 22 battalions deployed there had formed the *"corset rods"* of the Bulgarian front. The second reason was the demoralization of the Bulgarian troops, which we have already described. Had the German *"corset rods"* still been there, it can be assumed with a certain degree of certainty that the attack would have proceeded similar to the attacks in previous years. This also explains who was actually responsible for the defeat: it was Ludendorff who pulled the *"corset rods"* out of the Bulgarian front, in favor of his western offensive.

But even later, the Germans abandoned the

Bulgarians. As early as September 16th, i.e. on the second day of the battle, the Bulgarian commander in chief, General Georgi Todorov, citing Hindenburg's German weapons aid promised in December 1917, asked for rapid reinforcement of at least six divisions with the associated artillery. Hindenburg replied negatively on September 17th: the fighting in the west did not allow the surrender of troops to the east. On September 19th, Tsar Ferdinand again turned to the German OHL and asked for reinforcements. Hindenburg replied regretfully that at the moment he was unable to *"make stronger German forces available for the Macedonian front. The enemy onslaught on our western front has been going on for two months with the greatest effort of men and material. [...] The whole outcome of the war would be jeopardized if this front were weakened now. On the other hand, if the Bulgarian troops could not hold out in the current lines, the abandonment of further territory in Macedonia must be accepted in the interests of the greater whole."*[730]

On September 24th, General Todorov made a desperate plea for help to the OHL, which said: *"At the beginning of the offensive, fewer divisions were sufficient, but the situation is critical at the moment, [...] perhaps even ten divisions can no longer help."* Czar Ferdinand wired Wilhelm II on September 21th: *"Without rapid help, a catastrophe cannot be avoided."* On September 25th, he spoke of an impending doom for everyone if Macedonia was lost. The OHL finally began to rethink the situation, and on September 26th, two days before the Bulgarian surrender, three German divisions marched into the Balkan theater of war. Austria-Hungary agreed to send two divisions. Furthermore, the OHL suggested to the Czar that the general management of the Bulgarian troops should be transferred to General von der Scholtz. More cheap advice followed that had nothing to do with the reality on the ground. [731] In other words, the OHL simply abandoned a loyal ally. In addition, the German troops did not have to be taken from the western front, there were a large number of idle divisions in the east on the former Russian and Romanian fronts.

The success of the Allied offensive was only possible against this background. Guillaumat had planned minor advances. Franchet d'Espérey wanted to advance into southern Serbia and possibly occupy Bulgaria. When the armistice came in on September 30th, he wanted to go on defense first, and perhaps move forward later. It was the Serbian units that had made the first phase of the offensive a success. The Serbian armies had achieved the breakthrough. The French units followed as flank protection. The British had little success. And it was again the Serbs who pushed through to the north, to continue the offensive. They were highly motivated; after all they had liberated their own country. But because Franchet d'Espérey had led the offensive, the fame hung on the flags of the French army.

But even if the Macedonia front had held, this would have been irrelevant for the further course of the war. The war had already been decided and lost in the West, even if this was not yet completely apparent.

When the armistice was signed, Franchet d'Espérey ordered the 1st Greek Corps to occupy eastern Macedonia. Officers of the I. Corps contacted the commanders of the Bulgarian units in East Macedonia to persuade them to withdraw. But the Bulgarians were uncooperative. Thereupon an Evzones regiment quickly occupied Serres on October 4th. It was discovered that almost the entire male population between the ages of 14 and 60 had been deported to Bulgaria for forced labor and that the city itself had been devastated. A regiment of the 2nd division occupied Siderokastro and the Evzones regiment of the 13th division advanced to the village of Palaiokomi. On October 5th this regiment approached Kavala. On October 6th, regiments of this division occupied the village of Rodolivos and the city of Kavala. The villages and towns made a run-down impression. The residents were old people, women and children. Here, too, the men had been deported to Bulgaria for forced labor. Kavala, which had had 68,000 inhabitants when the war broke out, had just 9,000 left. Some villages had been completely destroyed.[732]

By October 10th, the Greek divisions occupied Macedonia as far as the Nestos River. On the same day, Franchet d'Espérey subordinated the 1st Corps to Milne in the event that it was ordered to make an advance on Constantinople. The French 122nd Division and the British 22nd Division advanced on the Turkish border in Thrace. But Turkey surrendered before there was an attack. The two allied divisions then occupied European Turkey and the Dardanelles together with the 28th British division.[733]

Interpretations and Legends Following the Bulgarian Capitulation

We have already reported on the creation of German legends by Hindenburg and Ludendorff, as well as the official historiography. While for the Germans the defeat of the Bulgarians in Macedonia was a welcome excuse for their own defeat on the Western Front, for the Allies it was a matter of talking up a failed military undertaking in the end and justifying it in retrospect.

It is interesting that Lloyd George adopted Ludendorff's argument in his memoirs: *"Ludendorff notes that on the evening of September 28th, the day the Bulgarian special envoy arrived in Saloniki, he and Hindenburg had decided that immediate action must be taken for a ceasefire and conditions for peace. The next day he instructed the Foreign Minister to take the necessary steps for this purpose."* Then, however, Lloyd George adds an argument from General Hermann von Kuhl, which the latter had brought before the Reichstag investigation committee after the war: After the collapse of Bulgaria, the Allies could have occupied Romania and thereby interrupted the Romanian oil supplies to Germany, whereby the German Armed forces would have been immobilized within two months. There is no doubt that this statement is correct, but there is no evidence that anyone thought of it at the time of the Bulgarian surrender – certainly not Hindenburg and Ludendorff – as Lloyd George assumes. Indirectly, he admits that the Allies did not come up with this idea either: *"If we had taken steps in 1915 to secure the Balkans, as we should have, the loss of oil supplies would have shortened the war by at least two years."*[734] Lloyd George attempted to justify the operation *ex post*.

Churchilll's verdict is astonishing: *"It was only indirectly as a result of the violent clash in the West that the last blow against the German resistance came about. The theater of war, where the war dragged on in a loss-making and senseless manner since the summer of 1915, the theater where initiatives were generally condemned by the highest Allied military commanders, was destined to make the final decision. The strength of a chain, even if it is strong, is always determined by the weakest link. The Bulgarian chain link was on the verge of rupture and with it the cohesion of the entire enemy coalition."*[735] So Churchill initially admits that the operation in Macedonia was pointless, but then finds a way to justify it. Bulgaria was the weakest link in the chain. More important, however, is his indirect statement: apparently he too believed that the key to victory in World War I had been hidden in the Balkans. And this has become a strategic credo for him.

But not only Churchill tried to justify the Macedonian defeat. Shortly after the end of the war, a well-known war correspondent published a book with the programmatic title *"Salonica and After"*. *"The Sideshow that Ended the War."*[736] In it he repeatedly quotes Ludendorff's memoirs in detail to prove that the activities of the Allies in Macedonia were decisive for the war. In the introduction to this book, the former commander of the British Army in Macedonia, Milne, described the Macedonia Front as the *"Achilles' heel of the Central Powers which was instrumental in the rapid collapse of the dramatic autumn of 1918."*[737] Referring to Ludendorff's testimony of the importance of the collapse of the Bulgarian front, Owen wrote in August 1919: *"The British Saloniki forces could not ask for more remarkable recognition for their long dedication and ultimate triumph than these few honest words from Ludendorff. Together with the important letter from Hindenburg reporting on the collapse of Bulgaria, in which he says: "We are no longer able to resist, we must ask for a ceasefire," they destroy all criticism that has been said so far regarding the value of the Saloniki Armed Forces.*[738]

Since no one contradicted or refuted these claims, they became an established truth, a strategic concept that carried within it a victory, and was applied again in World War II.

The Saloniki Front was one of the quietest in World War I. Until the summer of 1918 there were occasional small attacks, but these did not affect the course of the war at all. The Macedonia front did not even tie up German troops of any size or number worth mentioning. The Macedonia effort was as pointless as the Gallipoli effort. But where the latter was ceased when the senselessness was recognized, the former continued to the bitter end. Without the collapse of the war-weary Bulgarian army, there would have been no justification. The commentators therefore readily took up the Bulgarian collapse, as it provided the *ex post* justification for the pointless undertaking.

The author of the official British history of the Macedonia fighting, Captain Cyril Falls, comes to an appropriate conclusion when he writes: *"It has been three years since the Allies landed in Saloniki, almost three years since the Germans first made fun of the "Allied internment camps". Now all of this led to victory. Major events dwarfed this victory, for the entire three years they faded into the background. The scene aroused little interest in the people of the conquerors nor those of the conquered. There were no laurels for the Allied armies in the east."* The Macendonia Front was not even remembered at the great victory parade on July 14th, 1919 in Paris. But then Falls discovered some recognition after all*: "Perhaps the first true recognition came from an enemy, the commander in chief of the former grand coalition. It came from Field Marshal von Hindenburg and appeared in the war memoirs of his right-hand man General Ludendorff."* This was followed by the well-known quote from Ludendorff. But then Falls added a remark that shows that he knew how to differentiate: *"The situation was too serious for anything but the truth, yet it is possible that the Field Marshal tried, albeit unconsciously, to blame the German defeat in the west on the Bulgarian surrender in Macedonia."*[739] So Falls recognized that Hindenburg and Ludendorff were looking for excuses for the defeat, and they did not do this unconsciously, but rather fully consciously. But had Falls admitted that, he would have underlined the first sentences of his assessment.

For the Greeks, of course, their own contribution to the last battle for Macedonia is of central importance, as the General Staff Office notes: *"Without a doubt, the participation of Greece had a serious influence on the course of the war; one can argue that without its participation there might not have been a victory."* It is the well-known tendency to self-exaggeration, which can also be found in historical depictions of the Second World War.[740]

The Greek army contributed selectively to the Allied success, but the Serbs provided the real achievement. The final conclusion of the General Staff again follows the familiar pattern: *"Hence it is certain that the Allied victory on the Macedonia front hastened the end of the war. Without the win, the commitments would have continued through the following year, possibly even longer."*[741]

A more recent scientific study from the USA comes to the conclusion that *"The Battle of Dobro Pole was one of the most decisive battles in World War I. The breakthrough, achieved by the French and Serbs in three days, threw Bulgaria out of the war in less than two weeks and ended the heavy fighting in south-east Europe. The magnitude of this success is proof of the bravery of these soldiers as well as of the war-weariness of their Bulgarian opponents."*[742] Apart from the exaggeration in the first part of this text, this judgement is close to reality.

The Allied victory in Macedonia over the Bulgarians has the same status as that in Palestine over the Ottomans. They are local victories, the decision had been made on the western front, though, and for a while, defeat had only been a matter of time.

Macedonia in the Second World War

The region of Macedonia, which is strategically important for the Balkans, naturally played a role in the Second World War, simply because of its geostrategic location. In the following pages we want to briefly consider the existing parallels and differences.

After the end of the Polish campaign in the winter of 1939/40, the period known as the *sitting war* or *Drôle de guerre* began. The French sat on the Maginot Line and waited for the Germans to attack. They firmly believed that their line of bunkers was impenetrable. But they also believed this of the West Wall and therefore had refrained from attacking during the Polish campaign, although Hitler had pulled almost all good troops from the West. They became victims of their own wishful thinking. Paris firmly believed that the war would not be decided in the West but on the periphery in the Balkans. After all, this had been proven by General d'Esperey in the First World War. The new commander-in-chief of the French troops, General Gamelin, dreamed of building a huge eastern front from Saloniki to Poland.[743]

Consequently, the French had worked towards building this front since July 1939.[744] The French commanding general in the Levant, General Weygand, contacted the Greek and Turkish governments to discuss the matter with them. The Greeks declined a visit as they did not want to provoke the Italians. When Weygand insisted, London warned the French government that actions in the Balkans could lead Mussolini to join the war on the German side. So London rejected these French plans. But the French did not give up. On September 10th, Weygand met with Colonel Konstantinos Dovas, head of the Third Bureau of the Greek General Staff in Ankara. Weygand talked about his plans, and Dovas wanted to know how many troops the French would have in the Levant to carry them out. Weygand had to admit that he had only one division and not a single airplane. Dovas replied that the problem was no longer one or two divisions, but that the existing divisions of the Balkans had to be equipped with modern weapons and an air force had to be built up. If the French really wanted to build a beachhead at Saloniki, they should send strong air forces there and deliver the weapons that Athens had ordered a long time previous. Weygand reported back to Paris.[745]

The Italian-Greek summer crisis of 1939 ended on September 11th, and Paris was satisfied that this rapprochement would not hinder future Greek-French cooperation. Weygand continued his contacts with the Greek General Staff. But since the British did not want to upset the Italians, there were no concrete results. And when Weygand tried to win over the British Commander-in-Chief in the Middle East, General Archibald Wavell, for his ideas, he gave him the cold shoulder: The Allied forces in the region were far too weak and any transport to the Aegean would be attacked by the German Air Force.[746] With this, the topic of a Balkan Front was, for the moment, tabled.

The next time the subject came back into discussion, the British brought it up in May 1940 during the fighting in France. The Foreign Office suggested entering into an alliance with Greece. It could prevent the Greeks from reaching an agreement with Italy; it could keep Bulgaria neutral and induce Turkey to fulfill its contractual obligations towards Greece. But the British military were against an alliance: it would increase England's moral and military obligations to Greece; the latter would also be strategically unhealthy, since England could not help Greece materially.[747] At the end of May 1940, Wavell proposed landing in Greece and marching to and occupying the Ploësti oil fields in Romania. The chiefs of staff in London told him that they believed it wrong to repeat the mistakes of World War I when Britain backed small countries that were nevertheless overrun.[748] After the fiasco in Norway and the defeat in France, the chiefs of staff had even less appetite for peripheral operations.

But it was precisely this fiasco that fundamentally changed the situation. World War I "*Easterner*" Winston Churchill returned to the political scene as Prime Minister on May 10th. Until the Italian attack on Greece on October 28th, 1940, he was barely interested in Greece and the Balkans. That day, Secretary of War Anthony Eden announced that Britain was unable to assist Greece efficiently, either on the ground or in the air, and that it looked like another nation to which it had given a guarantee would fall to the Axis.[749] The British Cabinet agreed. But when Greece did not collapse but rather resisted, Churchill saw the chance to fight back. He ordered all conceivable help to be given to Greece.[750] Eden called this a "*strategic folly*"[751] but Churchill's order was a kind of reflex to a months-long frustration during which he could do nothing. Militarily seen, this decision to help Greece was as senseless as the one that set up the Saloniki Front during the First

World War.

But on November 8th, 1940, Churchill's eagerness to help Greece – at whatever cost – subsided when Eden informed him that Wavell was planning an offensive against the Italians in North Africa under the code name *Compass*. As he writes in his memoir, he purred like six cats, and gave *Compass* absolute priority.[752] Greece was almost forgotten. It only received a little help from a few RAF squadrons. The offensive in North Africa was extremely successful. British troops reached Tobruk within three weeks. They lost just over 600 men, but captured 38,000 Italians and 237 guns, 73 tanks and over 1,000 trucks.[753] If the British had pushed ahead with full force, they would probably have reached Tripoli and driven the Italians out of North Africa. The French colonies of Algeria and Tunisia would probably have switched sides and North Africa would have been in the hands of the Allies. There would never have been an Africa Corps. Allied bombers could have bombed Italy from there and the Italian fleet would have sought protection deep in the Adriatic Sea. Had *Compass* been pursued with all its might, the war would have seen a decisive turning point as early as 1941.

But the British stopped their successful advance and again offered help to the Greeks. The reason was not that they suddenly remembered their moral duty to honor the guarantee they had issued in 1939, but that they themselves urgently needed help. Since the beginning of the war, the British had bought huge amounts of war material from the US as part of the *cash and carry program*. But by the end of 1940 the British ran out of funds. On December 8th, Churchill wrote the infamous letter to the re-elected President Roosevelt asking for help. Roosevelt responded positively and brought the *lend lease bill* to Congress. But the legislative body was neutralist and hardly interventionist-minded. It quickly became clear that the law would only pass Congress if the American public sympathized with England. Any negative headline, such as England abandoning Greece in need, could stop it. So Churchill had to help Greece, or at least give the impression that he was doing that.

In mid-January 1941, the British offered their help to the Greek head of state Ioannis Metaxas. Metaxas, a former general, realized immediately that the help was totally inadequate and refused it. In his opinion, this aid would only provoke a German attack.[754] Churchill accepted this refusal and Wavell was able to continue his advance on Tripoli. But at the end of January Metaxas died. Churchill had followed the rather controversial Congressional debate with increasing unrest and believed he could influence it if Greece accepted British aid. To get the Greeks to accept the aid, he sent Foreign Minister Eden to Athens. He had two tasks: he was supposed to talk the Greeks into accepting the aid, and win the neighboring states as allies. The rest of the story of how Eden operated with fake numbers about British help and how he placed massive pressure on the new Greek Prime Minister to accept the aid is well-known.[755] But Churchill hadn't really intended to help Greece. His "*help*" was intended to influence American public opinion and, indirectly, Congress. Indeed, he had hoped that the Germans would overrun Greece before British aid could arrive.

But the timing of the whole operation got mixed up. The *Lend Lease Act* was passed by Congress in early March. But the long winter delayed the German attack until April. So the British were forced to keep their word. Churchill had to stop the attack in North Africa. He weakened the desert front by sending the only reserves, a tank brigade and an infantry division each from Australia and New Zealand, to Greece. The small chance of defending Greece with these forces was ruined by the British themselves when they initiated the coup in Yugoslavia. The unprepared Yugoslavs were not a serious opponent for the Wehrmacht, worse still: Yugoslavia's entry into the war allowed the Wehrmacht to launch a flank attack via Bitola into northern Greece. It was basically the same operation as the Allied attack in 1918, only in the opposite direction. The result was a disaster in every respect: Greece was overrun in a few days and occupied as far as the southern tip of the Peloponnese. At the end of May 1941, Crete was captured by an airborne operation.[756] The weakening of the front in north Africa allowed Rommel's counterattack, and his offensive toward Egypt.

All along Churchill knew he was playing *hazard*. But the risks to England were limited. A couple of tanks and the two divisions were at stake. If he won, unlimited American support and possibly even the United States' entry into the war beckoned. If he lost, the loss was manageable. As in the First World War in the case of Gallipoli or Macedonia, the consequences of this game were borne by the Australians, New Zealanders, Greeks and Serbs. The French weren't part of the action this time.[757]

Later in the war, Churchill sporadically reverted to the idea of a Balkan landing to open a second front there. Macedonia was mentioned, but Churchill spoke more of a landing in the north of the Adriatic. Apparently he had returned to his original opinion that Macedonia was not a successful concept as long as the enemy was strong enough to defend the mountain ranges in northern Greece. Churchill's renewed interest in a Balkan landing was not aimed at driving the Germans from the Balkans, but rather at preventing the Soviets from occupying Central Europe. The Americans saw his motives, but insisted on crossing the Channel (*Operation Overlord*), so Churchill had to change course yet again. From now on, instead of a military solution, he tried to find a political solution that would serve British interests alone and that would keep the Soviets away from the Mediterranean and thus from the "*life line*" of the British Empire from England to India and to oil from Mosul. In May 1944 he made a temporary arrangement with the Soviets on spheres of interest. On October 9th, this was converted into a permanent system in the so-called "*percentage agreement*". From then on the Balkans were divided into permanent spheres of interest.[758] In December 1944 *(Dekemvriana)* he even intervened militarily in Athens, in order to secure his control over Greece as a British protectorate.[759]

Macedonia played a role again in the Greek civil war when the Greek Communist Party (KKE) was ready to cede parts of the country to Bulgaria. But then, Macedonia became meaningless. Only when, in the wake of the breakup of Yugoslavia, the former Republic of Macedonia became a separate state and called itself "*Macedonia*", did a new dispute break out over the name. The Greeks believed that only their part of Macedonia had the right to call itself that. But that is another story for another day.

Part 2

Here we begin the attempt of Prof. Dr. Richter, in which he expanded the political dimension of his report to include the military's daily routine, from German sources. For this purpose, the authors would like to examine some excerpts of historical documents, because the process of recognition and acquisition of knowledge is based on attentive observation of the course of the war and the evaluation of existing documents. We start with the assessment of the *"Balkans as a secondary theater of war"*.

Secondary War Theater Balkans[760]

"Air and ground war conditions in the western theater of war have determined the organizational, tactical and technical course of development of the German air troops. The fronts in Russia, Macedonia, Romania and Asia Minor have had to adapt to this. But it would be remiss to ignore the achievements of the aviation units there, especially since they had to operate under the same, if not more severe, conditions.

Even though here, especially in Russia, the air situation and the defense against the ground fire may at times have been less threatening than in the west, the peculiarity of these theaters of war and the even less favorable balance of forces with the enemy - with the exception of Russia - made aviation sorties extremely difficult. In the Balkans [...] the opponent was six times superior.

Then there is also the big difference in the front widths, at the end of 1916:
- West: 8 kilometers
- East: 32 kilometers
- Balkans: 70 kilometers

While the tasks of aerial warfare had long been honed and specialized in the west, the units here (in the east) were used simultaneously for reconnaissance, infantry and artillery service, battle and bombing flights until 1917, and lacked adequate fighter protection. In the east there was only one fighter squadron on the whole front, in Romania and Palestine initially not a single fighter plane was available. In Macedonia, the "Fokkerstaffel Vardar" covered a front of 200 kilometers in length with four aircraft until the summer of 1916. In Thrace, until the end of 1916, Lt. von Eschwege fought a force ten times superior. When he fell on November 21st, 1917, he had become the legendary hero of a foreign people [Bulgaria], and his name was barely known in his homeland. Bulgaria solemnly erected a memorial for him in Drama. Only when, with his death, the aeronautical balance in the wide area from Vardar to Mesta, from Krusa-Balkans to the Aegean was finally and clearly lost, and the events in Monastir [today Bitola] forced the relocation of the only Balkan fighter unit to the Cerna-Arch, was a second squadron set up in Macedonia.

In the east, the cold and snow made it difficult to maintain the aircraft. [...] Wood for building barracks was in short supply. In addition, monotonous, often scarce food and an often unhealthy climate, with cases of dysentery, Pazatac and malaria. In addition, the enemy always equipped their pilots in the region with the most modern equipment, while the German units mostly had to be content with outdated types. In terms of flight technology, the vast swamp and steppe areas, the up to 3,000m-high snow peaks of the Dudika massif and the Albanian mountains demanded a great deal of the flight personnel. The formation of vortices in the cold mountain and warm sea winds on the Aegean or the Mediterranean often made the sure hand of a proven pilot fail, or made the aircraft insensitive to his control. An emergency landing in the gorges of the Balkans or the swamps was synonymous with certain death.

"The supply situation [to the Balkans] caused particular difficulties, because only the undisturbed operation on the single rail line [761] from Budapest via Belgrade to the Vardar valley was a matter of life and death for the units deployed in the southeast. Further transport by cars, horse or buffalo columns failed. They were not equipped for a mountain war. [...] Even after the front had solidified, the supply situation remained at risk, especially since there were still 200 km from the advanced depots in the Balkans to the war front. [...]

Illnesses in the Southeastern Area

Gerhard Fieseler reports about the difficult hospital conditions in Macedonia in his book „Meine Bahn am Himmel" ("My track in the sky").[762] He contracted typhus in the summer of 1917:

"I was taken to the disease hospital in Prilep, a former barracks that was overcrowded with

sick people. In addition, tents were also set up to accommodate everyone. The paramedics did what they could. I have never forgotten those terrible pictures there. Beside me men died a death against which they could not defend themselves. Alone, far from their relatives, they died in their prime. I spent three weeks here in a state of excruciating drowsiness and apathy. I had never before believed that the human body could deteriorate so quickly and relentlessly. I went through all stages of this insidious infectious disease and was only able to travel for a home leave three months after my admission."

Unit leader Oblt. Haupt-Heydemarck from FA 30 also reported on the insidious diseases in the summer of 1917. Due to illness Lt. von Wobeser, von Weppen's deputy officer, medical officer Dr. Woermann (to replace Dr. Frenzel), as well as Kuhlo and Stattaus. But Oblt. Haupt-Heydemarck also fell ill and had to give up FA 30 to Lt. König.

According to an English report, on the island of Thasos 18 of 20 English aviation teams were unavailable due to malaria.

Zeppelins on the Southeastern Front

It should also be mentioned here that army airships were also used in Macedonia. The first of these was LZ 81, beginning in November 1915. This airship was primarily intended to disrupt troop landings in Saloniki and destroy ammunition and supply stores. The performance of the engines was too weak to withstand the strong Mediterranean and Balkan winds. It was recalled.

The airship LZ 85 came as a replacement in mid-January 1916. On the night of January 31st to February 1st 1916 this one bombed the naval port of Saloniki with 2,000 kg bombs. The journey covered 1,450 km and lasted approx. 18 hours. Since the attack came as a surprise, no air defense was identified.

The second attack took place on the night of March 17th-18th, 1916. 1,850 km were covered in 26 hours.

The 3rd attack on Saloniki on May 4th and 5th, 1916 brought the end of LZ 85. It was hit numerous times and made an emergency landing in the swamps of the Vardar estuary, during which all crew members were captured.

SL 10, a Schütte-Lanz airship, came to the southeast as a replacement for the downed LZ 85. Since the long and moonlit nights with the increasing Allied air defense made a mission on Saloniki appear very questionable, it was used for naval reconnaissance over the Black Sea. There, it was lost with the entire crew on July 27th, 1916 during an attack on Sevastopol.

When more airships arrived, they were not used on the Macedonian front, but against Romania, Russia and Italy.

Captive Balloons in Action on the Macedonian Front, an Excerpt

The use of the observation balloons began with the advance in 1914 in the west. Before they were used, two years of war effort in the west were characterized by heavy losses, but the experiences of the combat procedures that were developed are now presented in extracts. The author is Captain and Army Airship Commander *Stegmann*, who compiled the report[763] to present to the air forces..

"Although the balloons had proved their value in defensive battles, it was questionable whether, because of sluggishness in their movements, they would be suitable for attack battles and for mobile warfare with far-reaching goals. [...]

No conclusion could yet be drawn from the war of aggression in Serbia in 1915. Two field airship detachments, each with a balloon, were set up here. [...]

After crossing the Danube at Cemendria, a storm set in that made it impossible to cross over, and the gas supply also was cut. It was still possible to catch up with the fighting troops, but then the balloons fared like most of the heavy artillery: they could not follow the foremost troops quickly enough on the groundless paths and were stopped. With double skins and very light vehicles it might have been possible to keep pace with the infantry for a few more days.

After a long period of rest and a six-week march through Serbia and Macedonia, the two field airship detachments arrived at the Greek border at the end of March 1916 and were deployed on both sides of the Vardar. Ascending barely two kilometers behind the foremost infantry line, they were able to observe and surveille the structure of the fortifications and the enemy artillery well. Climbing to altitudes of up to 1,700 m, excellent visibility of up to 30 km made the surveillance easier, at times even shipping traffic was observed in the port of Saloniki. In late April 1916, a French balloon also arrived at the front.

The heat that set in in June put a great strain on the teams. A portion of the group fell ill and the newly arriving replacements could not get used to the strong climatic changes at all. The losses of horses - mostly heavy horses - was extremely high. As a result of the strong solar radiation, the device also suffered, the balloon material quickly decomposed, the adhesions loosened and it was not uncommon for a balloon to disintegrate into its component parts.

Replacing the device (from Berlin) and gas (Fischamend near Vienna) was difficult because of the distance. In remote theaters of war, the use of balloons is only appropriate for a long time if the replacement of equipment and gas is guaranteed close behind the front. In August 1916 the balloons were withdrawn from Macedonia and used in Romania."

Combat Squadron 1 (Kampfgeschwader 1) in Macedonia

The designation "Kagohl" is the abreviation of "**Ka**mpf**g**eschwader der **O**bersten **H**eeres**l**eitung" ("Combat Squadron of the top military command").

Through the knowledge and evaluation of operational reports, the decision was made that "*worthwhile*" targets were mostly to be found in the enemy's hinterland. The bomb squadrons were set up in order to combat these sustainably, i.e. long-range targets on the English Channel and in Lorraine. At the end of 1915 the *carrier pigeon department Ostend* (BAO) became the Kagohl 1 and the *carrier pigeon department Metz* (BAM) became Kagohl 2.

Commander of KG 1 in Macedonia was the successor to the first commander of KG 1, Hptm. Ernst von Gersdorff,[764] Oblt. Hermann Kastner, who took over KG 1 and ran it in the Balkans. From KG 1 only squadrons 2, 3, and 5 were moved to the Balkans. The other three squadrons remained as *Halbkampfgeschwader 1* in the west. Oblt. Franz Kaestner was the leader of *Kasta 20* of Kagohl 4, which was subordinate to KG 1 in Macedonia. The *Kasta 20* was taken to Macedonia to try out the first *Gotha G.II* aircraft. In autumn 1916 Squadron 20 from Kagohl 4 with its 8 G.II aircraft was transferred to the Bulgarian Razgrad airfield, in the south-east, as a *half-squadron*. The *half-squadrons* were a necessity because the inadequate airfields in the south-east region could not accommodate an entire squadron.

In the west, the *Kasta 20* flew aircraft with the identification "20", but with different liveries. Single-engine Rumpler C.I biplanes without the identification "20" were on the railway transport wagons. These *Kasta 20* with Kaestner, Kempf, etc. then retrained to Gotha G.II in Bucharest-Pipera and flew missions with the Gothas.

The last air raid from KG 1 took place on May 1st, 1917 and then KG 1 moved back to the western front.

Field Aviation Unit 1 (Feldflieger-Abteilung 1)

In the Guard Corps, the Feldflieger-Abteilung 1 (FFA) (Field aviation Unit 1) had been in the Balkans since the end of April 1915 and stayed until the end of April 1916. On January 12th, 1916, Lieutenant Herbert von Chappuis and Georg Trenkmann were fatally shot down over Saloniki.

Letter from the Aircraft Mechanic Karl Hiemann of FFA1 to His Sister:

"*Xanthi*[765] , *January 14th, 1916*

My dear Rosa!

So far I haven't received any mail since we were transferred, who knows how long it will take to get regular mail again. This map is a view of this city where we find ourselves. You can see what huge mountains there are. I must hereby give you the sad news that I and my friend "Hörstel" have been abandoned for 2 days. Unfortunately, our officers[766] *have not yet returned from a long-distance flight. They were our best officers in the unit. Unfortunately, nobody knows yet what happened to them.*"

Aviation Unit 20 (Flieger-Abteilung 20)

The field aviation unit 51[767] was set up on November 28th, 1914, in Königsberg, through the Aviation Replacement Unit 3 as the second fortress aviation unit. On January 11th 1917 it was renamed to *Flieger-Abteilung 20*, headquartered in Schlettstedt in Lower Alsace. At the end of July 1917, the unit was transferred to Romania, and at the end of 1917 to the Macedonian war theater.

Aviation Unit 30 (Flieger-Abteilung 30)

Field Aviation Unit 30 was set up on August 1st, 1914 by the 1st Company of Aviation Battalion 1 in Döberitz. On January 11th, 1917, the name was changed to Aviation Unit 30. When the war broke out, the unit was on the Western Front, since November 1914 in Russia and since September 1915 in Macedonia.

SIn January 1917, Hptm. Von Blomberg, as commander of the aviators in Üsküb, introduced the squadron leader of the FA 30, Hptm. Haupt [Heydemarck], into his area of responsibility: "*Your squadron flies for the XX. Turkish Corps and for the 10th Bulgarian Division, headquarters in Drama, with Drama airfield.*

Front section: from Orljak airport (British) to the mouth of the Struma = 60 km as the crow flies, as well as the coastal section from Tschajagzi to the mouth of the Mesta = 100 km, Own forces: three reconnaissance aircraft and a combat single-seater [Lt. von Eschwege]." He took over the squadron from Oblt. Geisler.

Crew of B- or C-Aircraft:

The observer was usually an officer while the pilot was recruited from all ranks (troops, sergeants, officers).

At the turn of the year 1916/1917, all field (FFA) and artillery aviation departments (AFA) were reorganized into aviation units or aviation units (A) and the area of responsibility was only differentiated by location and target exploration, but the separation in long-range and local reconnaissance aircraft remained. The local scouts supplied divisions and general commands, the long-range scouts worked for the higher staffs. High-quality camera technology for series imaging were added.

Field pilot units (FFA) were reorganized and renamed into aviation units (Flg.-Abt. Or FA) at the turn of the year 1916/1917. The task of the aviation units was the close and long-range reconnaissance behind the enemy front on behalf of the division, the army corps or the army. Here, the reconnaissance with images was of particular importance. On important fronts, the army high command had a series imaging. All aircraft were two-seater biplanes (Type C) with machine-gun armament.

Stage-aircraft and Army Air Parks (AFP)

Each Army High Command formed a hinge for the subordinate aviation units between the front and the stages. Initially, these were immobile Etappen-Flugzeug-Parks (EFP) (stage aircraft parks), which were converted into mobile army aircraft parks in early 1915 and relocated near the front. The parks took care of the personnel and material supplies in the respective storage area. After initial improvisations, the situation changed suddenly with the establishment of the *"Chief of Field Aviation"* unit on March 11th, 1915. Major Hermann Thomsen created clear structures and set up an aviation staff officer (Stofl) at every Army High Command (AOK).

The army air parks gradually reached the size of small industrial enterprises with up to 1,000 soldiers. The parks created factory-similar facilities for equipping and flying in replacement aircraft arriving from home, for repairing damaged machines, overhauling engines and storing equipment, ammunition and consumables. At the same time, they served educational purposes.

The *"Army Air Park 11"* mentioned in this documentation was formed from the *"Army Air Park 13"* on September 24th, 1916 and was active in supplying all the 11th Army aviation units on the Macedonian theater of war.

The army air parks gradually reached the size of small industrial enterprises with up to 1,000 soldiers. The parks created factory-similar facilities for equipping and flying in replacement aircraft arriving from home, for repairing damaged machines, overhauling engines and storing equipment, ammunition and consumables. At the same time they served educational purposes. There were stage or army aviation parks with the armies, army groups and the army departments, or the large associations with names.

The *"Army Air Park 11"*, which is important in this documentation, was formed from *"Army Air Park 13"* on September 24th, 1916 and existed for the supply of airplanes active in the Macedonian area. until the end of the war.

Air Force News Bulletin, Individual Reports:

The *Air Force News Bulletin*[768] is used as a guide by date in this publication. It lists all aviation activities from all theaters of war. But since the 1st edition is from March 1917, the losses of the aviators in the south-east area are listed here from January 1916:

- BO Lt. d.R. Fritz Herrlich, born on April 16th, 1889 in Berlin, died on January 23rd 1916 in Demirkapu (Macedonia), FFA 1
- Offiz. Stellvertreter Otto Witt, born on June 9th, 1891 in Grimmau, died on January 23rd 1916 in Demirkapu (Macedonia), FA (A) 246
- BO Lt. Hellmut Camin, born on June 27th, 1892 in Lauenburg, died on March 27th, 1916 at Amatovo Lake, FFA 69
- FF Offiz. Stellvertreter Max Koestle, born on August 27th, 1898 in Saargemünd, died on March 27th, 1916 at Amatovo Lake, FFA 69
- Oblt. Hans Ludwig, born on October 30th, 1891 in Wittenberg, died on April 9th 1916 near Hudova (Dojran), FFA 30
- Lt. Richard Mengel, born on June 24th, 1895 in Köslin, died on April 9th, 1916 near Hudova (Dojran), FFA 30
- Flieger Erich Pilkenroth, born on March 15th, 1895 in Magdeburg, died on April 15thm 1916 in Gümuldzina (Bulgaria)
- Flieger Fritz Graichen, born on November 1st, 1896 in Leipzig, died on April 30th, 1916 in Üsküp (Skopje)
- Flieger Kurt Wagner, born on January 4th, 1887 in Freiberg (Saxony), died on June 9th in Üsküb (Skopje)
- Luftschiffer Franz Vostell, born on July 11th, 1879 in Enzen, died on June 29th, 1916 in Üsküb (Skopje)
- Gefreiter Harm Doejen, born on May 28th, 1893 in Norden-Süderneuland II, died on July 13th, 1916 in Üsküb (Skopje)
- Gefreiter Jürgen Rower, born on June 18th, 1891 in Mühlenbarbek, died on July 13th, 1916 in Hudova (Macedonia)
- Lt. d.R. Walter Strauß, born on December 10th, 1891 in Liverpool, died on July 18th, 1916 in Üsküb (Skopje)
- Uffz. Ernst Lohmann, born on January 10th, 1890 in Celle, died on July 31st, 1916 in Üsküp (Skopje)
- Luftschiffer Vinzent Jackowski, born on June 28th, 1876 in Obornik, died on August 12thm 1916 in Üsküb (Skopje)
- Luftschiffer Josef Machi, born on April 29th, 1876 in Kaltenhausen, died on August 20th, 1916 in Üsküb (Skopje)
- Gefreiter Franz Liecker, born on December 8th, 1886 in Linden, died on August 22nd, 1916 in Üsküb (Skopje)
- Hptm. Max Gissot, born on November 11th, 1883 in Bückeburg, died on August 31st, 1916 in Üsküb (Skopje), FAA 30
- Flieger Karl Salbreiter, born on January 3rd, 1875 in Struth, died on September 6th, 1916 in Üsküb (Skopje)
- Uffz. Johannes Ostenkötter, born on January 20th, 1893 in Berlin, died on Spetember 29th in in Piravo (Macedonia)
- Flieger Otto Sengewald, born on July 18th, 1881 in Schleiz, died on December 5th, 1916 in Nisch
- Oblt. a.D. Eckard, born on June 28th, 1886 in Kohlhof, died on December 8th, 1916 in Hudova
- Lt. d.R. Ernst Griesemann, born on December 2nd 1890 in Wolmirstedt, died on December 14thm 1916 in Drama, FFA 30
- BO Lt. Heinz Burkart, born on July 19th 1896 in Graudenz, died on January 1st, 1917 in Prilep, FA (A) 246
- FF Sergant Bruno Blume, born on March 27th, 1891 in Schneidemühl, died on January 1st, 1917 in Prilep, FA (A) 246
- FF Sergant Bruno Froese, born on March 27th, 1891 in Schneidemühl, died on January 1st, 1917 in Prilep, FA (A) 246
- Vzfw. Erwin Kernchen, born on January 20th, 1895 in Berlin, died on January 5th, 1917 in Muryas (Bulgaria), Jasta 25
- Offizier Stellvertreter Viktor Göldner, born on December 9th, 1888 in Wilhelmsbrück, died on January 13th, 1917 in in Buzau (Romania)

News Bulletin 01 from March 1917:

- In January 1917, the following aircraft were shot down in the southeast:
- January 9th, 1917, Drama, Farman D. D., Lt. von Eschwege, Art. Fl. Abt. 30
- January 15th, 1917, Smolari at Lake Dojran, double decker, type no longer able to be

determined, Oblt. Burckhard, Jasta 25
- Febrary 18th, 1917: Southeast front, near the army group Below, the enemy lost two aircraft, we lost one after an air fight [769] behind the front

News Bulletin 02 from March 1917:

- February, 22nd 1917: On February 22nd, 1917, the commander of Flieger 11 reported that reconnaissance aircraft of Flieger-Abteilung 30 observed small movable triangular objects on the water in the port of Saloniki in mid-February. To the east of Fort Karaburun there are clearly recognizable lines of buoys about 1,000 m long, in the water, some red, some white, running from north to south. The port can be entered approximately 1,000m west and north of the fort. The observations were made from an altitude of 1,000m.
- The buoys were captured in the photograph several times. They are apparently the suspension points of nets that are supposed to block the entrance to the port. The white lines observed between the buoys are wooden beams that create the connection between the individual buoys and help to carry the net.
- February, 22nd 1917: Report Commander of the aircraft near AOK 11: The English airport Samli has disappeared.
- February, 26th 1917: Southeast front: Kampfgescwader 1 bombed enemy airport Gorgop with 3,224kg bombs, airport Armutci with 3,150kg bombs.
- *Addendum* report Kampfgeschwader 1 on February 26th, 1917: Squadron attacked Gorgop airport in the morning with 3,224 kg bombs. Many hits were observed in the hangars and the aircraft in front of them, and confirmed by image reports.
- In the afternoon, Kampfgeschwader 1 took off again and attacked Armutci airport with 3,150 kg bombs. At least 12 direct hits in tents observed without any problems. The enemy anti-aircraft fire was very abundant and good, an aircraft landed in Warda soon after take-off due to an engine failure. Observer seriously injured, pilot and machine gunner slightly injured.
- February, 27th 1917: Southeast front: Kampfgeschwader 1 attacked a munitions and troop storage facility north of Saloniki with 3,500kg bombs. Good impact, apparently in an ammunition storage, observed.
- In February 1917, the following aircraft were shot down in the southeast:
- February 5th, 1917: northwest Moglia, 8km west of Monastir, Farman, Vfw. Könnecke, Jasta 25

- February 11th, 1917: near Balince (Vardar), A.W. D.D. 6166 or 97, Lt. d.R. Brauneck [770], Jasta 25
- February 17th, 1917: due to flak, AOK 11 a Farman D.D. 12 in the Orla-canyon
- February 18th,1917: Drama, B.E. D.D. Single-seat 5289, Lt. von Eschwege, Fl. Abt. 30
- [transcription error? February 12th, 1917? probably the 4th air victory from von Eschwege over Lt. Owen of the 16th Wing, started in Orljak]
- March 4th 1917: Southeast front: Kagohl 1 attacked a munitions storage facility north of Saloniki. Several fires and one explosion were noted. Unsuccessful attack by 12 enemy aircraft against the squadron's airport.

News Bulletin 03 from March 1917:

- March 7th, 1917: Kampfgeschwader 1 attacked the airport Gorgop with 3.000 kg bombs.

Kampfgeschwader 1 reports:

- March 12th, 1917: Kampfgeschwader 1 attacked the train station Vertekop in the early morning with 3.925 kg bombs. The numerous good hits and explosions, which could be seen from an altitude at 2.000 meters at 12 o'clock midday are confirmed through photographic reports.
- Addendum to the *"News bulletin 11"*:
- The air raid on Vertekop, carried out by KG 1 on March 12th, 1917, had a major impact, according to two Bulgarian soldiers who had escaped from captivity. Almost the entire ammunition depot blew up; explosions were still occurring four days later. The food and food stores were also burned, as were some wooden barracks. The human sacrifices were small.
- At 1 p.m. there was an enemy attack with 16 aircraft. The squadron started to defend itself, prevented the undisturbed implementation and shot down an aircraft. The crewmember is dead. No material damage was caused, only one man was slightly wounded.
- In the evening the squadron started again and successfully attacked Armutci airport with 3,425 kg bombs. Good hits on the airfield and the camp immediately adjacent were observed without any problems. Numerous dogfights took place over the target, in which the squadron retained the upper hand.

News Bulletin 04 from March 1917:

- March 18th, 1917: Kampfgeschwader 1 dropped 4,000 kg bombs on enemy troop storage. Vizefeldwebel Strey, Jasta 25, shot down an enemy aircraft behind the front near Monastir.
- March 20th, 1917: Kampfgeschwader 1 attacked

enemy troop storage south of Monastir, successful with 2,250 kg bombs.
- LZ 101 attacked Mudros on Lemnos on the night of March 20th to 21st, 1917 between 11 p.m. and 11.30 p.m. with 1,400 kg of explosive bombs. The airship landed smoothly at its home base on March 21st at 3:30 am.

News Bulletin 05 from March 1917:

- March 21st, 1917: In the Cerna-Arch an enemy tethered balloon was shot down through flak fire.
- March 22nd, 1917: Lt. von Eschwege, Fl. Abt. 30 shot down an enemy Nieuport near Drama; he has now shot down 5 aircraft.
- March 25th, 1917: Kampfgeschwader 1 attacked the airport and the storage facilities near Snevce with 2,050kg bombs. Excellent effect was observed on the storage facility.

Excerpt from "Army Day Order of the High Command of Army Group von Below" from March 16, 1917:

"I would like to express my appreciation to Kampfgeschwader 1 for its excellent work over the past few days. In the period from February 26th until March 12th, 1917 eight large squadron flights were carried out under the personal guidance of the commander, Captain Kastner [771], and despite considerable enemy counteraction, 20,000 kg bombs were dropped, with extraordinary success, on enemy ammunition piles, train stations and airfields. Towns and camps were boldly attacked with machine guns, and the enemy suffered heavy losses everywhere. The combat squadron can look back on these achievements with justified pride.
The commander in chief
signed von Below
General of the infantry."

- Order of the command of the Mediterranean Division in Constantinople dated March 8th, 1917:
- "The reports and recordings made by Aviation Units 30 and 34 have produced valuable material on the conditions in the Gulf of Saloniki. Thank you very much for your support of the naval warfare in the Aegean Sea.
- signed Souchon."
- Other personnel losses (incomplete) of flying units in the southeastern area for the month of March 1917:
- FF Flugzeugführer Matrose Walter Ludwig, born on October 11th, 1887 in Neu Langsow, died on March 9thm 1917 in Xanthi, Bulgaria, SFA Xanthi
- BO Lt.z.S. Julius Wild, born on August 9th, 1894 in Bayreuth, died on March 9thm, 1917 in Xanthi, Bulgaria, SFA Xanthi
- Gefr. Fritz Becker, born on July 29th, 1892 in Küblingen, died on March 20th, 1917 in Prileb
- Flieger Josef Jeziorowski, born on November 15th, 1887 in Berlin, died on March 20th in Prileb
- Flieger Adam Gasser, born on April 14th, 1875 in Guckheim, died on March 21st 1917 in Prileb, FA 38
- Flieger Paul Wollina, born on April 26th, 1884 in Wittstock, died on March 23rd, 1917 in Prileb
- Matrose Gottlieb Auer, born on May 15th, 1895 in Ohnastetten, died on March 25th, 1917 in Xanthi, Bulgaria, SFA Xanthi
- Vizeflugmeister Johannes Baldeweg, born on April 15th, 1892 in Bautzen, died on March 25th, 1917 in Xanthi, Bulgaria, SFA Xanthi
- Flieger Rudolf Michel, born on December 4th, 1888 in Frauwaldau, died on March 25th, 1917 in Prileb
- BO Lt. d.R. Johannes Schröder, born on November 12th, 1892 in Dresden, died on March 31st, 1917 in Patesti (Romania)

News Bulletin 06 from April 1917:

- March 30th, 1917: Kampfgeschwader 1 successfully attacked the camps south of Monastir with 3,125 kg bombs. Lt. von Eschwege shot down his 6th opponent on the Xanthi - Philippopel road. Lt. Degner brought down an enemy tethered balloon after several attacks.
- March 31st, 1917: Kampfgeschwader 1 occupied troop camps near Brod in the Cerna-Arch with 3,277 kg bombs. Excellent hits and ongoing fires within the camp were detected.
- South of Berkili, a tethered balloon was lit on fire and brought down. In addition, Lt. d. R. von Brauneck shot down his 3rd captive balloon in the Cerna-Arch.
- April 3rd, 1917: Kampfgeschwader 1 threw more than 3,500 kg of bombs at the railroad facilities at Vertekop. Several fires were detected through photographs. Nocturnal air raids against the squadron's airfield caused no damage.

Telephone message of the XX. Turkish Corps to the Advance Command of Fliegerabteilung 30:

"Before I leave Macedonia with the General Command, I would like to express my warmest thanks and appreciation to the advance command and its leader and to the Corps for their valuable

service and bravery. May God crown the work of the command with new victories.
General Abdul-Kerim.

News Bulletin 07 from April 1917:

- April 8th, 1917: KG 1 successfully bombed troop housing, railway buildings and airfields near Janes and Armutci (20km south of Lake Dojran) with 3,387 kg bombs. One aircraft of KG1 did not return.

Aviation activity near the Heeresgruppe von Below:

- The enemy air forces were somewhat reinforced on the Saloniki front in March. Two new airfields have been identified, two old ones seem to have disappeared. The English show a great deal of aggressiveness; they flew east of the Vardar. Despite the fierce fighting near Monastir, the French held back a lot. Five enemy planes were shot down, three of which are in our hands. One tethered balloon was brought down by airmen and one by artillery fire.
- The air raid on Dudular (north-east of Saloniki), which KG1 carried out on February 27th or March 4th, 1917 claimed 200 dead and wounded, according to an English defector.
- According to intelligence reports, an aircraft belonging to Fliegerabteilung 38, crewed by Vzfw. Gall and Lt. Heinrich, on March 26th, 1917 had to make an emergency landing at Grn. Gorica, west of Lake Prespa. After the crew set fire to the plane, despite the proximity of French soldiers, they were captured.
- The Jagdstaffel 35 has taken the impressive sum of 16,100 marks on the 6th war loan. Of this, NCOs and crews alone account for 15,200 marks.
- Army Aviation Park 11 takes part in the suppression of a gang uprising in Serbia.
- The following aircraft were shot down in the southwest in March 1917:
- March 22nd, 1917: Road Drama-Mavala, Nieuport 3182, Lt. von Eschwege,
- VorKdo. Fl.-Abt. 30
- March 30th,1917: Road Xanthi - Philippopel, Sopwith, Lt. von Eschwege,
- VorKdo. Fl.-Abt. 30
- March 31st, 1917: a ballon in the Cerna-Arch, Flakzug 11. Armee

News Bulletin 08 from April 1917:

- Other news:
- The impact of the German bombing raids on the Balkans has increased enormously in recent days, as numerous agent reports show. There are various reasons for this: the soldiers cannot build proper shelters due to the lack of wood; the houses are all weak and dry and therefore burn easily; they are mostly without basements. The frequent attacks have a depressing effect on the troops.
- According to a reliable agent in Zvezda (5 km south of Lake Prespa), the commanding general of the troops which had been standing between the lakes, and 14 people in his unit, were killed in his quarters during an aircraft bombing raid April 1st, 1917. In addition, 22 soldiers were killed or wounded in Zvezda, and 16 people and 11 soldiers killed or wounded in Korzica, on the same day.

News Bulletin 09 from April 1917:

- April 22nd, 1917: KG 1 dropped 3,500 kg bombs on artillery and infantry positions between the Varda valley and Lake Dojran.
- April 23rd, 1917: KG 1 attacked troop camps and enemy artillery and infantry positions south of Lake Dojran with machine guns and 3,200 kg bombs. The camps were hit, with subsequent fire and explosions.
- April 24th, 1917: KG 1 attacked the ammunition depot near Dobroveni with 3,297 kg bombs. Fires and explosions were seen.
- On the Macedonian front, an enemy aircraft crashed behind our lines and was destroyed by artillery fire.
- Other personnel losses (incomplete) of flying units in the southeastern area for the month of April 1917:
- Hptm. Kurt Rehdans, born on September 17th, 1890 in Graudenz, died on April 15th, 1917 in Batinesti (Romania)
- Flugzeugmeister Arnold Martin, born on May 27th, 1892 in Passau, died on April 29th, 1917 in Dünige (Romania)

News Bulletin 10 from May 1917:

- April 25th, 1917: enemy losses: two aircraft (one behind our lines), shot down by KG 2 during an attack on the Kolinova train station.
- April 27th, 1917: Two British aircraft shot down by flak.
- April 28th, 1917: KG 1 attacked storage and train station Kilindir with 2,500 kg bombs. Fire destruction was detected.
- April 29th, 1917: KG 1 successfully attacked the warehouses at Skocivir with 2,300 kg bombs. Two enemy aircraft were shot down in a dogfight.

- April 30th, 1917: KG 1 attacked the train stations Vertekop and Vodena with 2,297 kg Bombs. Hits were observed.
- May 1, 1917: In a closed squadron, KG 1 successfully attacked the greatly expanded aerial camp Bac, south of the Cerna-Arch. A warehouse exploded and caused a huge, fire that burned for a long time.

News Bulletin 11 from May 1917:

- May 5th, 1917: An enemy tethered balloon was shot down and burned in the Cerna-Arch.

News Bulletin 12 from May 1917:

- May 10th, 1917: Supplementary notification: near Porna, west of the Tahino Lake, B.E. single-seater, Lt. von Eschwege, Fl.-Abt. 30, eleventh aerial victory.
- May 16th, 1917: An aircraft was shot down in a dogfight beyond the enemy lines.

Army group command [no date]:

- *"To Kampfgeschwader 1. I would like to express my full appreciation and thanks to Kampfgeschwader 1 as I leave the Army Group, for the effective attacks against important installations behind the enemy front. May the Combat Squadron in the West enjoy similar, beautiful successes.*
- *The commander in chief*
- *signed von Scholtz "*

Activity of the Kampfgeschwader 1 in Macedonia:

- The difference between the French and English pilots is also obvious in this theater of war. Each time it attacked, the combat squadron had to reckon with stronger enemy influence. The enemy flak was numerous and fired remarkably well. Most of the enemy camps that had been bombed have been demolished and moved further back. This has undoubtedly influenced the general tactical situation.
- The enemy claims that bombs were thrown on hospitals several times. If this were the case, it only proves that the enemy hospitals are located in the immediate vicinity of important targets such as railway stations and ammunition depots. The number of camps on the enemy side under the protection of the Red Cross was so strikingly large that one did not always have the impression that the Red Cross was being used properly. Apparently all camps to which the enemy withdraws their troops for relaxation are considered eligible to enjoy the protection of the Red Cross.
- The strange weather conditions in the Macedonian theater have often made life difficult for the squadron. Rapidly changing weather conditions – previously unheard of, wind speeds of up to 30 m / sec. at 2,000 to 3,000 m altitude, sometimes forced aircraft to rest, while the most beautiful and calm weather prevailed on the ground.

News Bulletin 13 from May 1917:

- Southeast front: no news.

News Bulletin 14 from May 1917:

- May 28th, 1917: One aircraft shot down by MG fire from the ground.
- Enemy aviation near Salonika:
- A German sergeant who escaped from hostile captivity reports the following about the enemy aviation near Saloniki: at Cape Micra, about 6 km south of Saloniki, there is an English, a French and a Serbian air park, but the latter is one in name only. It is run by French and the mechanics are also French. But there was a three-month observer course for Serbs, mostly non-commissioned officers. The Serbian park included 15 regular officers, 200 French, 100 Serbs and 50 prisoners. From mid-February to mid-March, the number of aircraft, which had been around 15 until then, was increased by around 20 a week. The increase in aircraft was probably due to the bombing attacks of KG 1. During his stay at the park (from mid-December to early May) at least 10-12 planes crashed, in the English park around 5-6 planes. The French attribute the crashes to bad work. The French mechanics were very often drunk.
- Other personnel losses (incomplete) of flying units in the southeastern area for the month of May 1917:
- BO Lt. d.R. Walter Crome, born on August 13th, 1893 in Goslar, died on May 5th, 1916 in in Üsküb
- Flieger Karl Mayer, born on July 6th, 1895 in Hofen, died on May 5th, 1917 in Üsküb, AFP 11
- Flieger Paul Heil, born on November 25th, 1895 in Sontheim, died on May 13th, 1917 in Prilep
- Vzfw. Ludwig Gröppel, born on September 21st, 1894 in Leipzig, died on May 16th, 1917 in Üsküb, AFP 11
- Luftschiffer Karl Brüning, born on November 2nd, 1894 in Groß Strehlitz, died on May 17th, 1917 in Braila

News Bulletin 15 from June 1917:
- May 31st, 1917: one aircraft shot down.

Army day order f the Heeresgruppe Scholtz on May 25th, 1917:

"In close collaboration with other weapons, the aviation units on the Macedonian front rendered valuable services to the fighting troops through tireless reconnaissance, daring flights and fighting enemy targets from low altitudes.
In particular, the Aviation Department 30, which has a particularly difficult position as it is numerically very outnumbered by very active British aviators, distinguished itself. We can rightly be proud of the achievements of our aviators, who in every respect are more enterprising than the combined air power of our enemies. I express my full appreciation for this.
The commander in chief
von Scholtz
General of the Artillery"

News Bulletin 16 from June 1917:
- June 6th, 1917: English airfield Topci dismantled.
- June 12th, 1917: An enemy aircraft is shot down.

News Bulletin 17 from June 1917:
- June 18th, 1917: An enemy aircraft is shot down this side, due to flak.

News Bulletin 18 from June 1917:
- Southeast front: no news.
- Other personnel losses (incomplete) of flying units in the southeastern area for the month of June 1917:
- Lt. z.S. Heinrich Immisch, born on May 8th, 1895 in Leipzig, died on June 5th, 1917 in Xanthi, SFA Xanthi. Lt. Immisch drowned while swimming, and squadron leader Haupt-Heydemarck sent Lt. Greiff from Drama to replace him
- FF Lt. d.R. Horst Müller, born on November 3rd, 1891 in Neustettin, died on June 7th, 1917 in Mrawinka, FA 34
- Lt. d.R. Theodor Schultheis, born on February 23td, 1894 in Frankfurt am Main, died on June 25th, 1917 in Ciusiea (Romania)
- Flieger Walter Claus, born on September 25th, 1892, died on June 29th, 1917 in Prilep

News Bulletin 19 from July 1917:
- June 28th, 1917: An enemy aircraft is shot down in a dogfight.
- June 30th, 1917: An enemy aircraft is shot down in a dogfight.
- July 2nd, 1917: Jasta 25 bombed an enemy squadron of 14 aircraft, shot down two aircraft behind enemy lines and forced a third to land behind our lines.
- July 3rd, 1917: An enemy aircraft shot down behind our lines.

News Bulletin 20 from July 1917:
- July 4th, 1917: An own aircraft shot down by artillery on this side, crewmembers unharmed.
- July 5th, 1917: An enemy aircraft shot down in a dogfight.

Testimony of an English flying officer shot down by artillery near Livunovo (Macedonia) [undated]:
Prisoner belonged to the 17th Squadron, which was provided with 90 hp and 140 hp B.E. The 90 hp B.E. is used for squadron bombing flights, mostly without an observer, and climbs to 2,000 meters with two bombs of 50 kg and four of 8 kg, in 45 minutes. The three flights are separate.

Flak: The British are very uncomfortable with the flak south of Rupel and at Demir Hissar. The prisoner saw English flak on cars. He didn't know the caliber.

Various:
Greeks are to be trained as pilots in Saloniki. There has been much talk about the dismantling of the Saloniki companies. The French should be ready to do so, but the British are not. If the English gave up Saloniki, the Germans would immediately place a submarine station there and from there disrupt English shipping traffic in the Mediterranean.

- List from June 1st-30th, 1917: the following enemy aircraft were shot down in the southeast:
- June 6th, 1917: South of Cape Kojunnalla, Henry Farman, Lt. von Eschwege, Fl.-Abt. 30, 8th aerial victory
- June 11th, 1917: Ahinos, B.E., Lt. von Eschwege, Fl.-Abt. 30, 9th aerial victory.
- June 12th, 1917: At the moutn of the Mesta, Sopwith, Lt. von Eschwege, Fl.-Abt. 30,
- 10th aerial victory
- June 18th, 1917: 4 km south of Livunovo, B.E., Vzfw. Käppler, Flakzug 97e

News Bulletin 21 from July 1917:
- July 11th, 1917: The enemy airfield in Bac, south of the Cerna-Arch has been reduced by two hangars and five one-aircraft-tents. Thereby the capacity of the field has been reduced by 30 aircraft.
- List from June 16th-30th, the following enemy aircraft were shot down in the southeast:

- June 20th, 1917: Near Rastani, Sopwith single-seater, Lt. Rose, Jasta 25
- June 20th, 1917: Near Rastani, Sopwith single-seater, Vzfw. Treptow, Jasta 25
- June 28th, 1917: over Radulevo, Nieuport single-seater, Uffz. (later Vzfw.) Theodor van Ahlen, Fl.-Abt. 30

News Bulletin 22 from July 1917:

- Southeast front: no news.

News Bulletin 23 from July 1917:

- Southeast front: no news.
- List from July 1st-15th: the following enemy aircraft were shot down in the southeast:
- July 2nd, 1917: over the Cerna-swamps, Nieuport single-seater, Fw. Treptow, Jasta 25
- July 2nd, 1917: Puteros, Farman, Lt. Rose, Jasta 25
- July 5th, 1917: north of Monastir, Farman, Lt. Burckhardt, Jasta 25
- Other personnel losses (incomplete) of flying units in the southeastern area for the month of July 1917:
- Flieger Jakob Ewald, born on September 19th, 1897 in Griesheim, died on July 7th, 1917 in Braila
- Flieger Josef Grott, born on January 29th, 1885 in Filehne, died on July 12th, 1917 in Prilep
- Lt. Georg Engelmann, born on March 4th, 1895 in Dresden, died on July 20th, 1917 in Prileb, FA (A) 246
- Lt. d.R. Hieronymus Kistenfeger, born August 21st, 1890 in Altona, died on July 25th, 1917 in Romania
- Uffz. Adolf Scholz, born on December 2nd, 1893 in Hannover, died on July 25th, 1917 in Rotesti (Romania)

News Bulletin 24 from July 1917:

- August 2nd, 1917: Enemy bombing on the airfield of the Vorkommandos Fl.-Abt. 30 (11th Army, Struma-Front), one aircraft destroyed.
- August 3rd, 1917: Aircraft of the Vorkommandos Fl.-Abt. 30 took part in a retaliatory attack during the night of August 2nd-3rd on an enemy airfield on the island Thanos, with 300 kg explosive and burning bombs, and extensive MG fire. One hangar was destroyed, numerous explosions and long-burning fires were observed.

Report by Oblt. Scherzer about the war trip of the LZ 85 on May 4th / 5th, 1916:

- *"Mission: Attack on ships, port and port facilities in Saloniki.*
- *Course of the voyage: The strong risk of thunderstorms that had prevailed in the Serbian mountains for some time has decreased considerably or even completely disappeared. This good weather situation had to be taken advantage of, in order to fulfill the order.*
- *The departure took place at 5.30 p.m. The course was along the Morava valley via Nisch - Lescovac to the south to the Vardar, which was reached at Krivolac, at an average speed of 80 km / h. The trip went without any issues, the engines worked perfectly. Shortly before flying over the front, we climbed to an altitude of 3,000 m. As with the previous trips, the drop zone was again to be approached from the sea, since there was least likely to be a strong defense. The approach from the water turned out to be fatal, because it was here that the Entente, which had been notified of the airship's approach, had prepared a particularly strong defense.*
- *At 02.30 a.m. LZ 85 was clear to drop over the recognized targets when suddenly an extraordinarily strong artillery fire began, from all sides. From the south and south-east the fleets fired with heavy caliber, from the east, north-east and north, land batteries, from the west and from below - from the bay itself - special artillery with incendiary grenades. As I later observed, there were a number of small, fast-moving boats in the harbor, armed with searchlights and anti-aircraft guns. Certainly these boats also kept firing at us with incendiary grenades.*
- *As far as it was observed, one of the first volleys of the heavy ship guns was near the stern of LZ 85. The loss of gas was felt so quickly that an estimated 3 cells in the stern must have been torn open. In addition, since gasoline tanks and gasoline lines to the rear engines were shot up, they soon stopped and could no longer be restarted, 3/4 of the dynamic stroke failed, and one minute after the fire started, the airship turned to a steep incline and descended.*
- *Simultaneously, with the opening of enemy fire, bombs were dropped on ships sighted in the harbor. Accurate detection of the hits was not possible due to the strong detonations of the enemy projectiles in the immediate vicinity of the ship. According to reports from the Entente, the bombs detonated on the targets. Soon afterwards a message came through that the bombs had stopped falling. So the discharge cables seem to have been destroyed.*
- *In the meantime, I could see that it was no longer possible to return to our front with the*

airship, which was hardly controllable and still sinking rapidly. So I gave the order to turn south-west with the northeast wind in order to reach Greek territory.

- *The bombs that had not fallen were released by hand without arming the detonators, as the targets had already been flown over and the orders were to protect neutral property.*
- *In order to reach land, the front engine was used to its maximum limits. Any material that could be reached and easily extricated (MG, dropping device and radio) was thrown overboard. At 02.45 a.m. LZ 85 landed hard on the sea in the swamps of the Vardar estuary. Toward the end the ship sank at about 4–5 m / sec and had an incline of approximately 50°. Experience has shown that if it had landed on hard earth it would have exploded with a violent impact. The rear was sunk deep and had filled with water, the rear gondola was no longer accessible. The nose rose up about 15 m above the swamps. Under these circumstances it was not possible to saw off the rear gondola and then try to continue.*
- *The secret material was destroyed. After Oblt. Rippe and I were the last to leave the airship, I set it on fire at a distance of 30 m. The cartridge fired in two seconds. As a result of the explosion, Oblt. Rippe and I were thrown away quite violently and then had to walk another 50 m through the burning swamp. In similar cases, shooting at a distance of less than 100 m is hardly ever allowed.*
- *We marched in a south-southwest direction toward Greek territory, most of the time up to our chests through swamps or thistle fields. After dawn we hid in a grain field. At 9:30 am we were spotted by airmen who had been looking for us for hours and soon afterwards surrounded and captured by French and Serbian cavalry."*

News Bulletin 25 from August 1917:

- August 9th, 1917: Enemy squadron of 9 aircraft dropped 21 bombs on the airfield of Vorkommandos Fl.-Abt. 30 (Struma-Front) without effect. In the air war with three B.E. of this squadron, Lt. von Eschwege forced a B.E. straight down behind the enemy front. Two B.E. made emergency landings at enemy airfields and Lt. von Eschwege shot at them with MG from 800 meters' altitude.
- August 11th, 1917: At the airfield of Orljak five aircraft tents have disappeared.
- August 12th, 1917: English airfield Basimal reduced by two aircraft tents.
- August 13th, 1917: We successfully bombed the infantry camp near Ile Zvezda (north of Korca) and enemy aircraft on the island of Thasos.

News Bulletin 26 from August 1917:

- August 16th, 1917: The 1st Bulgarian Army observed numerous explosions and high fire towers while fighting enemy batteries southwest of Majadag.
- August 19th, 1917: We bombed the the pier in the harbor of Saloniki and enemy airfield on the island of Thasos.
- August 20th, 1917: Lt. von Eschwege shot down an enemy aircraft.
- August 21st, 1917: Enemy camps near Florina and Bodena were bombed.
- Four own losses: one aircaft in a dogfight, one through flak, two did not return.

Excerpt from the Army Day Order of the High Command of Army Group Scholtz dated August 11th, 1917:

- "On July 30th, 1917, the heavy 15 cm cannon battery M. 11 with air surveillance carried out a mission against enemy tethered balloons near Kalinowa and destroyed three tethered balloons in the process.
- My special appreciation goes to the battery and the flight crew, Lt. d. R. Wagner and Gefreiter Steinvorth from Flieger Abtteilung 30, who directed the shooting with energy, including 5 ½ hours of flight time, despite numerous dogfights and strong flak fire.
- The commander in chief von Scholtz
- General of the Artillery "

News Bulletin 27 from August 1917:

- August 25th, 1917: Successful bombing runs on Tepavci, camp near Grn. Gorcia and on enemy airfield on the island of Thasos.
- August 26th, 1917: We bombed and MG-fired Bukovo and the camps north of it.
- August 27th, 1917: 30 large ships sighted in the harbor of Saloniki.
- Ahead of the front of the 11th Army and the 1st Bulgarian Army enemy camps were successfully attacked with bombs and MG-fire.
- August 28th, 1917: Three large fires ensued through aircraft shooting into enemy batteries northwest of Monastir.

Assessment of the situation of the south-eastern front:

- Enemy airmen are active in front of the Scholtz Army Group. However, their bombs were always in the range of the enemy artillery. As a matter of principle, the enemy planes refused to engage in aerial battles on this side of the positions. On the other hand, our reconnaissance planes have to endure heavy dogfights beyond the positions. Nieuports and Spads (140 hp), in groups of three to six aircraft, have recently been attacking with great tenacity. In the last few days, Italian single-seaters have been reported for the first time.

August 1917, recognition by the Commanding General of the Air Force:

"To Kampfgeschwader 1.
After returning from the Macedonian theater of war, the squadron has shown the old aggressive spirit on numerous flights in Flanders. The use of attack aircraft shows understanding for new tasks. I am pleased to be able to express my special appreciation to the squadron.
The commanding general
signed by Hoeppner"

- Other personnel losses (incomplete) of flying units in the southeastern area for the month of August 1917:
- Lt. d.R. Willi Hotz, born on February 28th, 1892 in Truchlershaim, died on August 6th, 1917 in Romania.
- Lt. d.R. Walter Voigt, born on June 18th, 1893 in Goslar, died on August 14th, 1917 in Braila
- Lt. Friedrich Protzek, born on April 4th, 1898 in Rawitsch, died on August 16th, 1917 in Barlad (Romania)
- Lt. d.R. Hermann Steinbrück, born on October 16th, 1887 in Leipzig, died on August 16th, 1917 in Focsani
- Flieger Otto Oppermann, born on April 17th, 1890 Schladen, died on August 29th, 1917.

News Bulletin 28 from September 1917:

- August 29th, 1917: In front of the 11th Army we bombed Florina. In front of the 1st Bulgarian Army, we successfully bombed camps south of Fustani.
- September 1, 1917: Near the 11th Army our aircraft successfully entered the fighting. They bombed enemy trenches and batteries from a low altitude, with MG fire. An enemy aircraft was shot down behind the lines.

News Bulletin 29 from September 1917:

- September 4th, 1917: We successfully attacked the harbor infrastructure of Stavros with bombs.
- September 5th, 1917: We bombed the camps near Holoven and Cegel, as well as the heavily occupied Korica Basatz.
- September 7th, 1917: We successfully bombed enemy camps near Zvezda southeast of the Malik Lake, near Holoven south of Monastir, near Kapinjani 20 km north of Vodena and near Jajladzik, 30 km northwest of Saloniki. Near the 1st Bulgarian Army two enemy battalions were attacked with artillery, under observation, and resulted in fire and explosions.
- September 9th, 1918: We bombed Biklista und Skocivir. An enemy aircraft shot down behind the lines.
- September 11th, 1917: We attacked Korca and Bukovo successfully with bombs. Ahead of the 11th Army for the first time English aircraft were detected, in addition to the French, Italian and Serbian air forces.
- List of August 1st-15th 1917: the following enemy aircraft were shot down in the southeast:
- September 9th, 1917: Near Iveron, B.E. single seater, Lt. von Eschwege, Fl.-Abt. 30, 13th aerial victory

News Bulletin 30 from September 1917:

- September 12th, 1917: Enemy camps near Livornik and Florina successfully bombed and attacked with MG fire. One enemy aircraft shot down in a dogfight.
- September 13th, 1917: We bombed Korca and the enemy camps between Boresani and Holoven.
- September 15th, 1917: Monastir was bombed. Fire was observed.
- September 17th, 1917: Enemy camps near Zemlak and the train station of Ostrovo were bombed. One enemy aircraft shot down beind the enemy lines in the Cerna Arch.

September 1917, recognition by the Commanding General of the Air Force:

- *"To the commanders of the aviators in the east and south-east.*
- *The aviators successfully participated in the victories in Galicia and near Riga by supporting infantry and artillery, through surveillance and fighting. With a lively zest for action and a fresh spirit of attack, the pilots in the east are equals to their comrades in the west. I am happy to be able to express my appreciation and hope to see some of the units personally in the course of the year.*
- *The commanding general*
- *signed by Hoeppner."*

News Bulletin 31 from September 1917:

- September 18th, 1917: In the morning, Lt. d. R. Toppenthal and Lt. d. R. Lammering, Fl.-Abt. A 246, were engaged in heavy aerial fighting with four Nieuports, and brought one down northwest of Mesdzidli (south of Monastir) behind the front. One further Nieuport was forced to land near Bac (Cerna-Arch), on enemy territory.
- September 20th, 1917: Baracks near Ljubanista (southeast of the Ochrida Lake) and observer traffic on the road between Monastir and Trnova were shot at with MG fire. Numerous dogfights, one Nieuport was forced to land on enemy territory.
- September 21st, 1917: In the morning, enemy airfield Korca was successfully bombed wtih 200kg explosives by a squadron flight of vour aircraft of the Fl.-Abt. 34. Enemy retaliatory strike against the airfield of Fl.-Abt. 34 caused littel damage.
- September 22nd, 1917: Bomb attack on the camp north of Florina.
- September 23rd, 1917: Successful bomb attack on Banica, 15 km southeast of Florina.

Overview of the southeastern front:

- The enemy used strong air forces to support its ground attacks. Our planes intervened effectively in the fight. They shot at enemy trenches, troop camps and firing batteries from low altitudes with machine guns. Several batteries were silenced in this way. Successful own bombing attacks on battery positions near Monastir, heavily occupied towns and warehouses, and on ships in the port of Stavros.

Army day order of the Heeresgruppe Scholtz on September 21st, 1917:

During the last few weeks, the flying units have provided the command and troops with particularly valuable help in the fighting on the right flank of the army.
Facing an enemy far superior in number, they succeeded in bringing important insights from the rear enemy territory through their tireless reconnaissance work, in damaging the enemy troops in their camps through attacks with MG and bombs, as well as by gallant successful aerial battles to beat enemy airmen.
I address the aviation associations involved, especially Unit A 246 under the leadership of Lt. Thierfelder [772], my special recognition.
The commander in chief
signed von Scholtz
General of the Artillery "

- Other personnel losses (incomplete) of flying units in the southeastern area for the month of September 1917:
- FF Vzfw. Karl Ehlers, born on August 29th, 1892 in Celle, died on September 1st, 1917in Focsani
- Flieger Richard Swegat, born on September 8th, 1892 in Pillkallen, died on September 5th, 1917 in Üsküb
- Lt. d.R. Wilhelm Gröger, born on Setptemner 14th, 1890 in Hamburg, died on September 8th, 1917 in on the island of Imbros (Turkey)
- Uffz. Hermann Utsch, born on January 4thm 1896 in Ehrenbreitstein, died on September 24th, 1917 in Macedonia FA (A) 246
- Luftschiffer Hermann Braun, born on June 1st, 1892 in Sadlauken, died on September 27th, 1917 in Prilep
- Flugz. Ober-Matrose Walter Krüger, born on September 18th, 1891 in Düren, died on September 8th, 1917 on the island of Imbros (Turkey), SFA Xanthi

Overall successes of the German air forces in September 1917:
"In the month of September 1917 our opponents lost a total of 374 airplanes and 22 balloons to the action of our ordnance on all fronts. In contrast, we have lost 82 aircraft and 5 balloons. Of these, 38 aircraft remained beyond the lines while the other 44 were lost over our territory. Of the 374 enemy aircraft put out of action, 362 were on the western front alone, of the 82 Germans, 76.

- In detail, the sum of enemy losses is made up as follows:
- • 324 aircraft were in aerial combat
- 40 from anti-aircraft guns
- 6 shot down by infantry
- 4 involuntarily landed behind our lines
- Of these aircraft, 167 are in our possession and 207 on the other side of our lines clearly crashed. This number of kills is the highest achievement in a month so far. In addition, 23 aircraft were forced to land beyond our lines."

News Bulletin 32 from October 1917:

- September 25th, 1917: Nine aircraft attacked the enemy airfield Korca with 450kg in explosives in a squadron flight. In addition, the land infrastructure near Skala Panajia (Island of Thasos) and ships in the harbor were attacked with bombs.
- September 26th, 1917: Bomb attacks on camps near Zelova and Pisoderi.
- September 27th, 1917: Bomb attack on enemy

camps west of Florina Florina (hit on target) and on Monitor near Kartasli Derbend (Gulf of Orfano).
- October 1st, 1917: Tent camp and roads near Biklista attacked with MG from 250 meters' altitude.

Situation at the south-east front

- Newly observed aircraft types: one from a captured English aviator designated as B.E. 2E. The aircraft appears to correspond to the aircraft depicted in newsletter No.11. A single-handled Bristol single-seater with a rotary engine has been observed on Thasos. It climbs better than D. 3, but is slower.

News Bulletin 33 from October 11th, 1917:

- October 4th, 1917: Unsuccessful french squadron attack on the airfield of Rosna. We successfully bombed the airfield Korca
- October 5thm 1917: Two enemy aircraft shot down in a dogfight.

Report of the Vzfw. Wilhelm Grasmeher [773] about a surveillance flight with his observer, Lt. Beitter [774], Fl.-Abt. A 246, 11th Army, from October 5th 1917:

"On the approach, about 5 km before Korca, I saw a lot of smoke at the Korca airfield, and shortly afterwards three single-seaters took off. When my observer had finished his photography, the three enemy planes approached. Since all three attackers attacked at the same altitude and did not react to the fire of my observer, I turned sharply, briefly took one, then the second under fire, whereupon everyone dispersed. When I made a turn to fly towards Lake Prespa, the second attack came. At the same moment I saw two new opponents approaching us from the front. I immediately took one under fire, the other joined the three attackers in the back. My observer tried as best he could to fend them off, but I could already hear the projectiles pattering very close to me. In the mirror I saw that my observer was hit and sank back into the seat. I tried to reach the western edge of Lake Prespa, constantly attacking and turning, because the enemy kept attacking again. Suddenly, to my delight, I saw my observer come up again and take up the defense with the machine gun, but after firing a few shots he fell back bleeding profusely and then gradually sank down for the second time.

At the same time, bullets rattled into my seat, and gasoline was squirting out between my right arm and hip. Then there was another attack by all five opponents. I tried to escape by letting the plane slip, caught the plane again at an altitude of 2,800 m and flew on. Then, to my delight, I saw that three planes were leaving and only two opponents followed me. In order to shorten the line of attack, I climbed at maximum power and soon came to the same altitude as they were. When the planes came within range of me, I tossed my aircraft around and fired at the first one, then the second within 20 meters. It tried to escape by flying sharp curves, but my aim was sitting well. After firing 120 rounds, he fell straight down and rolled over and over. Point of shooting down: Lake bay near Laisica.

I wanted to keep watching him, but the projectiles were already rattling into my aircraft: the second enemy was sitting behind me. I made a short turn and shot him a load with the propeller-mounted machine. Suddenly my machine gun stopped! I reloaded, but to my horror I saw my ammunition had run out. Now I had another pursuer, whom I tried to escape by making false attacks, making sharp turns and banking my aircraft hard.

The latter took place 800 - 1,000 m above the lake. My last opponent pursued me to our position on the eastern edge of Lake Prespa, then abandoned the pursuit and flew over the Peristeri ridge in an easterly direction. Then the enemy flak grabbed me at Krani with well-aimed shots, because I was only 1,200 - 1,300 m high. I tried to reach Resna airfield as quickly as possible, so that I might be able to get medical help for my observer in good time. But death had already occurred."

Vzfw. Grasmeher was given the following recognition in the Army Daily Order:

"During a reconnaissance flight on October 5th, 1917, an aircraft from Unit A 246 had to endure heavy aerial combat far behind enemy lines with a numerically superior enemy. Although the aircraft observer was fatally wounded, the pilot, Vzfw. Grasmeher, managed to achieve superiority in the air by means of a gallant attack and to land his aircraft smoothly at the airfield. I express my special appreciation to Vzfw.Grasmeher for this excellent performance.
The commander in chief
signed von Scholtz
General of the Artillery"

List of September 1st-30th 1917: the following enemy aircraft were shot down in the southeast:

- September 1st, 1917: near Rastani north of Monastir, two Nieuport single seater,
- Vzfw. Fuckel [775] und Lt. Wunderlich [776], both Fl.-

Abt. 246.
- September 12th, 1917: at the mouth of the Struma, Sopwith, Lt. von Eschwege, Fl.-Abt. 30, 15th aerial victory
- September 18th, 1917: south of Monastir, Nieuport, Lt. Toppenthal and Lt. Lammering, Fl.-Abt. (A) 246

News Bulletin 34 from October 18th, 1917:
- October 12th, 1917: We used 1,240kg explosives to attack camps near Brusnik, Airfield Thasos and Airfield Monuhi. Hits on target. One enemy aircraft shot down with flak.

News Bulletin 35 from October 25th, 1917:
- October 15th, 1917: On the road to Florina, northwest of Kroca very strong column traffic, primarily toward the west. More and more camps between Armensco and Florina, by about 500 small tents. Constructino of a new railway from the train station of Gorazanli to the southerly bridge Karadzokoiej. We successfully bombed camps Budimirea and attacked trenches and camps on the Great Tomoros with MG fire.
- October 16th, 1917: New marching camp south of Biklista with about 160 tents, near Benznica about 200 tents. We bombed the camp Bistrica with 350 kg Bombs. Hits on target.
- October 17th, 1917: English monitor, that shot at Bulgarian positions at the moutn of the Struma, was attacked wthi 150kg explosives.
- October 18th, 1917: We attacked the troop camp in the Cerna-Arch with 300kg bombs. Hits on target.
- October 23rd, 1917: One enemy tethered balloon shot down.

Overview southeast front:
- "The aviation activity in the Monastir basin, in the Cerna-Arch, on Dobropolje and between Vardar and Lake Dojran showed the usual activity in good weather. It was moderate on the rest of the front, weak to the west and east of Lake Ochrida, and weak on the Struma front. Few bombs.
- During the attack by English companies on October 6th, 1917, between 5 and 7 a.m., on Bulgarian protected areas at Akindzali (east of Lake Dojran) - for the first time in the Macedonian theater of war - six English attack aircraft supported the infantry attack with machine-gun fire and 36 bombs.
- The consequences of the breakdown of the use of enemy aviation forces, probably to protect Saloniki and to prevent our long-range reconnaissance, can be seen on every flight to Saloniki. Already over the city and on the entire return flight, our planes have to carry out aerial combat with an opponent who is far superior in number (on October 5th, 1917 against 5 Spads alone).
- Reddish flak clouds were observed twice in the Cerna arch."

Army daily order of the 11th Army on October 19th, 1917:
*"After the bombing of Prilep by an enemy squadron on October 16th, 1917, it was then in turn attacked on the return flight by Jagdstaffel 25 in fresh attack. Here Vice Sergeant Treptow shot down a French Nieuport via Krusovica. Lt. Rose, Lt. Brockhoff and Vice Sergeant Bauhofer attacked two French observation planes from the squadron at the same time, one of which had to land so close to the front of the 2nd Bulgarian Division that it could be destroyed by the Bulgarian artillery.
I congratulate the Jagdstaffel 25 on these victorious aerial battles and I am convinced that wherever the enemy airmen continue to show up, they will be attacked in the same gallant manner. I would like to express my very special appreciation to the participants.
The commander in chief
signed von Steuben
General of the infantry. "*

News Bulletin 36 from November 1st, 1917:
- October 28th, 1917: Lt. von Eschwege shot down a tethered balloon.
- October 29th, 1917: Two enemy aircraft shot down.
-
- Other personnel losses (incomplete) of flying units in the southeastern area for the month of October 1917:
- Lt. d.R. Theodor Siebold, born on December 23rd, 1891 in Gütersloh, died on Octpober 1st, 1917 in Hudova (Macedonia), Jasta 38
- Flieger Paul Wißmann, born on January 18th, 1895 in Ossweil, died on October 1st, 1917 in Üsküb, FlBauKp 1/2
- Uffz. Artur Stede, born on May 18th, 1888 in Bahren, died on October 30th, 1917 in Hudova

Overall successes of the German air forces in October 1917

"In October, our opponents lost 244 aircraft and

9 balloons on all fronts due to the activity of our ordnance. In contrast, we have lost 67 aircraft and one balloon. Of these, 39 aircraft remained beyond the lines while the other 28 were lost over our territory. On the western front alone, of the 244 incapacitated enemy aircraft: 201, of the 67 German: 53. On the Italian front we shot down 35 enemy aircraft and lost 9 of our own. In detail, the sum of enemy losses is made up as follows:
- 207 aircraft were in aerial combat
- 22 by anti-aircraft guns
- 3 shot down by infantry
- 12 involuntarily landed behind our lines

- Of these aircraft, 149 are in our possession, a recognized 95 were shot down behind the enemy lines."
- List from October 1st-31st, 1917: the following enemy aircraft were shot down in the southeast:
- October 3rd, 1917: West of Sarmusakli, B.E., Lt. von Eschwege, Fl.-Abt. 30, 16th aerial victory
- October 5th, 1917: South edge of Lake Prespa, Type unknown, Vzfw. Grasmeher, Fl.-Abt. A 246
- October 16th, 1917: Krusevica, Nieuport Eins., Vzfw. Treptow, Jasta 25

News Bulletin 37 from November 8th, 1917:
- Southeast front: no news.

News Bulletin 38 from November 15th, 1917:
- Southeast front: no news.

News Bulletin 39 from November 22nd, 1917:
- November 15th, 1917: Lt. von Eschwege shot down a balloon and two aircraft.
- November 19th, 1917: Lt. von Eschwege, Fl.-Abt. 30, 19. shoot down
- The command of the Commander-in-Chief of November 18th, 1917 granted the aviation units of the Scholtz Army Group the full recognition of their superiority in the air, which they had demonstrated in numerous aerial victories during the months of October and November.

News Bulletin 40 from November 29th, 1917:
- November 21st, 1917: Three aircraft in dogfights, a balloon shot down by Lt. von Eschwege, 20th aerial victory. Lt. von Eschwege [777] falls.
- November 23rd, 1917: An enemy aircraft behind our lines shot down.
- November 26th, 1917: Lt. Splittgerber [778] shot down a balloon, Jasta 38.

On the death Lt. von Eschwege daily orders and letters:

"Lt. von Eschwege fell on the enemy side after a successful attack on a tethered balloon of the 11th Army. Lt. von Eschwege had 20 air victories. Kofl. 11 received the following radio message on this occasion:
With the comrades of the 11th Army, the air forces mourn the death of Lt. von Eschwege. He was deserved his decisive successes. May his memory live on in future successes in the Macedonian theater of war. I extend my warm condolences to Flieger-Abteilung 30.
The commanding general
signed von Hoeppner."

Upon the death of Lieutenant von Eschwege, the following radio messages were exchanged between the Bulgarian Supreme Army Command and the Bulgarian Naval Air Station Varna-Penardiek on the one hand, and the Commanding General of the Air Force on the other:
*"The heroic death of the courageous German aviator, Lieutenant von Eschwege, filled me and all members of the Bulgarian Supreme Army Command with sadness. I hasten to interpret the grief of all of us on the occasion of this new dear loss before Your Excellency. Through his glorious deeds, Lt. von Eschwege has made himself immortal. His name will live on in the history of Bulgaria.
Generalissimo Schekoff."*

*"Excellency Schekoff
Your Excellency, you have pronounced your condolences on the occasion of the death of Lt. From Eschwege to His Excellency General Field Marshal von Hindenburg from the Bulgarian O.H.L. The heartfelt words of appreciation from Your Excellency are new proof to me that Lt. von Eschwege fulfilled his duty. I hope with his excellency. that the memory of this brave officer will always stay alive in the new deeds of my aviators together with their Bulgarian comrades. On behalf of the Air Force, I thank your Excellency for your warm condolences. The Commanding General of the Air Force
signed von Hoeppner."*

"To the Commanding General of the Air Force. Officers and aviators from the Bulgarian naval air station Varna-Penardiek express their regret over the loss of Lt. von Eschwege.

Lieutenant Ivanov, station head."

"Bulgarian Naval Air Station Varna.
I thank the Naval Air Station for expressing heartfelt condolences upon the death of Lieutenant von Eschwege. May the Bulgarian comrades and my pilots be granted victories. The spirit of the fallen lieutenant von Eschwege will live on in these successes.
The Commanding General of the Air Force
signed von Hoeppner."

- Other personnel losses (incomplete) of flying units in the southeastern area for the month of November 1917:
- Oblt. Walter Thiemann, born on March 25th, 1891 in Berlin, died November 25th, 1917 in Braila

News Bulletin 41 from December 6th 1917:

- November 28th, 1917: One enemy aircraft shot down. One own aircraft shot down by MG fire from the ground.
- December 3rd, 1917: Two enemy aircraft emergency landed on own territory.
- Overview southeast front:
- "When the weather cleared, the enemy's activity in the air increased considerably. It was more than twice as strong as the previous week. After the death of Lieutenant von Eschwege, the enterprising spirit of the English aviators grew rapidly. After observing the ground, the pilot of a French single-seater that was brought down on this side of the front is said to have jumped out with a parachute [779] and pushed across the front by the wind."

News Bulletin 42 from December 13th 1917:

- Southeast front: no news

Overall successes of the German air forces in November 1917:

- "In November 1917 our opponents lost 205 airplanes and 22 balloons on all fronts as a result of the activity of our ordnance. In contrast, we have lost 60 aircraft and two balloons. Of these, 33 aircraft remained beyond the lines while the other 27 were lost over our territory. On the western front alone, of the 205 enemy aircraft that were put out of action: 167, of the 60 German: 44. On the Italian front we shot down 26 enemy aircraft and lost 10 of our own. In detail, the sum of enemy losses is made up as follows:
- .. 150 aircraft were in aerial combat
- .. 38 by anti-aircraft guns
- .. 4 shot down by infantry
- .. 13 involuntarily landed behind our lines

- Of these aircraft, 85 are in our possession, 120 crashed behind enemy lines."
- List from November 1st-30th, 1917: the following enemy aircraft were shot down in the southeast:
- November 16th, 1917: over Dedebal, Avro, Vzfw. Treptow, Jasta 25
- November 19th, 1917: Novak east of Monastir, Nieuport Eins., Lt. Lebram and Lt. Scheffler [780], Fl.-Abt. A 246

News Bulletin 43 from December 20th, 1917:

- December 10th, 1917: Two enemy aircraft behind our lines shot down in a dogfight.
- December 14th, 1917: One own aircraft did not return after flying over the enemy.

Daily order of the 11th army on December 15th, 1917:

"On December 10th, 1917, Lt. Werth from the Fl.-Dept. A 246 attacked four enemy Nieuports over Bitolj, of which forced one south of Kravari to crash, the other were forced to make an emergency landing.
To Lt. Werth, as well as his pilot Wiemer, I would like to express my congratulations and my special appreciation on his victory in the air.
The commander in chief
signed von Steuben
General of the infantry."

News Bulletin 44 from December 27th, 1917:

- Southeast front: no news.
- Overview southeast front
- "The aviation activity suffered from the unfavorable weather conditions. Noteworthy is an enemy bombing raid, which took place in the squadron formation with 21 aircraft. This consolidation was probably a consequence of the heavy losses in recent weeks. The zeroing activity with aerial observation was again limited this week, primarily to the area behind the Dojran."

Total successes of the German air forces in December 1917

"In the month of December 1917 our opponents lost 119 aircraft and 9 balloons on all fronts to the

activity of our ordnance. On the other hand, we lost 82 aircraft and two balloons. Of these, 57 aircraft remained behind the lines while the other 25 were lost over our territory. On the western front alone, 101 of the 119 incapacitated enemy aircraft are accounted for, 76 of the 82 German aircraft. On the Italian front, we shot down 17 enemy aircraft and lost 6 of our own. In detail, the sum of enemy losses is made up as follows:
- 83 aircraft in the air fighting
- 30 by anti-aircraft cannons
- 1 shot down by infantry
- 5 unwillingly landed behind our lines
- Of these aircraft, 47 are in our possession, 72 crashed behind the enemy lines."

Other personnel losses (incomplete) of flying units in the southeastern area for the month of December 1917:
- BO Lt. d.R. Ernst Loose, born on June 4th, 1893 in Chemnitz, died on December 15th, 1917 in Saloniki, FA 34
- FF Vzfw. Karl Twittmann, born on August 15th, 1891 in Hannover, died on December 15th in Saloniki, FA 34
- Flieger Robert Wieczorek, born on Jun 7th, 1872 in Gleiwitz, died on December 16th, 1917 in Prilep
- Gefreiter Friedrich Pasucha, born on April 10th, 1886 in Sensburg, died on Decembe r24th, 1917 in Butoin (Romania)
- Vzfw. Theodor Schüßler, born on October 3rd, 1892 in Eschwege, died on December 27th, 1917 in Braila

1918

News Bulletin 45 from January 3rd, 1918:
December 31st, 1917: Shot down an enemy aircraft in air battle.

News Bulletin 46 from January 10th, 1918:
January 8th, 1918: Shot down an enemy aircraft behind the lines.

News Bulletin 47 from January 17th, 1918:
Southeast front: no news.

Situation on the Southeast front:
Enemy aircraft used the few good days for extensive reconnaissance and observation. Compared to the previous weeks, the reconnaissance pilots were particularly active near the 1st and 2nd Bulgarian Army. This seems to have something to do with the fact that the enemy fears an offensive. While in the 11th Army the enemy aviators carried out almost only reconnaissance and observation flights, the enemy aviation activities in the 1st Bulgarian Army also included ….? [auch auf Sperre fliegen] *(could mean – "locking on targets…")*

News Bulletin 48 from January 24th, 1918:
Southeast front: no news.

News Bulletin 49 from January 31st, 1918:
- January 24th, 1918: Shot down an enemy aircraft. Lost an aircraft. [781]
- Other personnel losses (incomplete) from the flying units in the southeastern region for the month of January 1918:
- Flzg. Ober-Matrose Friedrich Gärtner, born on February 17th, 1893 in Landsberg (Warthe), died on January 17th, 1918 on the island of Imbros (Turkey), SFA Xanthi
- Flzg. Matrose Alfons Keppeler, born on December 7th, 1888 in Stetternich near Jülich, died on January 17th, 1918 on the island of Imbros (Turkey), SFA Xanthi

News Bulletin 50 from February 7th, 1918:
January 30th, 1918: Shot down an enemy aircraft behind the lines.

News Bulletin 51 from February 14th, 1918:
Southeast front: no news.

News Bulletin 52 from February 21st, 1918:
Southeast front: no news.

From here the numbering of the Air Force News

Bulletin changes. From February 28th, 1918 the editions are called "second year" and begin with the number 1 on February 28th, 1918.

News Bulletin 1 from February 28th, 1918:

Macedonian front: no news.

Other personnel losses (incomplete) from the flying units in the southeastern region for the month of February 1918:
- Pilot Andreas Soppa, born on November 17th, 1918 in Neustadt, Oberschlesien, died on February 1st, 1918 in Üsküb
- FF Lt. d.R. Friedrich Hermstedt, born on March 5th, 1892 in Erfurt, died on February 5th, 1918 in Topoljani near Üsküb, AFP 11
- BO Lt. Wilhelm Lenz, born on December 5th, 1895 in Berlin, died on February 5th, 1918 in Topoljani near Üsküb, AFP 11
- Pilot Eli Rützel, born on March 26th, 1899 in Kopenhagen, died on February 10th, 1918 in Üsküb

News Bulletin 02 from March 7th, 1918:

Macedonian front: no news.

The British Air Forces have renamed themselves Royal Air Forces (RAF) after the merger of the Royal Flying Corps and Royal Naval Air Service.

News Bulletin 03 from March 14th, 1918:

Macedonian front: no news.

Note: For military reasons, the kill lists will not be released until further notice.

News Bulletin 04 from March 21st, 1918:

- March 13th, 1918: Lost an aircraft in the air war.[782]
- March 15th, 1918: English single seat fighter emergency landing behind our lines and destroyed by English artillery. English tethered balloon crashed near Orljak, without our influence
- March 17th, 1918: English single seat fighter shot down over the Tahino Lake.
- March 18th, 1918: *"On the afternoon of March 18th, 1918, Vzfw. Bauhofer shot down a new enemy tethered balloon south of Opticar The commander in chief congratulated Vzfw. Bauhofer for this, his 3rd victory in the air. The Commander-in-Chief."*

March 1918, from a captured order of the English D / 295. Field Artillery Brigade:
"The Hun is preparing a large-scale offensive on the Western Front. If this should also be directed against our division front, we will have the best ever opportunity to *"slaughter the Huns"*."

News Bulletin 05 from March 28th, 1918:

- March 20th, 1918: A tethered balloon with two inflation balloons shot down by Lt. Hoesch.[783] An enemy aircraft burned and crashed without our influence.

Other personnel losses (incomplete) from the flying units in the southeastern region for the month of March 1918:
- Pilot Friedrich Kuhnt, born on April 8th, 1891 in Lauban, died on March 19th, 1918 in Nisch
- Lt. Hans Deetjen, born on August 19th, 1894 in Allenstein, died on March 29th, 1918 in Drama

News Bulletin 06 from April 4th, 1918:

Macedonian front: no news.

News Bulletin 07 from April 11th, 1918:

April 3rd, 1918: an enemy aircraft shot down by a Bulgarian anti-aircraft unit near the 11th Army.

Successes of German aviators in Macedonia:
- Vzfw. Fieseler, Jasta 25, shot down an enemy Nieuport single-seater near Caniste in a dogfight on April 5th, 1918 and thus achieved his third aerial victory.
- On April 12th, 1918, the pilot Uffz. Leu from the Fl.-Abt.38, although seriously wounded 20km behind enemy lines, brought his aircraft back to the airport in an almost 35 km long flight. The two pilots were recognized by the commander in chief of the army group.

News Bulletin 08 from April 18th, 1918:

Macedonian front: no news.

News Bulletin 09 from April 25th, 1918:

April 24th, 1918, Corps daily order
"On April 24th, 1918 an aircraft crew from the Fl.-Abt. 30 - Pilot Vzfw. Reichmuth, Observer Oblt. Wehmeyer, in a dogfight with three enemy aircraft

over Prespa Lake, shot down one enemy aircraft, and forced a second to break off the fight by going down in a steep glide.
I would like to express my congratulations and appreciation to the flight crew on their great success.
The commander in chief
signed Fleck
Lieutenant General."

Successes of German aviators in Macedonia in April 1918
In April 1918 our enemies lost 271 airplanes and 15 tethered balloons on all fronts thanks to the action of our ordnance. In contrast, we lost 123 aircraft and 14 balloons. Of these, 87 aircraft remained beyond the lines. In detail, the sum of enemy losses is made up as follows:
- 223 aircraft were in aerial combat
- 42 by anti-aircraft guns
- 6 shot down by gunfire"

Other personnel losses (incomplete) from the flying units in the southeastern region for the month of April 1918:
- Vzfw. Heinrich Schott, born on May 27th, 1891 in Eitra, died on April 25th, 1918 Opticar (Macedonia), Jasta 25

News Bulletin 10 from May 2nd, 1918:
Macedonian front: no news.

News Bulletin 11 from May 9th, 1918:
May 6th, 1918: Shot down an enemy aircraft.

Letter of appreciation:
"On the morning of May 4th, 1918, Vice Sergeant Bauhofer, Jasta 25, shot down an enemy two-seater at an altitude of 1,025 and thus achieved his 4th aerial victory.
I would like to offer my congratulations and appreciation to Vzfw. Bauhofer on his great success.
The commander in chief
signed von Steuben
General of the infantry "

News Bulletin 12 from May 16th, 1918:
May 7th, 1918: Shot down one enemy aircraft.

News Bulletin 13 from May 23rd, 1918:
May 14th, 1918: Shot down one enemy aircraft.

- On May 22nd, 1918 Lt. Koehler and Vzfw. Meyer from Jagdstaffel 25 each shot down an enemy aircraft close behind the enemy lines in the lake section. This was Lt. Köhler's first, Vzfw. Meyer's 3rd victory in the air. The victorious pilots received recognition from the Army Commander in Chief.
- "The Bulgarian aviation industry has recently been expanded based on the German model. The German air officers trained the Bulgarian air force. The leader of the German training command, Oblt. Walter, received a letter from the resigning chief of the technical department, Colonel Rankoff, in which Colonel Rankoff thanked him for his valuable support and expressed the hope that, with the help of Germany, the Bulgarian aviation industry would develop into a strong weapon."

News Bulletin 14 from May 30th, 1918:
Work of German aviators in Macedonia:
The enemy pilots generally showed a lively, aggressive spirit, but retreated in places, when German single-seater aircraft approached. Their frequent bombing raids on train stations and camps had little effect due to the fragmented operation.

The reconnaissance activity of our own aircraft was active. It discovered important findings about considerable regrouping and withdrawal of enemy troops, which apparently took place as a result of our victorious offensive in the west. Expansion of the enemy rail network was noted. The long-range reconnaissance carried out to Saloniki revealed important findings about shipping traffic in the port. Our reconnaissance and artillery pilots effectively supported a raid carried out by our infantry.

Other personnel losses (incomplete) from the flying units in the southeastern region for the month of May 1918:
- BO Lt. d.R. Artur Danneberg, born on October 19th, 1885 in Halle, died on May 4th, 1918 in Orljak (South Macedonia), FA 20
- FF Vzfw. Fritz Knauer, born on July 17th, 1888 Schlämpen, died on May 4th, 1918 in in Orljak (South Macedonia), FA 20

News Bulletin 15 from June 6th 1918:
Recognition of the achievements of an aviation unit on the Macedonian front:
"In the last few days, aircraft of the Fliegerabteilung 30 have undertaken several reconnaissance flights far into enemy territory in spite of bad weather,

and attacked enemy camps and vehicle columns with bombs and machine guns from low altitudes. I hereby express my appreciation to the aircraft crews involved in the flights.
The commander in chief
signed Fleck
Lieutenant General"

News Bulletin 16 from June 13th 1918:

Recognition of English air victories:
In order to be awarded an aerial victory, according to an English prisoner testimony, for *"ordinary"* aviators the confirmation of three witnesses is required. The *"tried and tested"* fighter pilots would be awarded a kill or an aerial victory on their own account. They do not need a triple confirmation. This would confirm many unfounded kill reports from the most famous French fighter pilots.

News Bulletin 17 from June 20th, 1918:

Macedonian front: no news.

From News Bulletin No. 21, Recognitions
"Army Daily Order:
A chain of three single-seat fighters from Jasta 25, with pilot Lt. Rose, Vzfw. Treptow and Vzfw. Fieseler, shot down four enemy single-seaters in dogfights in the last 15 days. All three fighter pilots were involved in the aerial battles.
I would like to express my full appreciation to the brave fighter pilots, and in particular to the proven Führer, for the proven aggressive spirit that has led to such beautiful successes.
The commander in chief
signed von Scholtz
General of the artillery."

Army daily order of June 21st, 1918:
"On June 20th, 1918, three Jasta 25 aircraft won a double victory in an aerial battle against the same number of enemy aircraft. Lt. Rose, Vzfw. Treptow and Vzfw. Fieseler shot down an enemy aircraft near Makovo, very near our front, while Vzfw. Fieseler crashed a second plane near Dunje and thus achieved his fourth aerial victory.
This successful flight bears witness to the brave and aggressive spirit of Jagdstaffel 25. Besides my best congratulations, I would like to express my fullest appreciation to those involved in this achievement.
The commander in chief
signed von Steuben
General of the infantry."

News Bulletin 18 from June 27th, 1918:

Macedonian front: no news.

Activities of the German air forces on the Macedonian theater of war from June 13th-19th, 1918:

- The French attacks in the Kamia area were halted. The enemy vacated its positions on the plain without a fight and retreated to the heights of Mahmudli - Bestavukmah.
- The weather was mostly clear with fresh westerly winds. Only on June 19th, 1918 was there some cloud cover. Enemy flight operations were only very active on the Vardar front and here they were superior to our armed forces. In addition to active front reconnaissance, the enemy explored Drenovo to Grabsko at various times. They carried out bombings mainly in larger squadrons without causing military damage.
- Through our own reconnaissance flights and photographs, earth huts for around 800 men were discovered in the Albanian section south of Lunga. In the Ronte-Vardar section, the enemy has begun digging 2,800m of trenches and expanding the conquered Bulgarian bases northwest of Ljumnica. Behind the 26th English division new accommodations, mostly marching camps, for 1,400 men have appeared.
- In the port of Brindisi 9 large, three medium-sized warships, 8 torpedo boats and 12 larger commercial steamers were moored. Five guard vehicles were observed in front of the open mine barrier. Three larger warships, four torpedo boats and five cargo steamers anchored in the port of Balona.
- Our artillery planes successfully zeroed in several times. Several hits were observed on the northern Struma Bridge, and larger explosions occurred on the banks. French camps and marching columns were harassed by frequent bombing and machine gun attacks.
- One of our own aircraft shot down in a dogfight, three enemy aircraft forced to land on the other side by anti-aircraft guns.

Total successes of the German air forces in June 1918:
- "In June 1918 our opponents lost 217 aircraft on all fronts as a result of the activity of our ordnance."
- Other personnel losses (incomplete) from airborne units in the south-east area for the

month of June 1918:
- Vzfw. Erich Dürre, born on October 15th, 1892 in Berlin, died on June 18th, 1918 in Stojakowo (Macedonia), Jasta 38

News Bulletin 19 from July 4th, 1918:
German losses:
- An English aviation expert commented to a journalist for the newspaper "*Globe*": "The figures that the Germans give about their aircraft losses are unreliable."
- In fact, all German information on aerial victories and losses is checked very carefully. The monthly list of casualties of the enemy aircraft that have remained in our possession is reprinted regularly, but without a word of comment. The English should therefore know that our claims are well-founded. As is well known, the English still work with the trick of reporting their losses only regarding aircraft that have not returned, but not regarding aircraft shot down over their own territory.

Army daily order Group Scholtz from July 2nd, 1918:
"From the reports available from the flak commander, I saw that in the month of June the Army Group's flak units shot down 6 enemy aircraft or forced them to land (Bulgarian three, German two, K.u.K. one). The attentive and reliable work of the personnel employed in the air detection service has contributed significantly to these successes. For this I express my full appreciation to all officers, NCOs and men involved.
The commander in chief
signed von Scholtz
General of the artillery."

News Bulletin 20 from July 11th, 1918:
Macedonian front: no news.

Activities of the German air forces in the Macedonian theater of war from June 27th to July 3rd, 1918:
- Smaller infantry attempts alternated with temporarily resurgent artillery fire. The favorable weather in the first half of the week was replaced by strong thunderstorms and rainfalls.
- The opposition's restrained activity west of the Vardar was brisk east of the Vardar. Numerous bomb attacks were aimed at the river valley itself. Enemy artillery pilots appeared in greater numbers. Our own surveillance revealed normal expansion of the position and enlargement of the camp by occupancy capacities between 305 and 3,656 men. Some artillery and balloon shootings were carried out with success. Our own fighter pilots disturbed the enemy several times as they intended to drop bombs, so that the squadron in question hurriedly and haphazardly dropped them. The flight crews suffered badly from Papataci fever, so that some of them are absent for the mission for days.

Weekly report (NB d. LSK No. 22)
Macedonian Front from July 4th to 10th, 1918:
Hostile aviation activity as usual. 16 dogfights, in which our own combat single-seaters shot down two enemy fighters.

News Bulletin 21 from July 18th, 1918:
Macedonian front: no news.

Activities of the German air forces in the Macedonian theater of war from June 27th-July 3rd, 1918:
Artillery activity was livelier at times, in connection with smaller infantry operations. Our own fighter pilots disturbed the enemy when they dropped bombs. Strong enemy bombing raids on the airports in the Varda Valley caused little damage. We attacked enemy camps, train stations and airfields.

News Bulletin 22 from July 25th, 1918:
July 17th, 1918: An enemy aircraft was shot down by flak.

Activities of the German air forces in the Macedonian theater of war from July 11th - July 17th 1918:
- When the sky was clear, the ground on the left flank and in the center of the front was generally calm, while on the right the Italians and French continued to advance. In addition to normal reconnaissance activities, the enemy showed particularly strong artillery activity. It paid for a major bomb attack in the squadron with the loss of two aircraft.
- Our own observation revealed small new camps in Albania, brisk boat traffic on Lake Prespa, a camp for 1,700 men near Zvezda, reduction of the camps on the Monastir front, greater expansion of the rear positions on the Struma front, and also the relocation of camps there.
- In Saloniki, the number of train orders increased from 350 to 750 wagons, and shipping increased from 10 ships on July 13th, 1918 to 27 ships on July 14th, 1918. Tank tracks were seen at a

training exercise north of Saloniki.
- Enemy losses: Two planes in aerial combat, one by flak, one made an emergency landing on this side, a tethered balloon.
- Own losses: an aircraft damaged by flak hit and destroyed during an emergency landing, crew uninjured. One officer wounded.

On the death of Lt. von Eschwege:
According to the newspaper *Guerre Aérienne* of July 25th, 1918, Lt. von Eschwege [died on November 21st, 1917] was the victim of a trap. After repeated successful attacks on English balloons in the Struma Valley, instead of an observer, a dummy that looked like an English officer was placed in the gondola of a balloon, which was filled with 350 kg of explosives. As Lt. von Eschwege attacked the balloon, there was a huge explosion to which he fell victim.

Army daily order of July 29, 1918 (NB d. LSK No. 25):
"On July 24th, 25th and 28th, 1918 Vzfw. Fieseler of Jagdstaffel 25 took down one enemy aircraft in dashing attacks on each of these days, one of which crashed behind our own lines, the other two on the other side of our front. Hereby, Vzfw. Fieseler achieved his 6th, 7th and 8th aerial victory. In the past five weeks, this successful fighter pilot has shot down five enemy planes. Proof of the prevailing aggressiveness of the Jasta 25.
I congratulate Vzfw. Fieseler and express my very special appreciation to him with the wish for further successes.
The commander in chief
signed von Steuben
General of the Infantry"

Activities of the German air forces in the Macedonian theater of war from July 18th to July 24th, 1918:
Little ground combat activity. Major regroupings have taken place on the Struma front in recent weeks. An increase by 9,000 is offset by a decrease of 15,000. The enemy air activity near the 11th and 1st Bulgarian Armies was greatly reduced, and on the Struma Front it was noticeably active.

Other personnel losses (incomplete) from airborne units in the south-east region for the month of July 1918:
- Aeronaut Karl Müller, born on September 29th, 1896 in Sprottau, died on July 11th, 1918 in Prilep

News Bulletin 23 from August 1st, 1918:

Macedonian front: no news.

Own and enemy air forces in Macedonia (according to Kofl's [784] monthly report):
Hostile flight activity:
1. French: Reconnaissance planes, fighter chains and bomb squadrons advanced at very high altitudes far behind our lines. Our fighters could barely come into combat contact with them. In two cases, enemy fighter planes had no insignia on the wings.
2. English: Superior in numbers and material, the English showed a lively aggressiveness towards the 1st Bulgarian Army. Bomb squadrons pushed far over the lines under the protection of single-seat fighters. The enemy showed a special attack tactic, attacking with chains of three single-seaters in such a way that two of its aircraft occupied ours from above without attacking at close range, while the third aircraft sat under the tail of the attacked aircraft.

Own flight activity:
It was possible to carry out our own reconnaissance everywhere, without gaps and to effectively hinder the enemy bomb squadrons from carrying out their bombing raids. Both could only be carried out with numerous heavy air battles and not inconsiderable heavy losses of our own. The establishment of a main flight time was carried out at the order of the aviators' commander in the area of the 1st Bulgarian Army, and each work plane was accompanied by a C-plane for protection. In the region of the 1st Bulgarian Army, the two fighter squadrons were brought together in the Varda Valley.

The accuracy of the enemy fighters, even during a fleeting attack, suggests the enemy's use of telescopic sights, and is also likely to be due to the MG on the upper deck of most enemy types.

News Bulletin 24 from August 8th, 1918:
Macedonian front: no news.

News Bulletin 25 from August 15th, 1918:
Macedonian front: no news.

Activities of the German air forces in the

Macedonian theater of war from July 25th to July 31st, 1918:
- In Albania the K.u.K. troops pushed the Italians back to the approximate line Belina on the Semeni, Remanica northwest of Berat and Krpica on the Tomorica.
- When the weather was not very favorable, the surveillance revealed nothing extraordinary. The hostile air activity in front of the 11th Army was normal, but very little in front of the other armies. Our own balloon on the Varda front was attacked again by enemy squadrons with bombs and machine guns, and damaged.

News Bulletin 26 from August 22nd, 1918:

Macedonian front: no news.

Activities of the German air forces in the Macedonian theater of war from August 1st to August 7th, 1918:
The enemy air activity was within the usual limits and only more active on the Struma, where two enemy bomb squadron attacks of 10 aircraft each took place on two days. Enemy airmen twice directed fire from monitors against Bulgarian positions at the mouth of the Struma. Two of our own planes each scored a bomb hit on a larger ship. Our own planes reconnoitered up as far as the Albanian Lakes area, went down to altitudes below 400 meters and found more traffic than usual. Otherwise nothing unusual.

Army daily order of August 7th, 1918 (NB d. LSK No. 27)
"After I was able to express my appreciation to Vice Sergeant Fieseler from Jagdstaffel 25 only a few days ago for his recent successes, today I am again able to congratulate him on his latest aerial victory. On the morning of August 4th, 1918, Vzfw. Fieseler hunted down his 9th opponent in an aerial battle. The enemy plane caught fire and fell behind our lines.
The commander in chief
signed von Steuben
General of the Infantry"

Attack of a fighter against ground targets, (NB d. LSK No. 27)
- On August 14th, 1918 at around 9 p.m. Lt. Thiede, leader of Jasta 38, attacked an enemy motorcade of about 12 cars, the first car of which was illuminated, from a height of 50 to 100 m with 600 MG rounds. The cars stopped immediately, the lights went out. People could be seen running away to the side. Observation revealed several violent explosions shortly after the attack.
- In the further flight, a lighted camp west of Dreveno was attacked with 200 rounds and a firing battery with 300 rounds. Immediately after the first attack, all lights in the Vardar Valley were extinguished. At all English airports, flares were constantly being fired. Apparently it was suspected that the aircraft lost its way since it was very dark and raining over the front. An airport that was brightly lit by parachute missiles and was bustling with activity was also attacked with machine guns on the return flight from an altitude of 30 m.

Army daily order of the 11th Army, (NB d. LSK No. 29)
*"Vice Sergeant Fieseler, Jagdstaffel 25, after crashing an enemy single-seater on August 5th and August 12th, 1918 each, achieved his 12th aerial victory on August 16th, 1918. During the attack behind enemy lines that day he shot down an enemy biplane.
I extend my congratulations to Vzfw. Fieseler and I am aware that, proud of what has been achieved so far, he will gain further fame and recognition for himself and his weapon in a bold attack.*

Also to Lt. Bender and Vzfw. Waldvogel of Jasta 25, I express my appreciation for the careful and energetic support in the course of these successful aerial battles, from which Vzfw. Fieseler emerged as the winner.
The Commander-in-Chief."

Army group order of August 19th, 1918 (NB d. LSK No. 27)
*"Lt. Ebert and Vzfw. Wilhelm of the Fliegerabteilung 20 brought back valuable coherent image reports despite the number of enemy airmen, and thus contributed significantly to clarifying the enemy situation.
I am pleased to be able to express my full appreciation to the crew.*
The commander in chief
signed von Scholtz
General of the artillery."

News Bulletin 27 from August 29th, 1918:

Macedonian front: no news.

Activities of the German air forces in the

Macedonian theater of war from August 22nd to August 28th, 1918:
- Active artillery activity on both sides, especially at Monastir and on both sides of the Vardar. Enemy reconnaissance was increased near the 11th and 1st Bulgarian Armies and restricted near the 2nd and 4th Bulgarian Armies.
- Four aircraft of our own supported the K.u.K. air forces in the fighting in Albania and took part in successful bombing raids against enemy airports near Valona. Troop movements have taken place in Albania in the past three weeks. Close to the west of the Vardar, the accommodation capacity was increased by over 6,700 men in July and August, and by 8,000 men in the Vardar - Dojran section. The large camps at Verria accommodate 11,200 men and consist almost exclusively of tents for the French infantry. On the Struma front, accommodation has been reduced to 7,000 men.

Other personnel losses (incomplete) from airborne units in the south-east region for the month of August 1918:
- Oblt. Kurt Graßhoff, born on January 24th, 1891 in Wesel, died on August 12th, 1918 in Macedonia, Jasta 38

Newsletter 28 of September 5th, 1918:

Macedonian Front: No news.

Activities of the German air forces in the Macedonian theater of war from August 28th to September 3rd, 1918:
"The K. u. K. XIX. Corps Command requests to accept its most binding thanks for the valuable support from German air forces and permits itself to announce that the aviators made a major contribution to the success of the difficult undertakings against Valona with the greatest bravery.
Colonel-General Baron Pflanzer-Baltin, Commander of the Austro-Hungarian Armed Forces in Albania."

News Bulletin 29 from September 12th, 1918:

Macedonian front: no news.

Activities of the German air forces in the Macedonian theater of war from September 4th to 10th 1918:

- Enemy activity had increased near the 11th and 1st Bulgarian armies. Our own explorations revealed nothing special.

News Bulletin 29 from September 12th, 1918:

September 12th, 1918: two enemy aircraft shot down
September 13th, 1918: one enemy aircraft shot down.

Daily orders of the 11th army from September 16th, 1918 (NB d. LSK No. 32)
"Vzfw. Fieseler, Jagdstaffel 25, on August 23rd and September 4th shot down his 13th and 14th opponent in aerial combat. He added another to these great successes by crashing an enemy single-seater right in front of our lines on September 13th, 1918. The number of his aerial victories has increased to 15. It is a pleasure for me to bring this outstanding achievement to the knowledge of my army. The aggressive spirit, which was always fresh and alive in the 25th squadron, guarantees further victories over the numerically superior opponent, who is provided with many means.

I offer my congratulations and once again, my appreciation, to Vzfw. Fieseler.
The Commander-in-Chief. "

Army Daily Order, (NB d. LSK No. 32)
"Vzfw. Fieseler, Jasta 25, achieved his 16th aerial victory on September 16th, 1918 by shooting down a French single-seater.
I express my appreciation and congratulations to the daring aviator, who was recently awarded the Military Cross of Merit by His Majesty, on the new victory.
The Commander-in-Chief. "

News Bulletin 31 from September 26th, 1918:

- September 18th, 1918: one enemy aircraft shot down.
- September 19th, 1918: four enemy aircraft shot down.
- September 20th, 1918: one enemy aircraft shot down.

Overall successes of the German air forces in September 1918:
"In September 1918, our opponents lost 788 aircraft on all fronts as a result of our ordnance. In contrast, we lost 107 aircraft and 103 balloons. In detail, the sum of enemy losses is made up as follows:

- 652 aircraft in air combat
- 125 shot down by anti-aircraft guns."

Other personnel losses (incomplete) from airborne units in the south-east region for the month of September 1918:
- BO Lt. d.R. Paul Fröhner, born on April 22nd, 1893 in Berlin, died on September 2nd, 1918, Tefik Bey
- FF Flieger Johannes Hardt, born on April 25th, 1896 in Kelsterbach, died on September 2nd 1918, Tefik Bey
- Lt. Joachim Patzer, born on February 14th, 1897 in Meiningen, died on September 8th, 1918 in Prilep, FA 38
- BO Lt. Alfred Ebert, born on January 5th, 1895 in Königsberg, died on September 16th, 1918 in Drama, FA 20
- FF Vzfw. Hans Kiselowski, born on January 17th, 1899 in Berlin, died on September 16th in Drama, FA 20
- Pilot Friedrich Ruttloff, born on December 29th, 1894 in Oberwiesa, died on September 16th, 1918 in Drama
- Gefreiter Wilhelm Pieper, born on November 2nd 1885 in Gladigau, died on September 21st, 1918, Pass on the Strumitza (Bulgaria)
- Gefreiter Anton Klötzer, born on September 10th, 1893 in Mannheim, died on September 27th, 1918 in Nisch.

Former Admiral A. Hopmann [785] reports in his book "The War Diary of a German Naval Officer" in 1926 about the fighting against the Allied offensive from Greece that began on September 15, 1918:

"In the meantime, the general war situation of the Central Powers had received a shock of the greatest importance due to the collapse of the Bulgarian front in Macedonia in mid-September [1918]. The Entente offensive starting in Saloniki no longer met with any resistance worth mentioning, and the Supreme Army Command's lukewarm support of the Bulgarian army was being avenged.

Later I heard from French officers that they were extremely astonished at the ease with which the attack succeeded, and that they had expected and feared strong resistance, including German troops, all the more so when taken into account that their own had been weakened, shaken in attitude and discipline (and were no longer particularly reliable) due to summer heat, inactivity and malaria fever. The sudden collapse of Bulgaria evidently came unexpectedly to our Supreme Army Command.

Only now were troop transports ordered to the mouth of the Danube and Varna. [...]"

News Bulletin 32 from October 3rd, 1918:
Macedonian front: no news.

News Bulletin 33 from October 10th, 1918:
Macedonian front: no news.

News Bulletin 34 from October 17th, 1918:
Macedonian front: no news.

News Bulletin 35 from October 24th, 1918:
Macedonian front: no news.

News Bulletin 33 from October 31st, 1918:
Macedonian front: no news.

Other personnel losses (incomplete) from airborne units in the south-east region for the month of October 1918:
- Pilot Willi Hackbusch, born on May 30th, 1897 in Röbel, died on October 7th, 1918 in Nisch, FA 20
- Pilot Hermann Paulitschke, born on April 17th, 1896 in Glatz – died on October 7th, 1918 in Nisch, Jasta 38
- Pilot Michael Wegerie, born on May 25th, 1897 in Nürnberg, died on October 8th, 1918 in Nisch
- BO Lt. d.R. Wilhelm Lehmann, born on June 27th, 1898 in Greiz, died on October 13th, 1918 in Macedonia, FA 22
- FF Uffz. Paul Krüger, born on July 31st, 1894 in Berlin, died on October 13th, 1918 in Macedonia, FA 22

Handwritten note on the last issue of the "Nachrichten der Luftstreitkräfte Nr. 36.": "In light of the ceasefire / revolution Nov. 18, this issue is no longer registered, and was not distributed. Some pages were not written."

Died in British captivity in Greece:
- Vzfw. Wilhelm Schmitz, born on February 12th, 1893 in Köln-Deutz, died on January 24th, 1919 in St. Martin de Re
- Lt. d.R. Wilhelm Lepper, born on October 31st, 1891, died on April 8th, 1919.

Part III

Appendix

Documents, Photos:

Following are the instructions for the ***"Funkverkehr des Artilleriefliegers"*** (radio operation for artillery pilots), dated June 25th, 1918, which are published for the first time. Since the artillery and reconnaissance aircraft are constantly accompanied in this publication, it is necessary to *"officially"* document his tasks and working methods. In addition, one must not forget that this observer was also responsible for defense in flight.

Personal descriptions and military resumes:

Fritz Wilhelm Hermann Kempf

Born on May 9th, 1894 in Freiburg im Breisgau

Father: Dr. h.c. Friedrich Wilhelm Kempf (born on October 26th, 1857, died on October 25th, 1932), Architect and Master Cathedral Constructor in Freiburg.

Mother: Adelgunde Kempf, born Huber (born on January 29th, 1865 – died on August 7th, 1943)

School education: 1900 – 1904 Primary school, September 11th, 1904 - July 31st, 1910 Middle and high school.

October 16th, 1910 – August 11th, 1913 Polytechnikum Mechanical engineering (August 11th, 1913 Engineers' exam).

Internships during his university studies: Firma Lanz A.G., Locomotive construction, Pippart-Noll, Aircraft construction, Mannheim.

Military experience prior to and during the First World War (1913–1918):

October 1st, 1913: Entry into service as one-year volunteer in the (5. Baden) Inf.-Reg. Nr. 113 in Freiburg im Breisgau
April 1st, 1914: named to surplus private
August 2nd, 1914: promoted to sergeant and sent to the west front with Inf.-Reg. Nr. 113. Participated in the battle in Lorraine (Mülhausen, Saarburg), 6th Army
September 7th, 1914: seriously wounded on his foot near Mesnil und in the hospital until February 7th, 1915. After ¾ year recovery, transferred to aviation unit.

Training to be a pilot at the Flieger-Ersatz-Abteilungen (FEA) 3 (Gotha) and FEA 9 (Darmstadt and Aviatik-Fliegerschule Freiburg).

January 4th, 1916: transferred to Kampfgeschwader 4, Staffel 20, *[Kasta 20]*
January 27th, 1916: promoted to vice-sergeant. Participation in the battle near Verdun (5th Armee)
September 1916: Transfer of the *Kagohl 4, Staffel 20* to the army group Mackensen in the Balkans. There it was renamed *Kampfgeschwader 1*. First served in the Balkans (battles in Dobruja), shot at by war ships in the Black Sea, emergency landing behind enemy lines, freed by Bulgarians.
September 9th, 1916: promoted to Leutnant d.R.
1917: Transfer to *Jagdstaffel Boelcke* [786] in the west *[no clear date]*
1917: Participation in the spring battle near Arras (6th army) and the battle in
Flanders (4th army)
April 30th, 1917: First air victory (BE2b) in the *Jasta Boelcke*
June 5th, 1917: Second air victory (Pup, 54 Squ RAF) in the *Jasta Boelcke*
October 20th, 1917: Third air victory (Camel, 70 Squ RAF) in the *Jasta Boelcke*
1918: Participation in the great battle in France (17th army, „Michael I"), the battle around teh Kemmel (4th army) and the battle near Soissons and Reims (7th army „Blücher")
May 8th, 1918: Fourth air victory (Camel) in the *Jasta Boelcke*
August 18th, 1918: Transferred to the Jagdstaffelschule I (Famars) as a trainer.

Awards during the first world war:
May 1915: EK II
July 22nd, 1915: Silver merit medal of Baden

June 21st, 1916: Pilots' badge
March 20th, 1917: EK I
June 10th, 1917: Bulgariab military award V. class, with war band
November 8th, 1917: K.u.K. field pilot badge
February 21st, 1918: Knight's Cross with Swords of the Order of the Zähringer Lion II class
August 16th, 1918: Wounded badge in black

Work following the end of the war 1918:
Dec. 1918 – March 1919 Pilot and technical head at Fliegerhorst Freiburg.

March 1919 – Sept. 1919 Pilot and technical head at Deutsche Luftreederei (DLR), Berlin.

Oct. 1919 – Sept. 1920 Cathedral building authority, Freiburg.

Dec. 1920 – Dec. 1921 Mechanical engineer at the auto construction company Ludwig Weber, Freiburg. (Ludwig Weber was also a fighter pilot during the war).

Jan. 1922 – March 1925 Mechanical engineer at the company Held & Francke, Bau A.G., Heidelberg (construction of the Neckar canal).

August 24th, 1925 Marriage to the photographer Hedwig Unsin (born on September 17th, 1900, died on January 13th, 1971), who also works for Held & Francke.

April 1925 – Dec. 1927 Hangar director of the Junkers Flugzeug- und Motorenwerft at the Vienna-Aspern airfield and technical expert in civil aviation at the federal ministry of transportation in Vienna. On July 19th, 1926, his daughter Ingeborg is born in Vienna. For this reason, he could not begin a job as an expert in China.

Jan. 1928 – March 1932 External agent of the Deutsche Versuchsanstalt für Luftfahrt (DVL) Berlin-Adlershof for south and west Germany. Construction supervisor and head of testing for the aircraft and engine companies in this area. (Leichtflugzeugbau Klemm, Hirth Motorenbau, Daimler Benz).

April 1933 – Jan. 1936 scientific assistant and head of aircraft operations at DVL in Berlin-Adlershof.

In Aug. 1933 took part in a multi-day Germany flight with a Klemm. Had to give up on the third day as the aircraft was damaged during a landing in thick fog on a cow pasture.

On June 15th, 1934, his son Jörg was born in Berlin.

Air Force and Second World War
Jan. 1936 – Nov. 1937 technical head of the Fliegerhorst Oldenburg. Jobs: Maintenance of aircraft and engines, technical advisor to the Commander.
September 1st – October 26th, 1936: As a Leutnant d.R. a.D. called up to an introductory training at Fliegerhorstkommandantur Oldenburg.
June 1937 Took over as aircraft staff engineer at the engineering corps of the air force. September 1st, 1937: As Leutnant d.R. began work at the Fliegerhorstkommandantur Delmenhorst.

Nov. 1937 – Oct. 1938 Field work leader of the Reichsluftfahrtministeriums (RLM). Introduction of new aircraft and engines of the aircraft units. Training of technical personnel.

April 1st, 1938: Transferred to Fliegerhorstkommandantur Aibling.
Oct. 1938 – Aug. 1939 Transferred as Chief engineer of the aircraft command in Munich zum. Technical advisor and assistant chief engineer.

Jan.1939 named as aircraft sergeant engineer (?) Flieger-Oberstabsingenieur.

In Aug. 1939 transferred to Wiesbaden as district air command engineer, to the Luftgaukommando XII, and then transferred to western France with the Luftgaukommando XII. Had his own Bf 110 until the end of the war, which he flew himself.

From 1941 until the end of the war he was a colonel-engineer and district air command engineer at the Luftgaukommando VII in Munich. Jobs: Technical advisor of the commander, director of the hangars, maintenance of aircraft and engines, etc. supply of aviation equipment and spare parts for front units and hangars.

Work after the war:
In 1945 he moved to Freilassing in Bavaria (near Salzburg in Austria). At first he was a machinist carpenter at the construction and furniture company Heinrich Feil and the woodworking shop of Joseph Schwarz und Sohn. Afterwards, he was an employee of an iron wholesaler and finally for an oil company in Hamburg.

Retired in 1950

Died on September 2nd, 1966 in Freilassing. He could still meet his grandson Per Fredrik, who was born on January 3rd, 1966 in Oslo.

Life Data Gerhard Fieseler [787]

Gerhard Fieseler was born in Glesch/Rheinland on April 15th, 1896. His father August built up his own print shop in Bonn, and Gerhard had to help out in the firm in addition to attending elementary and middle school. The father wanted him to be his successor in the print shop. So at 18, he successfully completed an apprenticeship as a bookbinder in his father's company. In the meantime, however, Gerhard had developed his passion for aviation and volunteered for the air force at the beginning of World War I. During the second test flight to become a field pilot, he crashed on November 11th, 1915, and suffered a double broken leg. After his recovery, in the summer of 1916, he went to Bucharest as an artillery pilot. He completed his training as a fighter pilot and was transferred to the Macedonian front as a deputy sergeant to Jasta 25 in May 1917. Vzfw. Fieseler always tried the surprise attack from behind and was able to shoot down a total of 19 enemy planes. But he was also one of the few who managed to force the enemy to land on the German side of the front, unharmed. This is what happened to the French fighter pilot Marie Eugen in a Nieuport. This process was rated as Fieseler's eighth kill. He then flew the captured aircraft several times to test its flight characteristics.

Gerhard Fieseler was promoted to Lieutenant by Kaiser Wilhelm II because of his bravery against the enemy, and received the *"Sergeants' Pour le Merite"*, the Prussian golden military cross of merit, on September 3rd, 1918. He was called *"Tiger"* by his squadron comrades.

After the First World War, Gerhard Fieseler became one of the most successful aviation pioneers as an aerobatic pilot, and from 1930 he built his own aircraft in Kassel. With the aircraft type Fi 156, the *"Fieseler Storch"*, he achieved his big breakthrough due to the aircraft's fantastic take-off and landing characteristics.

After the Second World War, in 1953, Gerhard Fieseler founded a company for the production of thermal windows in light metal composite design. Hospital beds and bed lamps were also made there. But economic success eluded him. In 1957 Fieseler ended his entrepreneurial activity. He then drew income from renting his own industrial and residential buildings. He had not lost interest in aviation. After he had regained the necessary safety while flying under the supervision of a flight instructor, he bought a sport airplane. He also oversaw aerobatic events. In 1980 his Gerhard Fieseler Foundation began its work. To date, his foundation has supported over 700 successful projects (welfare, care for the elderly, art and culture; sport) with a total of around 4.1 million euros.

Gerhard Fieseler died on September 1st, 1987 in Kassel.

Life Data of Johannes Keller

Stations from his pay book:
Johann Keller was born on October 6th, 1894 in Lindenfels (Hessen) and died on February 14th, 1940. His occupation was specified as *"Chauffeur and fitter"*. Should war break out, he was to report to the 1st Recruit Depot, Replacement Battalion, 32nd Infantry Regiment.

October 2nd, 1914 – transferred to 2. Kompanie Ersatz-Bataillon, Infanterie-Regiment 32 Meiningen
October 3rd, 1914 – Transfer as driver to head veterinarian Dr. Varst, Generalgouvernement Brussels
July 16th, 1915 – August 12th, 1915 transferred to FEA 9 (**F**lieger-**E**rsatz-**A**bteilung 9)
August 12th, 1915 – Transfer to FA 42
September 19th 1915 – August 12th 1915 in the reserve hospital Worms (got sick while on vacation)
December 3rd, 1915 – January 8th, 1916 at the FEA 10 Böblingen-Stuttgart
January 8th, 1916 – July 25th, 1916 at the Artillerie-Flieger-Abteilung 216
July 28th, 1916 – January 29th, 1917 at the FEA 5 Hannover, Hamburg-Fuhlsbüttel branch
October 19th, 1916 – January 24th 1917 at the FEA 5
January 24th, 1917 – March 23rd, 1917 Flieger-Kommando „Nord" in Flensburg
March 23rd, 1917 – March 25th, 1917 FEA 5
April 11th, 1917 – May 6th, 1917 Armee-Flugpark 11
May 6th, 1917 – May 10th, 1917 FA 34
May 10th, 1917 – July 2nd, 1917 Armee-Flugpark 11
July 2nd, 1917 – October 15th, 1917 FA 30, Vorkommando
October 16th, 1917 – December 9th, 1917 FA 30
December 9th, 1917 – January 15th, 1918 FA 20, Vorkommando
January 15th, 1918 – August 17th, 1918 : Flieger-Abteilung 20 (earlier Feld-Flieger-Abteilung 51),
August 18th, 1918 – Transfer to FEA 5 Hannover
September 30th, 1918 – October 20th, 1918 to FEA 9
February 14th, 1919 – August 7th, 1919: II. Landsturm-Infanterie-Ersatz-Bataillon Worms

Dismissed from army service in Lindenfels

near Worms on August 9th, 1919 due to the demobilization order. Johannes Keller joins the police force.

Promotions:
March 29th, 1916 named to private
July 24th, 1916 promoted to sergeant
December 1st, 1917 according to the command of Flieger 11, Oberst Scholtz, promoted to vice sergent (Vizefeldwebel) for excellence against the enemy

Medals and awards:
September 18th, 1917: EK II
September 25th, 1917: Pilots' insignia
December 7th, 1917: Hessian bravery medal
February 27th, 1918: Bulgarian bravery cross
April 18th, 1918: EK I
September 28th, 1918: Hessian warrior badge of honor

Participated in the following battles:
July 2nd, 1917 – October 16th, 1917: Battles of the 2nd Bulgarian army in East Macedonia
October 16th, 1917 – December 9th 1917: Fighting at the Greek border at the Vardar- and Dojran Lake
December 9th, 1917 – January 15th, 1918: Fighting of the 2nd Bulgarian army in east Macedonia
January 15th, 1918 – August 17th, 1918: Royal 2nd and 4th Bulgarian army in east Macedonia.

Life Data of Walter Schwabedissen

Walter Schwabedissen was born on June 16th, 1896. At the beginning of the First World War he joined the Lauenburg Field Artillery Regiment No. 45 on July 29th, 1914 as a flag junior. As early as June 1915 he began his aviation career with training as an observer in Jüterbog. He took up his first service as a lieutenant (observer) on September 1st, 1915 at Feldflieger Abteilung 11, where he stayed until February 1917. Via Army-Flugpark 9 and Flg.-Abt. 28 he came to the Flg.-Abt. 20 in Macedonia. At the end of February 1918 he was again transferred to Army Air Park 4 and to Aviation Dept. 3 (LB), where he ended the war as first lieutenant in 1918.

He had various positions in the Reichswehr until December 15, 1935, when he was appointed *"Major in the Reich Aviation Ministry, Adjutant of the Wehrmacht to the Führer and Reich Chancellor"*. He held this position until July 30th, 1935. After serving numerous roles in the Air Force, he was appointed General of the Air Force. He was released in 1947 after being a prisoner-of-war for two years. Walter Schwabedissen died on February 19th, 1989.

Life data of Walter Wehmeyer

First lieutenant (ret.) Walter Wehmeyer, born on July 7th, 1887 in Spandau, died on March 31st 1936 in Berlin.. Came from the Guards Foot Artillery Regiment. During the war as an observer at FEA 10, FAA 283, FA 20, FA 38, among others.

Bibliography

PRIMARY SOURCES
Documentation
Germany
- Hölzle, Erwin (ed.), *Quellen zur Entstehung des Ersten Weltkrieges. Internationale Dokumente 1901-1914* (Darmstadt: WBG, 1995)
- Karl Kautsky (ed.), *Die deutschen Dokumente zum Kriegsausbruch, Bände II und III* (Charlottenburg: Deutsche Verlagsgesellschaft, 1919),
- Auswärtiges Amt, *Dokumente zum Konflikt mit Jugoslawien und Griechenland* (Berlin: Zentralverlag der NSDAP, 1941)

Greece
- American Hellenic Society, *The Greek White Book Diplomatic Documents 1913–1917* (New York: Oxford UP, 1918)
- *The Greek White Book Supplementary Diplomatic Documents 1913–1917* (New York: Oxford UP, 1919)
- *Report of the Greek University Commission upon the Atrocities and Devastations Committed by the Bulgarians in Eastern Macedonia* (New York: Oxford UP, 1919)

Secondary Sources: Memoirs, Diaries, Memoirs:
- Churchill, Winston: *The World Crisis 1911–1918*, 2 Vols (London: Odhams Press, 1938)
- Churchill, Winston: *The Second World War, IV, The Commonwealth Alone* (London: Cassell, 1964)
- Conrad von Hötzendorf, Franz: *Aus meiner Dienstzeit 1906–1918*, Band 4
- (Wien: Rikola, 1925)
- Dartige du Fournet, Louis: *Souvenirs de guerre d'un amiral 1914–1916*
- (Paris: Plon, 1920)
- David, Robert: *Le drame ignoré de l'Armée d'Orient Dardanelles – Serbie – Salonique Athènes* (Paris: Plon, 1927)
- De Guingand, Francis: *Operation Victory* (London: Hodder & Stoughton, 1947)
- Deville, Gabriel: *L'Entente, la Grèce et la Bulgarie. Notes d'histoire et souvenirs* (Paris: Eugène Figuière, 1919)
- Eden: *The Anthony Eden Memoirs. The Reckoning* (London: Cassell, 1965)
- Falkenhausen, Ernst, v., *Erdrosselung Griechenlands* (Berlin: Ullstein, 1918)
- Hibben, Paxton: *Constantine I and the Greek People* (New York: Century Company, 1920); Reprintedition (Memphis: General Books, 2012)
- Hindenburg, Paul von: *Aus meinem Leben* (Leipzig: Hirzel, 1934)
- Lawson, John Cuthbert: *Tales of Aegean Intrigue* (London: Chatto & Windus, 1920)
- Lloyd George, David: *War Memoirs* Vol. I (London: Odham, 1936) – *War Memoirs* Vol. III (London: Nicholson & Watson, 1934)
- Ludendorff, Erich: *Meine Kriegserinnerungen 1914–1918* (Berlin: Mittler, 1919)
- Mackenzie, Compton: *Greece in My Life* (London: Chatto & Windus, 1960)
- Packer, Charles: *Return to Salonika* (London: Cassell, 1964)
- Pierrfeu, Jean de: *Trois ans au grand quartier général* (Paris: Éditions Française Illustrée, 1920)
- Radoslawoff, Vasil: *Bulgarien und die Weltkrise* (Berlin: Ullstein, 1923)
- Sarrail, Maurice: *Mon commandement en Orient (1916–1918)* (Paris: Flammarion, 1920) - „La Grèce venizéliste. Souvenirs vécus" Revue de Paris (15 Décembre 1919), pp. 685–706

SECONDARY LITERATURE

A
- Abbott, George Frederick: *Greece and the Allies* (Hamburg: Tredition, 2008) reprint von (London: Methuen, 1920) – *Turkey, Greece and the Great Powers: A Study in Friendship and Hate* (London: Robert Scott, 1916)
- Alexatos, Gerassimos: "Xairete: Ein griechisches Armeekorps in Görlitz" in: Wolfgang Schultheiß (ed.), *Meilensteine deutsch-griechischer Beziehungen* (Athen: Stiftung für Parlamentarismus und Demokratie des Hellenischen Parlamentes, 2010), pp. 185–200
- Angelow, Jürgen (ed.): *Der Erste Weltkrieg auf dem Balkan* (Berlin: Be.bra, 2011)

B
- Batowski, Henryk: "Proposals for a Second Front in the Balkans in September 1939" Balkan Studies 9 (1986), pp. 333–344.
- Breuer, Falk & Waiss, Walter (eds.): *Richthofen braucht Nachfolger – Das kurze Fliegerleben des Josef van Endert (1898–1918)* (Aachen: 2016)
- Breuer, Falk & Waiss, Walter (eds.): *Julius Withenius 1898–1918 – Ein Flieger aus dem Bergischen Land* (Aachen: 2018)

C

- Cabanes, Bruno und Anne Duménil (eds.): *Der Erste Weltkrieg. Eine europäische Katastrophe* (Darmstadt: WBG, 2013)
- Clews, Graham T.: *Churchill's Dilemma. The Real Story Behind the Origins of the Dardanelles Campaign* (Santa Barbara: Praeger, 2010)
- Collins, Robert J.: *Lord Wavell (1883–1941). A Military Biography* (London: Hodder & Stoughton, 1947)
- Cosmetatos, S. P. P.: *The Tragedy of Greece* (London: Kegan, 1928)

D

- Deville, Gabriel: *L'Entente, la Grèce et la Bulgarie. Notes d'histoire et souvenirs* (Paris: Eugène Figuière, 1919)
- Duppler, Jörg (ed.): *Kriegsende 1918, Ereignis, Wirkung, Nachwirkung* (München: Oldenbourg, 1999)

E

- Ehrke, Hans: *Makedonka. Ein Buch der Balkanfront* (Braunschweig: Westermann, 1938)

F

- Falls, Cyril: *Macedonia, 2 Vols.* (London: Imperial War Museum, 1932) – *The Great War* (New York: Putnam, 1959)
- Fyfe, Albert J.: *Understanding the First World War. Illusions and Realities* (New York: Peter Lang, 1988)

G

- Gauvain, Auguste: *L'affaire Grecque* (Paris: Bossard, 1918)
- Geniko Epiteleio Stratou – Hellenic Army General Staff: *A Concise History of the Participation of the Hellenic Army in the First World War 1914–1918* (Athens: Army History Directorate, 1999)
- *Epitomi istorias tis symmetochis tou Ellinikou stratou ston proto pankosmio polemou 1914-1918* (Athen: Diefthinsis Istorias Stratou, 1993)
- *I Ellas kai o polemos eis ta Valkania* (Athen: Diefhinsis Istorias Stratou, 2012)
- *I symmetochi tis Ellados eis ton polemon 1918* (Athen: Diefthinsis Istorias Stratou, 1961)
- Grigoriadis, Neokosmos: *I ethniki amyna Thessalonikis tou 1916* (Athen, 1960)

H

- Hall, Richard C.: *Balkan Break Through. The Battle of Dobro Pole 1918* (Bloomington: Indiana UP, 2010)
- *Handbuch der neuzeitlichen Wehrwissenschaften*, I (Berlin: de Gruyter, 1936)
- Higham, Robin: *Diary of a Disaster. British Aid to Greece 1940–1941* (Lexington: The University Press of Kentucky, 1986)
- Hopmann, Albert: *Das Kriegstagebuch eines deutschen Seeoffiziers* (Berlin, Scherl-Verlag, 1925)

J

- Joachim, Joachim: *Ioannis Metaxas. The Formative Years 1871–1922* (Mannheim: Bibliopolis, 2000)

K

- Keegan, John: *The First World War* (New York: Vintage Books, 2000)
- Koliopoulos, John S.: *Greece and the British Connection 1935–1941* (Oxford: Clarendon Press, 1977)
- Kralik, Richard: *Geschichte des Völkerkrieges (1914–1918)* (Graz: Styria, 1923)
- Krüger, Friederike: "Die Verantwortung der militärischen Führung 1918", in: Jörg Duppler & Gerhard P. Groß, (eds.): *Kriegsende 1918. Ereignis, Wirkung, Nachwirkung* (München: Oldenburg, 1999), pp. 378–394

L

- Leeds, Stanton: "The Greek King an dthe Present Crisis" *The Century Magazine* (April 1916), pp. 947–951
- Leon, George B.: *Greece and the Great Powers 1914-1917* (Thessaloniki: IMXA, 1974)
- Long, Gavin: *Greece, Crete and Syria* (Canberra: Australian War Memorial, 1953)

M–N

- Morélos, Yanis G.: "Français et Grecs pendant le Drôle de Guerre" *Balkan Studies* 29:1 (1988), pp. 99–142
- Mückler, Jörg: *Deutsche Bomber im Ersten Weltkrieg* (Stuttgart: 2017)
- Mückler, Jörg: Deutsche Flugzeuge im Ersten Weltkrieg (Stuttgart: 2013)
- Mühlmann, Carl: *Oberste Heeresleitung und Balkan im Weltkrieg 1914/1918* (Berlin: Wilhelm Limpert, 1942), „Der Eintritt Griechenlands in den Weltkrieg Sommer 1917" Berliner Monatshefte (1937), pp. 515–524
- Münkler, Herfried: *Der große Krieg. Die Welt 1914–1918* (Berlin: Rowohlt, 2013)
- Misha, Glenny: *The Balkan 1804–2012* (London: Granta, 2012)

- Napp, Dr. Niklas: *Die deutschen Luftstreitkräfte im ersten Weltkrieg* (Paderborn: Verlag Schöningh, 2017)
- Neumann, Georg Paul: *Die gesamten deutschen Luftstreitkräfte im 1. Weltkrieg* (Berlin, Mittle & Sohn, 1920)
- McClymont, W. G.: *Official History of New Zealand in the Second World War 1939–45. To Greece.* (Wellington, New Zealand: Department of Internal Affairs, War History Branch, 1959)

O

- Österreichisches Bundesministerium für Heerwesen (ed.): Österreich-Ungarns letzter Krieg, Band 1. Das Kriegsjahr 1914 (Wien: Militärwissenschaftliche Mitteilungen, 1930)
- Ortner, Christian: "Die Feldzüge gegen Serbien in den Jahren 1914 und 1915", in: Jürgen Angelow (ed.), *Der Erste Weltkrieg auf dem Balkan* (Berlin: be.bra, 2011), pp. 123–142
- Owen, H.: *Collison Salonica and After. The Sideshow that Ended the War* (London: Hodder & Stoughton, 1919)

P

- Palmer, Alan: *The Gardeners of Salonika. The Macedonian Campaign 1915–1918* (London: Faber & Faber2009)
- Porte, Rémy: "*Commen faire plier un neutre? L'action politique et militaire de la France en Grèce (1915–1917)*" Cahiers de la Méditerranée 81 (2010), pp. 45–62
- Price, G. Ward: *The Story of the Salonica Army* (New York: Edward J. Clode, 1918)

R

- Reichsarchiv (ed.): *Der Weltkrieg 1914 bis 1918, Band II, Die Befreiung Ostpreußens* (Berlin: Mittler, 1925)
  - *Der Weltkrieg 1914 bis 1918, Band IX, Die Operationen des Jahres 1915* (Berlin: Mittler, 1933)
  - *Herbstschlacht in Macedonien Cernabogen 1916* (Oldenburg: Stalling, 1925)
  - *Das Weltkriegsende an der mazedonischen Front* (Oldenburg: Stalling, 1925)
- Reichskriegsministerium (ed.):
  - *Der Weltkrieg 1914–1918, Band X, Die Operationen des Jahres 1916* (Berlin: Mittler, 1938)
  - *Der Weltkrieg 1914–1918, Band XI, Die Kriegführung im Herbst 1916 und im Winter 1916/17*(Berlin: Mittler, 1938)
- Richter, Heinz: *Griechenland zwischen Revolution und Konterrevolution (1936–1946)* (Frankfurt: EVA, 1973)
  - *Griechenland im Zweiten Weltkrieg 1939–1941* (Bodenheim: Syndikat, 1997)
  - "*Die Auswirkungen der Operationen 'Marita' und 'Merkur' auf den Beginn des Unternehmens 'Barbarossa'*. THETIS 2 (1995), pp. 203–216
  - "The Impact of Operations Marita and Merkur on Barbarossa. The six missing weeks in front of Moscow. Myth or historical truth?" in: Institute for Balkan Studies (ed.), *Macedonia and Thrace, 1941–1944. Occupation – Resistance – Liberation* (Thessaloniki: IMXA, 1998), pp. 15–32
  - *Operation Merkur: Die Eroberung der Insel Kreta im Mai 1941* (Ruhpolding: Rutzen 2011)
  - *British Intervention in Greece. From Varkiza to Civil War.* (London: Merlin Press, 1986)
- Roudometof, Nikolas (ed.): *Tetradia voulgarikis katochis, II, Anatoliki Makedonia 1916–18* (Kavala: Istoriko & Logotechniko Archeio Kavalas, 2008)

S

- Schmidt-Annaberg, Hans: *Bilder vom Mazedonischen Kriegsschauplatz* (München: Kastner & Callwey, ca. 1917)
- Smith, Michael: *Llewellyn Ionian Vision. Greece in Asia Minor 1919–1922* (London: Allen Lane, 1973)
- Stavrianos, Leften S.: *The Balkans since 1453* (New York: Holt, Rinehart and Winston, 1965)
- Stegemann, Bernd: „Die italienisch-deutsche Kriegsführung im Mittelmeer und Afrika", in: Militärgeschichtliches Forschungsamt, (ed.): *Das Deutsche Reich und der Zweite Weltkrieg. Vol.III Der Mittelmeerraum und Südosteuropa* (Stuttgart: DVA, 1984)

T

- Theodoulou, Christos A.: *Greece and the Entente August 1, 1914 – September 25, 1916* (Thessaloniki: IMXA, 1971
- Tietz, Klaus-Dieter: „Griechen in Görlitz" Hellenika N. F. 5 (2010), pp. 60–71

U–Z

- Ventiris, Georgios: Η Ελλάς του *1910–1920* (Athen: Ikaros, 1970)
- Veremis, Thanos: *Eleftherios Venizelos. A Biography* (New York: Pella, 2010)
- Veremis Thanos & Helen Gardikas-Katsiadakis: „Protagonist in Politics, 1912–20", in: Paschalis Kitromilides (ed.), *Eleftherios Venizelos. The Trials of Statemenship* (Edinburg: UP, 2006), pp.

- 115–133
- Vinogradov, V. N.: „Romania in the First World War: The Years of Neutrality, 1914-1916" *The International History Review*, 14:3 (August 1992), p. 450f
- Wakefield, Alan & Simon Moody: *Under the Devil's Eye. The British Military experience in Macedonia 1915–1918* (Barnsley: Sword & Pen, 2004)
- Wentscher, Bruno: *Deutsche Luftfahrt* (Berlin, 1925)
- Willmore, John Selden: *The Story of King Constantine as Revealed in the Greek White Book* (New York: Longmans, Green & Co., 1919)
- Wolf, Klaus: *Gallipoli 1915. Das deutsch-türkische Militärbündnis im Ersten Weltkrieg* (Sulzbach: Report Verlag, 2008)
- Woodward, David R.: *Field Marshal Sir William Robertson*. (London: Praeger, 1998)
- Yokell, Matthew A.: *"Sold to the Highest Bidder? An Investigation of the Diplomacy Regarding Bulgaria's Entry into World War I"*, MA-Thesis (Richmond, 2010)

Endnotes

1. Romania had been a secret member of the triumvirate since 1886.
2. Matthew A. Yokell, "Sold to the Highest Bidder? An Investigation of the Diplomacy Regarding Bulgaria's Entry into World War I", MA-Thesis (Richmond, 2010), p. 42.
3. Stavrianos, *op. cit.*, p. 560.
4. Yokell, *op. cit*, p. 45f.
5. *Ibidem*, p. 47f.
6. *Ibidem*, pp. 47-49.
7. *Ibidem*, p. 50f.
8. *Ibidem*, p. 57.
9. *Ibidem*, p. 65.
10. Winston Churchill, *The World Crisis 1911–1918*, Vol. 1. (London: Odhams Press, 1938), pp. 600–602.
11. Yokell, *op. cit.*, pp. 67–81.
12. *Ibidem*, p. 82f.
13. *Ibidem*, pp. 84–86.
14. http://en.wikipedia.org/wiki/Bulgaria_during_World_War_I#.22The_Bulgarian_Summer.22_of_1915. [Stand vom April 2014].
15. Stavrianos, *op. cit.*, p. 561.
16. V. N. Vinogradov, "Romania in the First World War: The Years of Neutrality, 1914–1916" *The International History Review*, 14:3 (August 1992), p. 453. The description follows this, where it is not noted otherwise.
17. Geniko Epiteleio Stratou - Hellenic Army General Staff, *A Concise History of the Participation of the Hellenic Army in the First World War 1914–1918* (Athens: Army History Directorate, 1999), p. 7f.
18. George Frederick Abbott, *Greece and the Allies 1914–1922* (Hamburg: Tredition, 2013), p. 27.
19. American Hellenic Society, *The Greek White Book Diplomatic Documents 1913-1917* (New York: Oxford UP, 1919), p. 43; Document No. 12.. Ab hier zitiert als *Greek White Book*.
20. *Ibidem*, p. 45f; Documents Nos. 14 & 15.
21. *Ibidem*, p. 48; Document. No. 17.
22. Abott, *op.cit.*, pp. 19-22. Venizelos told the American war correspondent Hibben that *"the Greco-Serbian treaty foresaw only the possibility of a Balkan-War."* Hibben, *op. cit.*, p. 17.
23. Karl Kautsky (ed.), *Die deutschen Dokumente zum Kriegsausbruch* Band II (Charlottenburg: Deutsche Verlagsgesellschaft, 1919), p. 186f, Dokument Nr. 466 und Band III, p. 21, Dokument Nr. 504.
24. *Ibidem*, III, p. 163f.
25. *Ibidem*, p. 164.
26. *Greek White Book*,, Document No. 19: *"The Mediterranean is at the mercy of the united fleets of England and France. They would destroy our fleet and our merchant marine, occupy our islands [...] I am of opinion that neutrality is imposed upon us."*
27. Ernst v. Falkenhausen, *Erdrosselung Griechenlands* (Berlin: Ullstein, 1918), p. 20f; Hibben, *op. cit.*, p. 16.
28. Falkenhausen, *op. cit.*, p. 23; Carl Mühlmann, "Der Eintritt Griechenlands in den Weltkrieg Sommer 1917", *Berliner Monatshefte* (1937), p. 517.
29. Abbott, *op. cit,,* p. 16f.

30 *Ibidem*, p. 17.
31 Hibben, *op. cit.*, p. 107.
32 Abbott, *op.cit.*, pp. 23–24; Cosmetatos, *op. cit.*, p. 7.
33 *Ibidem*, p. 24f.
34 Stavrianos, *op. cit.*, p. 566.
35 Martin Gilbert, *Winston Churchill*, III, *1914–1916* (London: Heineman, 1971), p. 200f.
36 Churchill, *op. cit.*, I, p. 440f.
37 Churchill, *op. cit.*, I, p. 441. David Lloyd George, *War Memoirs*, Vol. I (London: Odham, 1936), p. "*When the Greeks offered to join the Allies [...] they were prepared to send an adequate contingent to occupy the Gallipoli Peninsula. Had they done so the whole story of the Dardanelles would have been different [...] But for some inscrutible reason Sir Edward Grey rejected Greek overtures.*"
38 Abbott, *op. cit.*, pp. 25–27.
39 Cosmetatos, *op. cit.* p. 8.
40 Abbott, *op. cit.*, p. 33.
41 Gabriel Deville, *L'Entente, la Grèce et la Bulgarie. Notes d'histoire et souvenirs* (Paris: Eugène Figuière, 1919), p. 136f; Cosmetatos, *op. cit.*, p. 8f.
42 Greek General Staff, *op. cit.*, p. 19.
43 Die Darstellung folgt Clews, *op. cit*, p. 137f.
44 Graham T. Clews, *Churchill's Dilemma. The Real Story Behind the Origins of the Dardanelles Campaign* (Santa Barbara: Praeger, 2010), p. 99.
45 Hellenic Army General Staff, *A Concise History*, p. 19f.
46 Abbott, *op. cit.*, p. 34f; Georgios Ventiris, *Η Ελλάς του 1910–1920* (Athen: Ikaros, 1970), p. 270.
47 zu der Schreibweise: *Dojran* is slavic, and *Doiran* is Greek.
48 *Ibidem*, p. 35f.
49 *Ibidem*, p. 36f.
50 Michael Llewellyn Smith, *Ionian Vision. Greece in Asia Minor 1919–1922* (London: Allen Lane, 1973), p. 54.
51 Abbott, *op. cit.*, p. 37; Joachim Joachim, *Ioannis Metaxas. The Formative Years 1871–1922* (Mannheim: Bibliopolis, 2000), p. 193.
52 Hellenic Army General Staff, *A Concise History*, p. 20.
53 Abbott, *op. cit.*, p. 37.
54 Hellenic Army General Staff, *A Concise History*, p. 21.
55 Abbott, *op. cit.*, p. 38.
56 Hellenic Army General Staff, *A Concise History*, p. 25.
57 Abbott, *op. cit.*, p. 39.
58 Churchill, *op. cit.*, I, p. 619.
59 *Ibidem*, p. 620.
60 *Ibidem*, p. 622; Gilbert, *op. cit*, III, p. 328f.
61 Abbott, *op. cit.*, p. 40.
62 Hellenic Army General Staff, *A Concise History*, p. 25.
63 Hellenic Army General Staff, *A Concise History*, p. 27; Abbott, *op. cit.*, p. 40f.
64 Cosmetatos, *op. cit.*, p. 36.
65 Abbott, *op. cit.*, p. 44.
66 *Ibidem*, p. 45.
67 Army General Staff, *A Concise History*, p. 37; Abbott, *op. cit.*, p. 46; Cosmetatos, *op. cit.*, p. 23.
68 Abbott, *op. cit.*, p. 47; Cosmetatos, *op. cit.*, p. 23.
69 Abbott, *op. cit.*, p. 48f.
70 Cosmetatos, *op. cit.*, p. 31.
71 *Ibidem*, p. 36f.
72 Abbott, *op. cit.* p. 51.
73 Hellenic Army General Staff, *A Concise History*, p. 37; Abbott, *op. cit.*, p. 52f.
74 Abbott, *op. cit.*, p. 63.
75 *Ibidem*, p. 64.
76 Abbott, *op. cit.*, pp. 64–66; This is at the same time the beginning of the accusation that is always heard later, of the Entente propaganda that Constantine rules like an absolute monarch and doesn't care about

the constitution.
77 Hellenic Army General Staff, *A Concise History*, p. 38; Abbott, *op. cit.*, p. 66;
78 Cosmetatos, *op. cit.*, p. 50.
79 Abbott, *op. cit.*, p. 66.
80 Cosmetatos, *op. cit.*, p. 50.
81 Abbott, *op. cit*, p. 66.
82 Cosmetatos, *op. cit.*, p. 51.
83 Article 99 says that without an express decision in Parliament, no foreign army may cross or be stationed on Greek territory. *Cosemtatos op.cit.*, p. 50.
84 *Ibidem*, p. 51.
85 *Journal Officiel* (28 Octobre 1919), p. 77 zitiert n. Cosmetatos, *op. cit.*, p. 51.
86 Cosmetatos, *op. cit.*, p. 52.
87 Hellenic Army General Staff, *A Concise History*, p. 41.
88 Abbott, *op. cit.*, p. 66f.
89 Cyril Falls, *Military Operations Macedonia from the Outbreak of the War to the Spring of 1917* (London: Imperial War Museum, 1932), p. 39.
90 *Ibidem*, p. 39f.
91 Hellenic Army General Staff, *A Concise History*, p. 41.
92 Abbott, *op. cit.*, p. 68.
93 Falls, *op. cit.*, p. 39f; Abbott, *op. cit.*, p. 67.
94 Hellenic Army General Staff, *A Concise History*, p. 43.
95 Cosmetatos, *op. cit.*, p. 52.
96 Text in Abbott, *op. cit.*, pp. 68-70.
97 *Ibidem*, p. 68. A slightly different version in Cosmetatos, *op. cit.*, p. 52f.
98 Falls, *op. cit.*, p. 41.
99 Hellenic Army General Staff, *A Concise History*, p. 43f.
100 Cosmetatos, *op. cit.*, p. 53f.
101 Abbott, *op. cit.*, p. 71.
102 Joachim, *op. cit.*, p. 223.
103 Cosmetatos, *op. cit.*, p. 54.
104 Joachim, *op. cit.*, p. 223.
105 Cosmetatos, *op. cit.*, p. 54f.
106 *Ibidem*, p. 58.
107 *Ibidem*, p. 56.
108 Abbott, op. cit., p. 73; Falls, op. cit., p. 41.
109 Paul Maurice Emmanuel Sarrail, *06.04.1856 – † 23.03.1929
110 George B. Leon, Greece and the Great Powers 1914–1917 (Thessaloniki: IMXA, 1974), p. 264f; Hibben, op. cit., p. 15: "General Sarrail felt he was being sent to Saloniki to get him out of France."
111 Falls, op. cit., p. 37f.
112 Ibidem, p. 42f.
113 Leon, op. cit., p. 259.
114 Ibidem, p. 260; Lloyd George, op. cit. , I, p. 294f.
115 Ibidem, p. 261f; Lloyd George, op. cit. , I, p. 296f.
116 Cosmetatos, op. cit., p. 78.
117 Leon, op. cit., p. 266f.
118 Falls, op. cit., p. 43f.
119 Ibidem, p. 44.
120 Ibidem, p. 45f.
121 Ibidem, p. 46.
122 Hibben, op. cit., p. 17.
123 Abbott, op. cit., p. 79f.
124 Hellenic Army General Staff, A Concise History, p. 51; Abbott, op. cit., p. 77f.
125 Leon, op. cit. , p. 249.
126 Ibidem, p. 251.
127 Hellenic Army General Staff, A Concise History, p. 52; Hibben, op. cit., p. 19.

128 Ibidem.
129 Falls, op. cit., p. 46f.
130 Hibben, op. cit., p. 20.
131 Hellenic Army General Staff, A Concise History, p. 52.
132 Hibben, op. cit., p. 20.
133 Ibidem.
134 Falls, op. cit., p. 47; during the subsequent discussion with Skouloudis, he gave Kitchener a similar assurance. Joachim, op. it., p. 233
135 Hibben, op. it.p. 20.
136 Reichsarchiv, Weltkrieg, IX, p. 300.
137 Joachim, op. cit., p. 234.
138 Hellenic Army General Staff, A Concise History, p. 52.
139 Falls, op. cit., p. 47f; Hibben, op. cit., p. 21; Falkenhausen, op. cit., p. 64. After him Skouloudis is alledged to have commented "These are strengths!"
140 Hellenic Army General Staff, A Concise History, p. 53f.
141 Lloyd George, op. cit., I, p. 314.
142 Leon, op. cit., p. 300; Falls, op. cit., p. 48f
143 Keegan, op. cit., p. 254f.
144 Leon, op. cit., p. 302.
145 Abbott, op. cit., p. 88; Leon, op. cit., p. 303.
146 Text des Statements in Falls, op. cit., p. 63.
147 Lloyd George, op. cit., I, p. 315.
148 Falls, op. cit, 49f.
149 Hibben, op. cit., p. 21.
150 Leon, op. cit., p. 309f.
151 Hibben, op. cit., p. 23.
152 Ibidem.
153 Ibidem, p. 29.
154 Reichsarchiv, Weltkrieg, IX, p. 196f.
155 Ibidem, pp. 197-200.
156 Ibidem, pp. 207f, 231.
157 Österreichisches Bundesministerium für Heerwesen (ed.), Österreich-Ungarns letzter Krieg, Band 3, 2. Teil: Das Kriegsjahr 1915. Von der Einnahme von Brest-Litowsk bis zu Jahreswende (Wien: Militärwissenschaftlichen Mitteilungen, 1932), p. 3.
158 Reichsarchiv, Weltkrieg, IX, p. 287.
159 Ibidem, p. 231f
160 Ibidem, p. 286f.
161 Ibidem, p. 256.
162 Ibidem, p. 292f.
163 Ibidem, p. 299f.
164 Ibidem, p. 304.
165 Ibidem, p. 305.
166 Ibidem, p. 321.
167 Leon, op. cit., p.257.
168 Falls, op. cit., pp. 50–52.
169 Ibidem, p. 52.
170 Ibidem, p. 53.
171 Ibidem, pp. 53–55.
172 Falls, op. cit, pp. 56–64; Reichsarchiv, Weltkrieg, IX, p 285.
173 Abbott, op. cit., p. 107f; Hellenic Army General Staff, A Concise History, p. 55f.
174 Ibidem, p. 108.
175 Falls, op. cit., pp. 65–81.
176 Ibidem, p. 82.
177 Hibben, op. cit., p. 21.
178 Cosmetatos, op. cit., p. 68.

179 Falls, op. cit., p.85; Hellenic Army General Staff, A Concise History, p. 55.
180 Alan Palmer, *The Gardeners of Salonika. The Macedonian Campaign 1915–1918* (London: Faber & Faber 2009), p. 92.
181 Rémy Porte, "Commen faire plier un neutre? L'action politique et militaire de la France en Grèce (1915–1917)" Cahiers de la Méditerranée 81 (2010), p. 4f. The online version is quoted http://cdlm.revues.org/5461.
182 Admiral Louis Dartige du Fournet, *02.03.1856 - † 16.02.1940, evacuated more than 4,000 threatened Armenians on his ships and saved their lives from the Turks.
183 Dartige du Fournet, op. cit., p. 115.
184 Abbott, op. cit., p. 97f; Leon, op.vcit., p. 315f.
185 Falkenhausen, op. cit., p. 73ff.
186 Hellenic Army General Staff, A Concise History, p. 60.
187 Abbott, op. cit., p. 97f; http://en.wikipedia.org/wiki/Kastellorizo#History.
188 Hellenic Army General Staff, A Concise History, p. 57.
189 Cosmetatos, op. cit., p. 118f. 5 Ibidem, p. 119.
190 Ibidem, p. 119
191 Dartige du Fournet, op. cit., p. 304.
192 Cosmettatos, op. cit., p. 144
193 Falls, op. cit., p. 94, Leon, op. cit., p. 316; Sarrail, op. cit., p. 94f
194 Hibben, op. cit., p. 42.
195 Alan Palmer, *The Gardeners of Salonika* (London: Andre Deutsch, 1965). p. 55.
196 Falls, op. cit., p. 100; Leon, op. cit., p. 316f; Palmer, op. cit., p. 55f
197 Leon, op. cit., p. 317.
198 Ibidem, p. 318f.
199 Ibidem, p. 320f.
200 Ibidem, p. 321.
201 Ibidem, p. 324f.
202 Ibidem, p. 325f
203 Abbott, op. cit., p. 99f.
204 Ibidem, p. 100f.
205 Ibidem, p. 98f; Cosmetatos, op. cit., p. 159f.
206 Hellenic Army General Staff, A Concise History, p. 61; Falles, op. cit., p. 120.
207 Falls, op. cit., p. 99.
208 Maurice Sarrail, *Mon commandement en Orient (1916–1918)* (Paris: Flammarion, 1920), p. 74.
209 Hibben, op. cit., p. 29.
210 Hellenic Army General Staff, A Concise History, p. 57; Palmer, op. cit., p. 54.
211 Leon, op. cit., p. 314.
212 Falls, op. cit, p. 99.
213 Abbott, op. cit., pp. 98, 105f.
214 Richard Kralik, Geschichte des Völkerkrieges (1914–1918) (Graz: Styria,, 1923), p. 262; this is almost exactly Hibben's acount, word for word, op. cit., p. 32, zitierten Aussage.
215 Hibben, op. cit., p. 32: "I have tried every way I know how to get fair play from the British and French press and a fair hearing by the British and French public. No sooner has a British newspaper attacked Greece with the most amazing perversions of fact and misrepresentations of motives than I have called its correspondent and given him face to face a full statement of Greece's position. I have given the frankest statement to the French press through one of the newspapers most bitterly attacking Greece. Its publication was not permitted by the French censor."
216 Falkenhausen, op. cit., p. 77; Hibben, op. cit., p. 32.
217 Hibben, op. cit., p. 33.
218 Ibidem, p. 34.
219 Leon, op. cit., p. 362.
220 Leon, op. cit., p. 363.
221 Sarrail, op. cit., pp. 107, 153, 354–356.
222 Hibben, op. cit., p. 25.
223 Ventiris, op. cit., II, p. 211.

224 Neokosmos Grigoriadis, I ethniki amyna Thessalonikis tou 1916 (Athen, 1960), p. 18.
225 Thanos Veremis & Helen Gardikas-Katsiadakis, "Protagonist in Politics, 1912–20", in: Paschalis Kitromilides (.), Eleftherios Venizelos. The Trials of Statemenship (Edinburgh: UP, 2006),p. 123. Die Bibliographie dieses Aufsatzes besteht nur aus griechischen und englischen Titeln.
226 Unterdrückung und Zwang
227 Falls, op. cit., p. 130.
228 Hibben, op. cit., p. 42.
229 Hellenic Army General Staff, A Concise History, p. 68. In a part of the French press the imposition of the siege of a part of a neutral state was described as:"La proclamation de l'état de siège ne constitue pas une brutale mainmise sur l'autorité grecque et ne lèse pas les droits souverains de l'État hellénique, mais elle institue un contrôle rigoureux et efficace des ports, des douanes et des communications." Porte, op. cit., p. 9,
230 Palmer, op. cit. p. 67; Sarrail, op. cit., p. 108; Hibben, op. cit., p. 45: The Venizelist Zymprakakis became new police chief.
231 Palmer, op. cit,, p. 68, Leon, op. cit., pp. 367f–371.
232 Ibidem, p. 68f.
233 Leon, op. cit., p. 374.
234 Leon, op. cit., p. 364.
235 Falkenhausen, op. cit., p. 87f.
236 Ibidem, p. 69.
237 Stavrianos, op. cit., p. 567f.
238 Leon, op. cit., p. 70.
239 Ibidem.
240 Hellenic Army General Staff, A Concise History, p. 75.
241 Abbott, op. cit., p. 121.
242 Hellenic Army General Staff, A Concise History, p. 70
243 Ibidem, p. 76.
244 Hibben, op. cit., p. 27.
245 Palmer, op. cit., p. 94.
246 Hellenic Army General Staff, A Concise History, p. 76.
247 Palmer, op. cit, pp. 77–79.
248 Abbott, op. cit., p. 122.
249 Leon, op. cit., p. 375f. In the allied press the formation of the Lega was portrayed as work of the Germans. Falkenhausen, op. cit., p. 91f.
250 Palmer, op. cit., p. 97.
251 Ibidem.
252 Falls, op. cit., p.85; Hellenic Army General Staff, A Concise History, p. 55.
253 Alan Palmer, The Gardeners of Salonika. The Macedonian Campaign 1915–1918 (London: Faber & Faber 2009), p. 92.
254 Rémy Porte, "Commen faire plier un neutre? L'action politique et militaire de la France en Grèce (1915–1917)" Cahiers de la Méditerranée 81 (2010), p. 4f. The online version is quoted http://cdlm.revues.org/5461.
255 Admiral Louis Dartige du Fournet, *02.03.1856 – † 16.02.1940, evacuated more than 4,000 threatened Armenians on his ships and saved their lives from the Turks.
256 Dartige du Fournet, op. cit., p. 115.
257 Abbott, op. cit., p. 97f; Leon, op.vcit., p. 315f.
258 Falkenhausen, op. cit., p. 73ff.
259 Hellenic Army General Staff, A Concise History, p. 60.
260 Abbott, op. cit., p. 97f; http://en.wikipedia.org/wiki/Kastellorizo#History.
261 Hellenic Army General Staff, A Concise History, p. 57.
262 Cosmetatos, op. cit., p. 118f. 5 Ibidem, p. 119.
263 Ibidem, p. 119
264 Dartige du Fournet, op. cit., p. 304.
265 Cosmettatos, op. cit., p. 144
266 Falls, op. cit., p. 94, Leon, op. cit., p. 316; Sarrail, op. cit., p. 94f

267 Hibben, op. cit., p. 42.
268 Alan Palmer, *The Gardeners of Salonika* (London: Andre Deutsch, 1965). p. 55.
269 Falls, op. cit., p. 100; Leon, op. cit., p. 316f; Palmer, op. cit., p. 55f
270 Leon, op. cit., p. 317.
271 Ibidem, p. 318f.
272 Ibidem, p. 320f.
273 Ibidem, p. 321.
274 Ibidem, p. 324f.
275 Ibidem, p. 325f
276 Abbott, op. cit., p. 99f.
277 Ibidem, p. 100f.
278 Ibidem, p. 98f; Cosmetatos, op. cit., p. 159f.
279 Hellenic Army General Staff, A Concise History, p. 61; Falles, op. cit., p. 120.
280 Falls, op. cit., p. 99.
281 Maurice Sarrail, Mon commandement en Orient (1916-1918) (Paris: Flammarion, 1920), p. 74.
282 Hibben, op. cit., p. 29.
283 Hellenic Army General Staff, A Concise History, p. 57; Palmer, op. cit., p. 54.
284 Leon, op. cit., p. 314.
285 Falls, op. cit, p. 99.
286 Abbott, op. cit., pp. 98, 105f.
287 Richard Kralik, *Geschichte des Völkerkrieges (1914–1918)* (Graz: Styria,, 1923), p. 262; this is almost exactly Hibben's acount, word for word, op. cit., p. 32, zitierten Aussage.
288 Hibben, op. cit., p. 32: "I have tried every way I know how to get fair play from the British and French press and a fair hearing by the British and French public. No sooner has a British newspaper attacked Greece with the most amazing perversions of fact and misrepresentations of motives than I have called its correspondent and given him face to face a full statement of Greece's position. I have given the frankest statement to the French press through one of the newspapers most bitterly attacking Greece. Its publication was not permitted by the French censor."
289 Falkenhausen, op. cit., p. 77; Hibben, op. cit., p. 32.
290 Hibben, op. cit., p. 33.
291 Ibidem, p. 34.
292 Leon, op. cit., p. 362.
293 Leon, op. cit., p. 363.
294 Sarrail, op. cit., pp. 107, 153, 354–356.
295 Hibben, op. cit., p. 25.
296 Ventiris, op. cit., II, p. 211.
297 Neokosmos Grigoriadis, I ethniki amyna Thessalonikis tou 1916 (Athen, 1960), p. 18.
298 Thanos Veremis & Helen Gardikas-Katsiadakis, "Protagonist in Politics, 1912-20", in: Paschalis Kitromilides (.), Eleftherios Venizelos. The Trials of Statemenship (Edinburgh: UP, 2006),p. 123. Die Bibliographie dieses Aufsatzes besteht nur aus griechischen und englischen Titeln.
299 Unterdrückung und Zwang
300 Falls, op. cit., p. 130.
301 Hibben, op. cit., p. 42.
302 Hellenic Army General Staff, A Concise History, p. 68. In a part of the French press the imposition of the siege of a part of a neutral state was described as:"La proclamation de l'état de siège ne constitue pas une brutale mainmise sur l'autorité grecque et ne lèse pas les droits souverains de l'État hellénique, mais elle institue un contrôle rigoureux et efficace des ports, des douanes et des communications." Porte, op. cit., p. 9,
303 Palmer, op. cit. p. 67; Sarrail, op. cit., p. 108; Hibben, op. cit., p. 45: The Venizelist Zymprakakis became new police chief.
304 Palmer, op. cit,, p. 68, Leon, op. cit., pp. 367f–371.
305 Ibidem, p. 68f.
306 Leon, op. cit., p. 374.
307 Leon, op. cit., p. 364.
308 Falkenhausen, op. cit., p. 87f.

309 Ibidem, p. 69.
310 Stavrianos, op. cit., p. 567f.
311 Leon, op. cit., p. 70.
312 Ibidem.
313 Hellenic Army General Staff, A Concise History, p. 75.
314 Abbott, op. cit., p. 121.
315 Hellenic Army General Staff, A Concise History, p. 70
316 Ibidem, p. 76.
317 Hibben, op. cit., p. 27.
318 Palmer, op. cit., p. 94.
319 Hellenic Army General Staff, A Concise History, p. 76.
320 Palmer, op. cit, pp. 77-79.
321 Abbott, op. cit., p. 122.
322 Leon, op. cit., p. 375f. In the allied press the formation of the Lega was portrayed as work of the Germans. Falkenhausen, op. cit., p. 91f.
323 Palmer, op. cit., p. 97.
324 Ibidem.
325 Ibidem, p. 51.
326 Hellenic Army General Staff, A Concise History, p. 55.
327 Ibidem.
328 Falls, op. cit., pp. 85–88.
329 Keegan, op. cit., p. 255f; Palmer, op. cit., p. 62.
330 Lloyd George, op. cit., I, p. 317f.
331 Hellenic Army General Staff, A Concise History, p. 62.
332 American Hellenic Society, The Greek White Book Supplementary Diplomatic Documents1913–1917 (New York: Oxford UP, 1919), p. 63
333 Hellenic Army General Staff, A Concise History,, p. 63; Abbott, op. cit., p. 109.
334 The American Hellenic Society, The Greek White Book Diplomatic Documents 1913–1917 (New York: Oxford UP, 1918), pp. 76–79; from here quoted as Greek White Book I; Hellenic Army General Staff, A Concise History, p. 64; Abbott, op. cit., p. 119.
335 Greek White Book I, pp. 79–81.
336 Vasil Radoslawoff, Bulgarien und die Weltkrise (Berlin: Ullstein, 1923), p. 197.
337 Hellenic Army General Staff, A Concise History, p. 65.
338 Ibidem, p. 66f.
339 Falls, op. cit., p. 124.
340 Leon, op. cit., p. 361.
341 Cosmetatos, op. cit., p. 177f.
342 Palmer, op. cit., p. 62, Falls, op. cit., 116f
343 Palmer, op. cit., p. 319f; Falls, op. cit., p. 136.
344 Leon. op. cit., p. 365.
345 Palmer, op. cit., p. 64f; Falls, op. cit., p. 115.
346 Palmer, op. cit., p. 67.
347 Ibidem, p. 69.
348 Ibidem.
349 Ibidem, p. 70
350 Cyril Falls, *The Great War* (New York: Putnam, 1959), p. 139.
351 Hellenic Army General Staff, A Concise History, p. 80.
352 Ibidem, pp. 81–85
353 American Hellenic Society, The Greek White Book Supplementary Diplomatic Documents 1913–1917 (New York: Oxford UP, 1919), p. 41f. The *Komitatschis* had a Bulgarian and Turkish minority in West-Macedonia
354 Hellenic Army General Staff, A Concise History, p. 86.
355 Leon, op. cit., p. 383f
356 Keegan, op. cit., p. 306.
357 Falls, op. cit., p. 139f.

358 Winston Churchill, The World Crisis 1911–1918 Vol. I (London: Odhams Press, 1938), p. 1098f.
359 Keegan, op. cit., p. 307f.
360 http://de.wikipedia.org/wiki/Erich_von_Falkenhayn#F.C3.BChrungsschw.C3.A4chen_an_der_Ostfront.
361 Hellenic Army General Staff, A Concise History, p. 86.
362 Ibidem, p. 87.
363 Ibidem: Leon, op. cit., p. 399.
364 Ibidem, p. 87.
365 Ibidem, 88; Abbott, op. cit., p. 130f.
366 Abbott, op. cit., p. 131.
367 Hellenic Army General Staff, A Concise History, p. 88; Abbott, op. cit., p. 131.
368 Hellenic Army General Staff, A Concise History, p. 89; Abbott, op. cit., p. 131.
369 Hellenic Army General Staff, A Concise History, p. 89f.
370 Klaus-Dieter Tietz, „Griechen in Görlitz" Hellenika N. F. 5 (2010), p. 60.
371 Abbott, op. cit., p. 131f; Cosmetatos, op. cit., p. 204 explains that the French press published the number 40,000 and the transport to Görlitz represented a new treason of Greece.
372 Ibidem, p. 132.
373 Leon, op. cit., p. 401.
374 American Hellenic Society, Report of the Greek University Commission upon the Atrocities and Devastations Committed by the Bulgarians in Eastern Macedonia (New York: Oxford UP, 1919)
375 Tietz, op. cit., p. 63f.
376 Ibidem, p. 64f
377 Elisabeth Logothetopoulou said this in discussions with the author. The author got to know some of these women, who stayed in Greece, in 1967/68
378 Ibidem, p. 65f.
379 Ibidem, p. 66.
380 Gerassimos Alexatos, "Xairete: Ein griechisches Armeekorps in Görlitz." in: Wolfgang Schultheiß (ed.), Meilensteine deutsch-griechischer Beziehungen (Athen: Stiftung für Parlamentarismus und Demokratie des Hellenischen Parlamentes, 2010), p. 192.
381 Tietz, op. cit., p. 67.
382 Ibidem, p. 68f.
383 Leon, op. cit., p. 383.
384 Falls, op. cit., p. 211.
385 Leon, op. cit,, p. 390f; Falls, op. cit., p. 213.
386 Ibidem, p. 391f.
387 Hibben, op. cit., p. 53.
388 Hellenic Army General Staff, A Concise History, p. 106. Four German and three Austrian merchant ships. The crews were taken prisoner and locked up on French ships. Hibben, op. cit., p. 54.
389 Hibben, op. cit., p. 53.
390 Palmer, op. cit., p. 101.
391 Leon, op. cit, p. 392f.
392 Hibben, op. cit., p. 54: "A campaign in France and England had long been in progress against [...] Schenck, the head of the German propaganda in Greece. Rather an insignificant figure in fact, the baron had been raised to a pinnacle of diabolical cunning and almost superhuman influence by the more sensational British and French newspapers."
393 Leon, op. cit., p. 393f.
394 Ibidem, p. 394; Hibben, op. cit., p. 57.
395 Hellenic Army General Staff, A Concise History, p. 106. This was even confirmed by a court. Hibben, op. cit., p. 57; Cosmetatos, op. cit., p. 210.
396 Falkenhausen, op. cit., p. 110.
397 Hibben, op. cit., p. 57.
398 Leon, op. cit, p. 395.
399 Hibben, op. cit., p. 58.
400 Ibidem, p. 59.
401 Hellenic Army General Staff, A Concise History, p. 107.

402   Abbott, op. cit., p. 135f.
403   Hibben, op. cit., p. 59f.
404   Ibidem, p. 60.
405   Abbott, op. cit., p. 136.
406   Leon, op. cit., p. 404. Es handelte sich um Lysander Kaftanzoglou, Dimitrios Voktopoulos und Loukas Roufos. Sie waren gegen einen Kriegseintritt. Hibben, op. cit., p. 59.
407   Abbott, op. cit., p. 137.
408   Ibidem.
409   Falls, op. cit., p. 214f; Abbott, op. cit., p. 137; Constantine fully supported this suggestion. Hibben, op. cit, p. 64.
410   Falls, op. cit., p. 216; Hibben, op. cit., p. 65 Politis "had striven without avail to persuade Sir Francis Elliot that the Government was honestly in favor of leaving neutrality on the side of the Entente. [...] He had done all he could to convince the Allied diplomatists that they were committing the greatest political blunder of the war in boycotting the Calogueropoulos cabinet. When Sir Francis and M. Guilemin remained obdurate, he finally gave up his attempt and went to Saloniki, where he joined Venizelos as minister for foreign affairs."
411   Leon, op. cit., p. 403f.
412   Ibidem, p. 404.
413   Hibben, op. cit., p. 64.
414   Ibidem.
415   Ibidem, p. 61.
416   Leon, op. cit., p. 405f.
417   Ibidem, p. 407f.
418   Hibben, op. cit., p. 66.
419   Abbott, op. cit., p. 139; Details about the conspiracy on Crete in:: John Cuthbert Lawson, Tales of Aegean Intrigue (London: Chatto & Windus, 1920), passim.
420   Hibben, op. cit., p. 61.
421   Ibidem, p. 62.
422   Palmer, op. cit., p. 97f.
423   Leon, op. cit., p. 409.
424   Palmer, op. cit., p. 98.
425   Abbott, op. cit., p. 141; Leon, op. cit., p. 409; Palmer, op. cit., p. 98.
426   Leon, op. cit., p. 410f.
427   Ibidem, p. 411f.
428   Leon, op. cit., p. 413; Palmer, op. cit., p. 101.
429   Leon, op. cit., p. 414; Hellenic Army General Staff, A Concise History, p. 107; Palmer, op. cit., p. 102.
430   Abbott, op. cit., p. 152f.
431   Leon, op. cit p. 414f; Hellenic Army General Staff, A Concise History, p. 108.
432   Abbott, op. cit., p. 154.
433   Gardener, op. cit., p. 102.
434   Hellenic Army General Staff, A Concise History, p. 108, Leon, op. cit., p. 421.
435   Leon, op. cit., p. 415.
436   Falls, op. cit., p. 219.
437   Leon, op. cit., p. 422.
438   Ibidem, p. 422f.
439   Hellenic Army General Staff, A Concise History, p. 108f; Palmer, op. cit., p. 102.
440   Hibben, op. cit., p. 71f.
441   Hellenic Army General Staff, A Concise History, p. 109f.
442   Ibidem, p. 110.
443   Ibidem.
444   Ibidem, p. 111.
445   Hibben, op. cit., p. 74 gives the Details: "Sixteen batteries of field artillery, with 1000 rounds of ammunition for each gun; 16 that is, 6 in addition to the 10 already mentioned batteries of mountain artillery, with 100 rounds for each gun; 40,000 Mannlicher rifles, with 8,800,000 rounds of rifle ammunition; 140 machine guns, with a proportionate quantity of ammunition; and 50 military trucks.

Save in the matter of the machine guns and rifles, this was virtually the entire available equipment of the Hellenic Army."

446 Ibidem.
447 Ibidem, p. 75.
448 Ibidem, p. 76.
449 Ibidem, p. 77.
450 Hellenic Army General Staff, A Concise History, p. 112; Palmer, op. cit., p. 102; Falls, op. cit., p. 222; bei Falkenhausen, op. cit., pp. 133-140 findet sich eine ausführliche Schilderung.
451 Ibidem, p. 112f.
452 Hibben, op. cit., p. 79; *Le Temps* had listed all of Constantine's "sins" in detail on the first page, but not one of them corresponded to the truth.
453 Robert David, Le drame ignoré de l'Armée d'Orient Dardanelles - Serbie - Salonique - Athènes (Paris: Plon, 1927), p. 173. Ähnlich propagandistisch verfälscht ist die Darstellung bei: Auguste Gauvain, L'affaire Grecque (Paris: Bossard, 1918), p. 151. Unter anderem schreibt er: "Aussitôt les royalistes postés sur les emplacements préparés se mirent à tirer à coup de mitrailleuses non seulment sur les détachements alliés, mais encore sur les Français cantonné au Zappeion et sur l'annexe de la légation d'Angleterre, centre de la police anglo-française."
454 Hibben, op. cit., p. 80f; Hellenic Army General Staff, A Concise History, p. 113.
455 Ibidem, p. 114
456 Ibidem, p. 115
457 Cosmetatos, op.vcit., p. 254.
458 Hibben, op. cit., p. 81f.
459 Leon, op.vcit., p. 435.
460 Hibben, op. cit., p. 82.
461 Ibidem.
462 Hellenic Army General Staff, A Concise History, pp. 115-119. Where not otherwise noted, the presentation is from this source and Abbott, op. cit., pp. 169-171, as well as Hibben, op. cit., p. 84f.
463 Cosmetatos, op. cit., p. 257f.
464 Abbott, op. cit.,p.171; du Founet, p. 210; Compton Mackenzie, Greece in My Life (London: Chatto & Windus, 1960), p. 73.
465 Abbott, op. cit., p. 171; Hibben, op. cit., p. 87 nennt die Details: "*Allein in Venizelos Haus wurden 66 Gewehre, 6000 Schuss Munition, 49 Revolver mit Patronen, 2500 Dynamit-Sprengkapseln mit 36,5 Meter Zündschnur und 15 Handgranaten gefunden.*"
466 Gauvain, op. cit., p. 155 schreibt: "Les venizélistes notoires d'Athènes et du Perée furent massacrés, tortués, emprisonnés. On pilla leurs maison de fond en comble. On détruisit les bureaux et les imprimeries des journaux libéraux ..."
467 Ibidem.
468 Leon, op. cit., p. 437.
469 Hibben, op. cit., p. 96.
470 Abbott, op. cit., p. 178f.
471 Hibben, op. cit., p. 107f. "*Upon publication of a copy in Athens, Mr. Venizelos declared the above letter as a fake. I presented the letter to a dozen of his closest friends, including men who worked for him as his Secretary, and who were familiar with his handwriting. All of them emphasized that the veracity was unquestioned.*"
472 Ibidem, p. 89.
473 Ibidem, p. 89f.
474 Ibidem, p. 91.
475 Falls, op. cit.,p. 224.
476 Dartige du Fournet, op. cit., p. 231, 245.
477 Falls, op. cit., p. 223; Porte, op. cit., p. 10.
478 Palmer, op. cit., p. 106f.
479 Hibben, op. cit., p. 90.
480 Abbott, op. cit., p. 177.
481 Ibidem., p. 173; Hibben, op. cit., p. 91.
482 Hellenic Army General Staff, A Concise History, p. 120.

483 Cosmetatos, op. cit., pp. 259-262.
484 Hellenic Army General Staff, A Concise History, p. 120.
485 Ibidem, p. 120f.
486 Ibidem,, p. 121.
487 Hellenic Army General Staff, A Concise History, p. 102.
488 Ibidem, p. 419f.
489 Palmer, op. cit., pp. 92, 93f.
490 Leon, op. cit., p. 427.
491 Ibidem, 427-430.
492 Ibidem, p. 431f.
493 Hellenic Army General Staff, A Concise History, p. 102. The *Berliner Tageblatt* announced the declaration of war on November 27th 1916, without commenting on it, but with the following headline: *"War declaration of the venizelist government on Bulgaria and Germany"*. Neither on this day, nor later, did the newspaper publish a reaction of the German government.
494 The *Berliner Tageblatt* did not mention this new declaration of war with a single word.
495 Eyewitness account of this process with Hibben op. cit., p. 93.
496 Palmer, op. cit., p. 100. Falkenhausen writes about the coercive recruiting methods, op. cit., p. 107: "Since no one came, they moved to coersion. Thousands of militarily capable citizens of Saloniki and the Chalkidiki fled to Ancient Greece. In order to force them to return, they threw their relatives into jail. They torched the houses of those who had fled. In some villages of the Chalkidiki, those who refused shortly thereafter ended up in the gallows."
497 Hibben, op. cit., p. 104.
498 Hellenic Army General Staff, A Concise History, pp. 103–105.
499 Hibben, op. cit.,p. 62f.
500 Ibidem, p. 63.
501 Falls, op. cit., p. 225.
502 David R. Woodward, Field Marshal Sir William Robertson. (London: Praeger,1998), pp. 30-3, 66-7.
503 Falls, op. cit., pp. 226-230.
504 Hibben, op. cit., p. 96.
505 http://en.wikipedia.org/wiki/Autonomous_Republic_of_Northern_Epirus.
506 Ibidem.
507 Hellenic Army General Staff, A Concise History, p. 91f.
508 Ibidem, p. 92–94.
509 http://en.wikipedia.org/wiki/Northern_Epirus.
510 Hellenic Army General Staff, A Concise History, p. 71f
511 Ibidem, p. 72f.
512 Book by Haupt-Heydemarck: Hptm. von Blomberg, Januar 1917 Kommandeur der Flieger in Uesküb: Einweisung der Staffel: "Your squadron flies for the XX. Turkish Corps and for the 10th Bulgarian Division, headquarters in Drama with an airfield. Front section: from Orljak airport (British) to the mouth of the Struma = 60 km as the crow flies, as well as the coastal section from Tschajagzi to the mouth of the Mesta = 100 km, own forces: 3 reconnaissance aircraft and a combat single-seater [Lt. von Eschwege]."
513 Ibidem, p. 73f.
514 Ibidem, p. 97.
515 Falls, op. cit., pp. 152-169.
516 Reichsarchiv (ed.), Herbstschlacht in Macedonien Cernabogen 1916 (Oldenburg: Stalling, 1925), pp. 21–26.
517 Ibidem, pp. 22–26.
518 Reichskriegsministerium (ed.), Der Weltkrieg 1914-1918, Band XI, Die Kriegsführung im Herbst 1916 und im Winter 1916/17 (Berlin: Mittler, 1938), 338–343; Reichsarchiv (ed.), Cernabogen, pp. 27–33, 41–46.
519 Reichsarchiv (ed.), Cernabogen, pp. 73–109.
520 Palmer, op. cit., p. 81f; Sarrail, op. cit., p. 162 "Le Général Cordonnier [...] était à Boresnica qu'il croyait être Florina, et où il voulait faire une belle entrée. Mais Florina était occupé."
521 Sarrail, op. cit., p. 371.
522 Palmer, op. cit, p. 83f.
523 Ibidem, p. 84.

524 Ibidem.
525 Ibidem, p. 85.
526 Sarrail, op. cit., p. 168.
527 Reichsarchiv (ed.), Cernabogen, p. 112; Jean de Pierrfeu, Trois ans au grand quartier général (Paris: Éditions Françaises Illustrées, 1920), p. 198.
528 Palmer, op .cit., p. 85
529 Hellenic Army General Staff, A Concise History, p. 98.
530 Palmer, op. cit., p. 90.
531 "*Allied strength reached 670,000 in 1917, a third of the forces were British. During the three-year stay in Saloniki, nearly half a million Brits reported sick and only 18,187 were treated for wounds.*" Albert J. Fyfe, Understanding the First World War. Illusions and Realities (New York: Peter Lang, 1988), p. 103f.
532 Falls , op. cit., p.240.
533 Hellenic Army General Staff, A Concise History, p. 120; Leon, op. cit., p. 445.
534 Lloyd George, op. cit., I, p. 549f.
535 Ibidem, 550.
536 Ibidem, p. 821f
537 Ibidem, p. 559.
538 Palmer, op. cit., p. 108f.
539 Falls, op. cit., p. 254f.
540 Palmer, op. cit., p. 110.
541 Falls, op. cit., p. 254.
542 Falls, op. cit., p. 254f; Palmer, op. cit., p. 111.
543 Palmer, op. cit, p. 110.
544 Ibidem, p. 44.
545 Lloyd George, War Memoirs, III, (London: Nicholson & Watson, 1934), p. 1426f. chapter xlvii
546 Palmer, op. cit., p.112.
547 Lloyd George, War Memoirs, III, p. 1428.
548 Hellenic Army General Staff, A Concise History, p. 121f; Falls, op. cit., p. 231.
549 Leon, op. cit., p. 450.
550 Lloyd George, War Memoirs, III, p. 1446.
551 Abbott, op. cit., p. 180.
552 Ibidem, p. 181.
553 Joachim, op. cit., p. 260f.
554 Palmer, op. cit., p. 108f.
555 Palmer, op. cit., p. 108f.
556 Heinz Richter, *Griechenland zwischen Revolution und Konterrevolution (1936–1946)*, (Frankfurt: EVA, 1973), pp. 138–148.
557 Abbott, op. cit., p. 184.
558 Ibidem.
559 Ibidem, p. 185ff.
560 Ibidem, p. 188.
561 Ibidem, p. 189.
562 Ibidem
563 Louis Dartige du Fournet, "*Souvenirs de guerre d'un amiral 1914-1916*" (Paris: Plon, 1920), p. 116.
564 Abbott, op. cit., p. 190.
565 Ibidem, p. 190f.
566 Falls, op. cit.. I, p. 348.
567 Rapprochement = Annäherung
568 Abbott, op. cit., p. 191.
569 Palmer, op. cit., p.138.
570 Leon, op. cit., p. 473f.
571 Falls, op. cit., p. 348.
572 Falls, op. cit., p. 349; Abbott, op. cit., p. 192.
573 Falls, op. cit., p. 349; Abbott, op. cit., p. 192.

574 Falls, op. cit., p. 349.
575 Ibidem.
576 Ibidem, p. 350.
577 Abbott, op. cit., p. 192.
578 Ibidem, p. 191.
579 Falls, op. cit., p. 352.
580 Hibben, op. cit., p. 102f.
581 Falls, op. cit, p. 317f.
582 Ibidem, p. 352.
583 Abbott, op. cit., p. 196.
584 Lloyd George, *War Memoirs*, Vol. III, p. 790.
585 Ibidem, p. 197f.
586 Ibidem, p. 199.
587 Hellenic Army General Staff, A Concise History, p. 133.
588 Falls, op. cit., p. 354.
589 Hellenic Army General Staff, A Concise History, p. 134.
590 Falls, op. cit., p. 354; Abbott, op. cit., p. 200.
591 Hellenic Army General Staff, A Concise History, p. 133f; Lloyd George, op. cit., VI. p. 3.202f.
592 Falls, op. cit., p. 355.
593 Hellenic Army General Staff, A Concise History, p. 134.
594 Abbott, op. cit., p. 200.
595 Falls, op. cit., p. 361.
596 Abbott, op. cit., p. 201; weitere Versionen in Joachim, op. cit., p. 269.
597 Ibidem, p. 202f.
598 Falls, op. cit., p. 362.
599 Abbott, op. cit., p. 203f.
600 Ibidem, p. 205f.
601 Hibben, op. cit., p. 4.
602 Mühlmann, Eintritt Griechenlands, p. 524.
603 Abbott, op. cit., p. 210; Falls, op. cit., p. 356.
604 Falkenhausen, op. cit., p. 178.
605 Hellenic Army General Staff, A Concise History, pp. 135-138.
606 Falls, op. cit., p. 356.
607 Hellenic Army General Staff, A Concise History, p. 139.
608 Ibidem, p. 140f.
609 Ibidem, p. 143.
610 Hibben, op. cit., p. 4.
611 Abbott, op. cit., p. 223.
612 Joachim, op. cit., p. 273.
613 Abott, op. cit. p. 211.
614 Joachim, op. cit., p. 274.
615 Abbott, op. cit., p. 212f.
616 Hellenic Army General Staff, A Concise History, p. 143f.
617 Abbott, op. cit., p. 214.
618 Ibidem, p. 215.
619 The opposition named this assembly the *Parliament of the Lazarusse*. Thanos Veremis, Eleftherios Venizelos. A Biography (New York: Pella, 2010), p. 62.
620 Abbott, op. cit., p. 217.
621 Ibidem.
622 Ibidem, p. 218.
623 Ibidem, p. 218f.
624 Ibidem, p. 220.
625 Ibidem, p. 221.
626 Ibidem, p. 222.
627 Ibidem, p. 223.

628	Joachim, op. cit., p. 282.
629	Hellenic Army General Staff, A Concise History, p. 123.
630	Ibidem, p. 125.
631	Ibidem, p. 124.
632	About the fighting, Falls, op. cit., pp. 294-343.
633	Ibidem, p. 345.
634	Hellenic Army General Staff, A Concise History, p. 145f
635	Ibidem, p. 147.
636	Ibidem.
637	Veremis, op. cit., p. 63.
638	Hellenic Army General Staff, A Concise History, p. 148f
639	Ibidem, pp. 154–159.
640	Falkenhausen, op. cit., p. 182f.
641	Palmer, op. cit., p. 140f.
642	Hellenic Army General Staff, A Concise History, p. 149.
643	Palmer, op. cit., p. 152
644	Ibidem, p. 148f.
645	Ibidem, pp. 151–153.
646	Palmer, op. cit., p. 165.
647	Ibidem., pp. 160–164.
648	Carl Mühlmann, Oberste Heeresleitung und Balkan im Weltkrieg 1914/1918 (Berlin: Wilhelm Limpert, 1942), p. 143.
649	Ibidem, p. 144.
650	Ibidem, pp. 151–154.
651	Ibidem, p. 155.
652	Ibidem.
653	Ibidem, p. 157f.
654	Ibidem, pp. 176-178.
655	American Hellenic Society, The Greek White Book Supplementary Diplomatic Documents 1913–1917 (New York: Oxford UP, 1919), p. 84, No. 75.
656	Mühlmann, Oberste Heeresleitung , p. 183f.
657	Ibidem, p. 184f.
658	Ibidem, p. 185f.
659	American Hellenic Society, The Greek White Book Supplementary Diplomatic Documents 1913–1917 (New York: Oxford UP, 1919), p. 87, No. 77.
660	Ibidem, p. 89f, No. 79.
661	Mühlmann, op. cit., p. 186.
662	The only text of Sophie's in German is the one that Mühlmann published. Nowhere else have the originals been published. Only French translations of the telegrams can be found in the Greek archives, which were the ones prepared for the white paper. It is unknown if the translators had access to the German originals. Thankfully, copies of three of the French translations were acquired for me from the Greek national archive by my colleague Dimitrios Thrasyvoulou from Samos.
663	John Selden Willmore, The Story of King Constantine as Revealed in the Greek White Book (New York: Longmans, Green & Co., 1919). The enclosed quotes from the white paper are not exact, but rather linguistically changed.
664	G. Ward Price, The Story of the Salonica Army (New York: Edward J. Clode, 1918), p. 217f.
665	Falls, op. cit., II, p. 48; Palmer, op. cit., p. 166.
666	Ibidem, p. 49f.
667	Hellenic Army General Staff, A Concise History, p. 168f.
668	Falls, op. cit., II, pp. 70–74; Hellenic Army General Staff, A Concise History, p. 171f.
669	Falls, op. cit., II, p. 76f; Hellenic Army General Staff, A Concise History, p. 172.
670	Falls, op. cit., II, p. 77.
671	Ibidem.
672	Hellenic Army General Staff, A Concise History, p. 173.
673	Ibidem, pp. 175–179. The conquest of the village of Skra is described on page 179–189.

674 Lloyd George, op. cit., II, p. 1.915.
675 Hellenic Army General Staff, A Concise History, p. 175.
676 Falls, op. cit., II, p. 101f; Palmer, op. cit., p. 180f
677 Ibidem, p. 79.
678 Ibidem, p. 64.
679 Ibidem, p. 65f.
680 Ibidem, p. 66f.
681 Ibidem, p. 68.
682 Ibidem, p. 69.
683 Hellenic Army General Staff, A Concise History, p. 190f.
684 Erich Ludendorfff, Meine Kriegserinnerungen 1914–1918 (Berlin: Mittler, 1919), p. 547.
685 Friederike Krüger, "*Die Verantwortung der militärischen Führung 1918*", in: Jörg Duppler & Gerhard P. Groß, (eds.), Kriegsende 1918. Ereignis, Wirkung, Nachwirkung (München: Oldenbourg, 1999), p. 390.
686 Falls, op. cit., p. 104.
687 Ibidem.
688 Ibidem, p. 106f.
689 Ibidem, p. 108f.
690 Ibidem, p. 110f.
691 Ibidem, p. 111; Lloyd George, op. cit., II, p. 1.917f.
692 Ibidem, p. 112.
693 Herfried Münkler, Der große Krieg. Die Welt 1914–1918 (Berlin: Rowohlt, 2013), pp.704, 706.
694 Ludendorff, op. cit., p. 551.
695 Paul von Hindenburg, Aus meinem Leben (Leipzig: Hirzel, 1934), p. 284.
696 http://de.wikipedia.org/wiki/Erster_Weltkrieg#Kriegsjahr_1918.
697 Münkler, op. cit., p. 722.
698 Hindenburg, op. cit., p. 303.
699 In the Spring of 1917 Kaiser Karl tried to kickstart peace talks with Prince Sixtus von Bourbon-Parma and Franz Xaver von Bourbon-Parma, the two brothers of his wife Zita. In March 1918, the affair became public.http://de.wikipedia.org/wiki/Sixtus-Aff%C3%A4re.
700 http://de.wikipedia.org/wiki/Erster_Weltkrieg#Kriegsjahr_1918.
701 Hindenburg, op. cit,, p. 298.
702 http://de.wikipedia.org/wiki/Waffenstillstand_von_Moudros
703 Keegan, op. cit., p. 412.
704 Keegan, op. cit., p. 412; eine geschönte Version in Ludendorff, op. cit., p. 582: "*The situation could only get worse because of the conditions on the Balkans, even if we held the western fron*»
705 Münkler, op. cit., p. 722f.
706 Ludendorff, op. cit., p. 589.
707 Hindenburg, op. cit., p. 286.
708 Ludendorff, op. cit., p. 547.
709 Reichsarchiv, Weltkriegsende, p. 179. The official Handbuch der neuzeitlichen Wehrwissenschaften, I (Berlin: de Gruyter, 1936), p. 454 has a rather objective description. In it, the Macedonian front is depicted as a secondary front.
710 Reichsarchiv, Weltkriegsende, p. 17.
711 Falls, op. cit., II, p. 125f; Hellenic Army General Staff, A Concise History, p. 195f
712 Falls, op. cit., II, p. 128.
713 Ibidem, p. 129f.
714 Ibidem, p. 131.
715 Ibidem, p. 132f; Hellenic Army General Staff, A Concise History, p. 202f.
716 Richard C. Hall, Balkan Break Through. The Battle of Dobro Pole 1918 (Bloomington: Indiana UP, 2010), p. 127f.
717 Falls, op. cit., II, p. 135; http://de.wikipedia.org/wiki/Aleksandar_Malinow#Ende_des_Ersten_Weltkrieges_und_Minis terpr.C3.A4sident_1918.
718 Hellenic Army General Staff, A Concise History, p. 209f.
719 Ibidem, p. 211.
720 Ibidem, 212.

721 Ibidem, p. 212f.
722 Ibidem, p. 213f.
723 Hellenic Army General Staff, A Concise History, pp. 214-220; Details about the battles on the Dojran-front in: Charles Packer, Return to Salonika (London: Cassell, 1964), passim; Falls, op. cit., II, pp. 159–192, 202-204.
724 Ibidem, pp. 221–223.
725 Ibidem, pp. 224–232.
726 Ibidem, p. 234f.
727 Ibidem, p. 236; Text of the armistice agreement in Falls, op. cit., II, p. 252f.
728 Ibidem, p. 237.
729 Ibidem, p. 238.
730 Mühlmann, Oberste Heeresleitung, p. 230f.
731 Ibidem, p. 234f.
732 Hellenic Army General Staff, A Concise History, p. 209f.
733 Ibidem, p. 244
734 Lloyd George, War Memoirs, II, p. 1.920f.
735 Winston Churchill, The World Crisis 1916–1918, II (London: Butterworth, 1927), p. 536.
736 H. Collison Owen, Salonica and After. The Sideshow that Ended the War (London: Hodder & Stoughton, 1919). Saloniki und danach. Der Nebenschauplatz, der den Krieg beendete.
737 Ibidem, p. IV
738 Ibidem, p. VI.
739 Falls, op. cit., II, p. 283f.
740 Heinz A. Richter, "Die Auswirkungen der Operationen 'Marita' und 'Merkur' auf den Beginn des Unternehmens
'Barbarossa'. THETIS 2 (1995), pp. 203–216; idem, "The Impact of Operations Marita and Merkur on Barbarossa.
The six missing weeks in front of Moscow. Myth or historical truth?" in: Institute for Balkan Studies (ed.), Macedonia and Thrace, 1941–1944. Occupation - Resistance - Liberation (Thessaloniki: IMXA, 1998), pp. 15–32.
741 Hellenic Army General Staff, A Concise History, p. 256.
742 Hall, op. cit., p. 173f.
743 Henryk Batowski, "Proposals for a Second Front in the Balkans in September 1939" Balkan Studies 9 (1986), pp. 335–344.
744 Heinz A. Richter, Griechenland im Zweiten Weltkrieg 1939–1941 (Bodenheim: Syndikat, 1997), p. 22f; Yanis G. Morélos, "Français et Grecs pendant le Drôle de Guerre" Balkan Studies 29:1 (1988), p. 106f.
745 Deutschland Auswärtiges Amt, Dokumente zum Konflikt mit Jugoslawien und Griechenland (Berlin: Zentralverlag der NSDAP, 1941), p. 151.
746 Robert J. Collins, Lord Wavell (1883–1941). A Military Biography (London: Hodder & Stoughton, 1947), p. 325; Yannis G. Mourélos, "Français et Grecs pendant la drôle de guerre" Balkan Studies 29: (1988), p. 116.
747 John S. Koliopoulos, Greece and the British Connection 1935–1941 (Oxford: Clarendon Press, 1977), p. 134f.
748 Robin Higham, Diary of a Disaster. British Aid to Greece 1940–1941 (Lexington: The University Press of Kentucky, 1986), p. 6.
749 Koliopoulos, op. cit., p. 174.
750 Winston S. Churchill, The Second World War, IV, The Commonwealth Alone (London: Cassell, 1964), p. 194f.
751 Anthony Eden, The Eden Memoirs. The Reckoning (London: Cassell, 1965), p. 169, „strategische Torheit"
752 Churchill, op. cit., p. 200; Richter, Griechenland im Zweiten Weltkrieg, pp. 164–167.
753 Bernd Stegemann, "Die italienische-deutsche Kriegsführung im Mittelmeer und Afrika", in: Militärgeschichtliches Forschungsamt, (ed.), Das Deutsche Reich und der Zweite Weltkrieg. Vol.III Der Mittelmeerraum und Südosteuropa (Stuttgart: DVA, 1984), p. 595ff.
754 Richter, Griechenland im Zweiten Weltkrieg, p. 216ff.
755 Francis De Guingand, Operation Victory (London: Hodder & Stoughton, 1947), p. 57; Richter,

Griechenland im Zweiten Weltkrieg, pp. 217-243.
756   Richter, Griechenland im Zweiten Weltkrieg, p. 216ff.
757   The Australian and New Zealand views can be found in: Gavin Long, Greece, Crete and Syria (Canberra: Australian War Memorial, 1953); W. G. McClymont, Official History of New Zealand in the Second World War 1939–45. To Greece. (Wellington, New Zealand: Department of Internal Affairs, War History Branch, 1959)
758   Richter, Revolution und Konterrevolution, pp. 460–466, 517-547.
759   Heinz A. Richter, British Intervention in Greece. From Varkiza to Civil War. (London: Merlin Press, 1986) passim.
760   Wentschler; Deutsche Luftfahrt, (Berlin: 1925)
761   As comparison: In the fall of 1914 the average speed of a military vehicle was about 15 km/h. This rose to 30 km/h by the end of the war in 1918.
762   Thanks to the *Fieseler-Archive/ Kassel*, especially to Mr. Rolf Nagel.
763   Neumann, Georg Paul: *Die gesamten deutschen Luftstreitkräfte*. Berlin 1920,
764   Hptm. Ernst von Gersdorff, *25.05.1878 in Straßburg – † 16.06.1916 in Neuburg, Lothringen
765   For information: The naval pilot station Xanthi was on Lake Buru, about 80km from Drama.
766   Oblt. von Chappius and Oblt. Trenkmann were shot down over Saloniki on January 12th, 1916.
767   Borlinghaus: German field mail, undated
768   Borlinghaus: German field mail, undated
769   FF Oblt. Bodo Freiherr von Lyncker, born on October 22nd, 1894 in Berlin, died on February 18th, 1917 in Piravo (Macedonia), Jasta 25
770   Book Haupt-Heydemarck: (born on February 27th, 1896 Sulzbach/ Saar , died July 26th, 1917 near Zonnebeke) [in 1916 he flew for the FFA 69 on the Macedonian front, then with the Jasta 25, also in Macedonia. Brauneck wanted to switch to Jasta 11 together with Bodo Freiherr von Lyncker, but he fell before that. From April he flew with the Richthofen squadron, until he died in a dogfight over Flanders.] Brauneck shot down the commander of the English 47th squadron, Major Black].
771   Hermann Kastner, born on July 20th, 1884 in Myslowitz, died on November 28th, 1960, leads the KG1 from June 1916 until Dezember 1917. Info: Jörg Mückler
772   Lt. d.R. Paul Thierfelder, born on August 17th, 1884 in Auerbach, died on September 24th, 1917 in Macedonia, FA (A) 246
773   Vzfw. Wilhelm Grasmeher, born on May 31st, 1888 in Stieden, Kreis Oberlahn, took on Polish citizenship after the end of the war.
774   Lt. d.R. Eduard Beitter, born on December 6th, 1885 in Wildenstein, died on October 5th, 1917 in Resna near Prilep
775   FF Vzfw. Theodor Fuckel, born on January 2nd, 1892 in Erfurt, died on November 27th, 1917 in Bratindol (Bratin Dol, Macedonia), FA (A) 246.
776   BO Lt. d.R. Otto Wunderlich, born on September 25th, 1888 in Falkenberg, died on November 27th, 1917 in Bratindol (Bratin Dol, Macedonia), FA (A) 246.
777   FF Lt. Rudolf von Eschwege, born on February 27th, 1895 in Homburg v.d.H, died on November 21st in Orljak, FA 30.
778   Lt.d.L. Otto Splittgerber, born on August 26th, 1883 in Schwaben, died on March 13th, 1918 in Hudova (Macedonia), FA (A) 246.
779   On the western front the French aviators were not permitted to fly with a parachute.
780   Lt. d.R. Fritz Scheffler, born on September 13th, 1886 in Mohrungen, died on June 25th, 1918 in Prilep, FA (A) 246.
781   Pilot Uffz. Heinrich Naegele, born on June 14th, 1891, died on January 24th, 1918 in Hudova (Macedonia), FA (A) 246.
782   Lt.d.L. Otto Splittgerber, born on August 26th, 1883 in Schwaben, died on March 13th, 1918 in Hudova (Macedonia), FA (A) 246
783   On March 20th, 1918, Lt. Hoesch crashed as a victim of a similar balloon trap – similar to von Eschwege. According to other information (J. Mückler) he flew with Jasta 38 in May 1918.
784   Commander of aviators
785   Albert Julius Emil Hopman (born on April 30th, 1865 in Olpe, died on March 14th, 1942 in Berlin).
786   Gastreich/Waiss: "*Jagdstaffel Boelcke*", Helios-Verlag, Aachen 2016
787   Source Book: *Rolf Nagel, "Kassel und die Luftfahrtindustrie seit 1923", Melsungen 2015*

Printed in Great Britain
by Amazon